Blackwell, N.A.
9 Feb 81
$8.75

Combinatorial Optimization

Combinatorial Optimization

Edited by

NICOS CHRISTOFIDES
Imperial College of Science and Technology, London

ARISTIDE MINGOZZI
SOGESTA, Urbino, Italy

PAOLO TOTH
University of Bologna, Italy

and

CLAUDIO SANDI
IBM Scientific Centre, Pisa, Italy

A Wiley–Interscience Publication

Wingate College Library

JOHN WILEY & SONS
Chichester · New York · Brisbane · Toronto

Copyright © 1979 by John Wiley & Sons, Ltd.

All rights reserved.

No part of this book may be reproduced by any means, nor transmitted, nor translated into a machine language without the written permission of the publisher

Library of Congress Cataloging in Publication Data:

Main entry under title:

Combinatorial optimization.

'A Wiley–Interscience publication.'
'Based on ... lectures given at the summer school in combinatorial optimization held in SOGESTA, Urbino, Italy from 30th May to 11th June 1977.'
Includes index.
1. Mathematical optimization. 2. Combinatorial analysis. I. Christofides, Nicos.
QA402.5.C545   511'.6   78-11131
ISBN 0 471 99749 8

Typeset in Northern Ireland at The Universities Press, Belfast and printed in Great Britain at The Pitman Press, Bath

Foreword

This book is based on a series of lectures given at the Summer School in Combinatorial Optimization held in SOGESTA, Urbino, Italy from 30th May to 11th June 1977. It was the purpose of this School to review the state of the art in the formulation and solution of practical problems as Combinatorial Optimization programs in the general area of Operations Research.

Real-world problems have in the past been approximated and dealt with mainly by the use of linear programming, since this was in many cases the only means by which practical-size problems could be solved. Attempts to represent reality more closely, however, lead directly to mathematical programs with integer restrictions. Because of the very general applicability and economic importance of these models, their study—under the general term of 'Combinatorial Optimization'—has been pursued at an ever-increasing rate in recent years.

This volume is a further contribution in this direction and is an additional step in the general effort to increase the size of combinatorial optimization problems that can be solved and which will enable the potential economic benefits to be realized.

G. Puppi, Bologna, 1978
(Chairman of SOGESTA)

Contents

Preface

1. **Branch and Bound Methods for Integer Programming** . . . 1
   R. S. Garfinkel

2. **The Theory of Cutting-Planes** 21
   R. Jeroslow

3. **Subgradient Optimization** 73
   C. Sandi

4. **A Partial Order in the Solution Space of Bivalent Programs** 94
   P. L. Hammer and S. Nguyen

5. **The Complexity of Combinatorial Optimization Algorithms and the Challenge of Heuristics** 107
   F. Maffioli

6. **The Travelling Salesman Problem** 131
   N. Christofides

7. **Set Partitioning—A Survey** 151
   E. Balas and M. W. Padberg

8. **The Graph-Colouring Problem** 211
   S. M. Korman

9. **The 0–1 Knapsack Problem** 237
   S. Martello and P. Toth

10. **Complexity and Efficiency in Minimax Network Location** . . 281
    G. Y. Handler

11. **The Vehicle Routing Problem** 315
    N. Christofides, A. Mingozzi, and P. Toth

12. **Loading Problems** 339
    N. Christofides, A. Mingozzi, and P. Toth

13. **Minimizing Maximum Lateness on One Machine: Algorithms and Applications** 371
    B. Lageweg, J. K. Lenstra, and A. H. G. Rinnooy Kan

14	**The Crew Scheduling Problem: A Travelling Salesman Approach**	**389**
	F. Giannessi and B. Nicoletti	
15	**Graph Theoretic Approaches to Foreign Exchange Operations**	**409**
	N. Christofides, R. D. Hewins, and G. R. Salkin	
Index		**421**

Preface

'Combinatorial Optimization' is a term that has emerged in recent years to describe those areas of mathematical programming that are concerned with the solution of optimization problems having a pronounced combinatorial or discrete structure. The title is used as a unifying term covering Integer Programming, Graph Theory, parts of Dynamic Programming, etc., and the areas covered are of increasing importance because of the large number of practical problems that can be formulated and solved as combinatorial optimization problems.

The chapters of this book are derived from lectures given at the Summer School in Combinatorial Optimization, held in Urbino, Italy from 30th May to 11th June, 1977. The chapters are of three types. Chapters 1 to 5 describe methodologies and results of general applicability to combinatorial optimization. Chapters 6 to 10 describe some of the best-known pure combinatorial problems and which often form the central core of larger and more complex practical problems. Chapters 11 to 15 are also concerned with problems with structure but are less 'pure' and are closer abstractions of problems as they appear in reality.

Chapter 1 introduces the concepts of branch and bound as the most important technique for the solution of combinatorial optimization problems.

Chapter 2 discusses the general theory of cutting planes and how these can be used to solve integer programming problems.

Chapter 3 describes an iterative technique—known as subgradient optimization—for the optimization of continuous but not everywhere differentiable functions. This technique is extremely useful, when used in conjunction with lagrangian relaxation, for deriving bounds to be used in branch and bound algorithms.

Chapter 4 derives some order relations in the solution space of linear 0–1 programs which can be used to obtain information on problem infeasibility, and the fixing of variables. A resulting algorithm and its computational performance are investigated.

Chapter 5 is an introduction to the field of computational complexity and the role that heuristic methods can play in solving hard combinatorial problems.

Chapter 6 gives a unified treatment of the travelling salesman problem—probably the most celebrated combinatorial optimization problem—from the viewpoint of lagrangian relaxation.

Chapter 7 is an extensive survey of the set partitioning problem: a mathematical model that is applicable to a very large number of practical situations. Theoretical, algorithmic, and application aspects are discussed.

Chapter 8 surveys the graph colouring problem and its applications to timetabling from the algorithmic point of view.

Chapter 9 deals with the 0–1 knapsack problem—the simplest of the integer programming problems. Algorithms for the solution of this problem are described and tested computationally.

Chapter 10 describes problems of optimally locating facilities on a network with minimax objectives such as those found in the practical location of emergency facilities.

Chapter 11 describes the problem of routing vehicles to supply customers from a central facility—a frequently encountered practical problem. Various heuristics are described and evaluated.

Chapter 12 deals with a new class of problems that involve the loading and unloading of different types of liquids into tanks. Solution procedures are described and evaluated.

Chapter 13 is concerned with a one-machine scheduling problem where the objective is to minimize the maximum lateness. Various algorithms are described and tested.

Chapter 14 is concerned with the problem of crew scheduling, a problem of considerable economic importance for the efficient operation of an airline.

Chapter 15 considers foreign exchange operations from a graph theoretic viewpoint and makes use of graph theoretic techniques for the solution of foreign exchange problems.

The editors wish to express their sincere thanks to Professor Puppi and Dott. Ing. Fossa (Chairman and Director of SOGESTA, respectively) and to the Directors of IBM (Italy) for the sponsoring of the Summer School in Combinatorial Optimization that was the catalyst without which this book would have never been produced.

N. CHRISTOFIDES
A. MINGOZZI
P. TOTH
C. SANDI

June 1978

CHAPTER 1

Branch and Bound Methods for Integer Programming

ROBERT S. GARFINKEL
Management Science Program, University of Tennessee

1.1 Introduction

In this paper we consider enumerative (tree search) approaches to the solution of the mixed integer linear program (MILP)

$$\max z = cx$$
s.t.
$$Ax \le b$$
$$x \ge 0$$
$$x_j \text{ integer}, j \in I.$$
(1.1)

The 'data' matrices c, A, and b in (1.1) are of dimension $(1 \times n)$, $(m \times n)$, and $(m \times 1)$ respectively. The 'variable' matrix x has dimension $(n \times 1)$. In the case $I = \{1, \ldots, n\}$, (1.1) reduces to an integer linear program (ILP). We assume $I \ne \emptyset$, for otherwise (1.1) is a linear program (LP) which can be solved by the simplex method.

Throughout this paper we will speak of solving LP's by *the* simplex method, as though there were only one simplex method. There are, of course, many variants of the simplex algorithm. For initial optimization one uses the primal simplex, generally in revised form with product form of the inverse. Most problems will have bounds on the variables to begin with, and additional bounds will be imposed by the enumeration process. These bounds are handled simply and directly by modern LP codes, as are generalized upper bound (GUB) constraints. Finally, addition of new bounds and constraints to an optimal simplex tableau requires reoptimization. This can be done by using the dual simplex algorithm or by parametric analysis. We will simply refer the reader to the many papers which relate the details of branch and bound codes based on linear programming to see how the linear programs are solved.

Although much of the research on (1.1) has been concentrated on the ILP, there is a general consensus (see, for example, Land and Powell, 1977) that in the majority of applications I is only a small subset of $N = \{1, \ldots, n\}$. That is, (1.1) is fundamentally a large LP with a few integer variables. For such problems, enumerative techniques based on linear programming constitute very appealing solution procedures.

Even for the ILP, no other solution technique has been able to approach the computational success of 'branch and bound'. To the best of our knowledge, every commercially available computer code uses some variation of this approach. Before giving a general description of the branch and bound technique, we note some of its inherent features which make it particularly appealing. For the reader interested in other approaches to (1.1), recommended references include Geoffrion and Marsten (1972), Garfinkel and Nemhauser (1972a), and Jeroslow (1977).

Branch and bound approaches are appropriate for virtually all combinatorial problems. In fact there is no necessity for the problem in question to be posable in a mathematical form such as (1.1) for the MILP. If there exists a (finite) set of objects whose total enumeration is possible, and if there is a well-defined objective function value associated with each member of the set, then a (finite) branch and bound algorithm can be developed for finding an optimal solution. (The reader is directed to Lawler and Wood, 1966; Agin, 1966; Mitten, 1970; or Balas, 1968 for general discussions of branch and bound techniques). Of course, for the algorithm to consistently do better than total enumeration there must be some structure which allows for computational shortcuts. Such structure abounds for the general MILP (1.1) and even more so for a number of specializations of (1.1) incuding set covering problems, location problems, scheduling problems, etc. We will restrict ourselves to discussion of the general problem (1.1) and to codes which can solve it.

The fundamental point to be made here is that the basic branch and bound concept is so simple that it allows ample room for the inclusion of many heuristic rules which can be devised for any particular instance of a combinatorial problem. These rules tend to be of little theoretical interest, but along with the cleverness of the implementation they are often the key elements in determining success or failure of these techniques. A large portion of this chapter will be devoted to discussion of rules which have been proposed for the MILP and the reasons for their success or failure.

In most commercial codes, the details of the algorithm are adjustable. That is, there are options for performing many of the steps. The options used in solving a particular MILP may be chosen internally by the program or externally by the user. The way in which these matters are handled by various codes is mentioned throughout the paper. The reader is directed to Land and Powell (1977) for more details.

Branch and bound algorithms are almost always 'primal' in the sense that they proceed from one feasible solution to another until optimality is verified. In fact they often find optimal or near optimal solutions early in the enumeration process and spend the majority of the time verifying optimal-

Branch and Bound Methods for Integer Programming

ity. Thus the user can be comforted by the expectation that termination before proof of optimality will likely yield a very good solution if not an optimal one. This contrasts to other approaches, for example Gomory's fractional cutting plane algorithm (Gomory, 1958), where a feasible solution is not found until termination.

Branch and bound was first proposed for MILP's by Land and Doig (1960). Since then a vast number of papers and books have appeared on the subject. Some other surveys include: Tomlin (1970), Garfinkel and Nemhauser (1972a), Geoffrion and Marsten (1972), and Land and Powell (1977). In Land and Powell (1977), eleven commercial codes are analysed with respect to specific techniques used and options available. Some of this work is briefly summarized in Section 1.13. The codes are numbered in Table 1.1 and will be referred to by these numbers throughout the text.

The special branch and bound methods which have been developed for binary ILP's by Balas, Geoffrion, Glover, and others and are often termed 'implicit enumeration' will not be discussed in this paper. In general, these methods are so specialized that they are often distinguished from branch and bound. The reader is referred to Breu and Burdet (1974) for a computational study of these problems.

1.2 Elements of Branch and Bound Algorithms

In this section we distinguish four fundamental elements of a branch and bound algorithm. These are: separation (partition) into subproblems; relaxation (upper bounding); fathoming of subproblems (lower bounding); and selection of subproblems (branching). Each of these elements is explored in detail in a later section. Here we give an overview of the elements and the relation between them.

Denote the set of feasible solutions to (1.1) by

$$S = \{x \mid Ax \leq b, x \geq 0, x_j \text{ integer}, j \in I\}.$$

Instead of attempting to solve the problem directly over S, the set is successively divided into smaller and smaller sets which have the property that any optimal solution must be in at least one of the sets. This is called separation and is often illustrated by an 'enumeration tree' such as that of Figure 1.1 (see Section 1.6 for an alternative representation of the tree).

Each node of the enumeration tree corresponds to a subproblem of (1.1). That is, node k is the problem

$$\max z, x \in S_k \qquad (1.1_k)$$

where $S_k \subseteq S$. As the enumeration proceeds farther down the tree, the sets S_k become progressively smaller until it finally becomes possible to solve

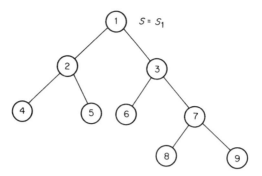

Figure 1.1

(1.1_k) or at least to determine whether or not it contains a potentially optimal solution. The enumeration process will be sucessful if this happens before the tree becomes impossibly large.

Upper bounds $\bar{z}_k$ on z_k^*, the maximum value of z_0 over S_k, will be calculated. As will be seen below, the quality of these bounds will be critical. By the quality of $\bar{z}_k$, we mean its closeness to z_k. $\bar{z}_k$ will generally be calculated by solving a 'relaxation' of (1.1_k). That is, a problem of the form

$$\max w(x), x \in T_k \qquad (1.2_k)$$

where $S_k \subseteq T_k$ and $w(x) \geq cx$ for all $x \in S_k$. Typically (1.2_k) will be a linear program (LP).

Subproblems are discarded or 'fathomed' when (1.1_k) is solved, or when it is determined that S_k does not contain a solution which can improve on the best known solution to (1.1). Denoting the latter as the 'incumbent', and its value by $\underline{z}$, we can certainly fathom S_k if $z_k^* \leq \underline{z}$.

If z_k^* is unknown, as is generally the case, it suffices to have

$$\bar{z}_k \leq \underline{z} \qquad (1.3)$$

for fathoming S_k. Thus it becomes clear that not only is the quality of the bound $\bar{z}_k$ of critical importance, but also the quality of $\underline{z}$. More will be said about achieving good lower bounds later.

Clearly node k can also be fathomed if (1.2_k) is infeasible ($T_k = \emptyset$) since $S_k \subseteq T_k$. This can be included as a special case of (1.3) by setting $\bar{z}_k = -\infty$ when $T_k = \emptyset$. It also follows directly from (1.3) that if the optimal solution to (1.2_k) is in S_k then node k is fathomed.

Loosely speaking, 'branching' is nothing more than the process of deciding which node to consider next. By 'considering' node k we may mean: solving (1.2_k); separating S_k; or possibly attempting to fathom by improving the bound $\bar{z}_k$. In the next eight sections we elaborate on the ways that these steps may be performed.

1.3 Relaxation (Upper Bounding)

Clearly, to be of any value, (1.2_k) must be considerably easier to solve than (1.1_k). In fact it is usually the case that the tighter the bound, the harder it is to calculate. The tightest bound would, of course, be achieved by letting $(1.2_k) = (1.1_k)$, and the easiest calculated would set $\bar{z}_k = \infty$.

By far the most popular choice of relaxation is to let (1.2_k) be the *linear program* (LP) obtained by relaxing the integrality constraint on x in (1.1_k). This LP relaxation (LPR) satisfies the criterion of relatively easy computability. (Codes 1-11 of Table 1.1 all use this relaxation.) In many cases it also gives a good approximation to the ILP solution. In fact it is demonstrated in Rardin and Lin (1977) that the most important parameter in the success of branch and bound codes for ILP's is the 'distance' from the LP optimum to the MILP optimum. Except for the remainder of this section, this chapter will deal exclusively with algorithms based on LPR. However, other relaxations have been proposed and in some cases implemented. Two of these are mentioned below.

Lagrangian relaxation has been proposed for MILP algorithms by a number of authors. The reader is referred in particular to Geoffrion (1974) and Shapiro (1971). For MILP's of the form

$$\max cx$$
$$Ax \leq b$$
$$Dx \leq d \qquad (1.4)$$
$$x \geq 0, \ x_j \text{ integer}, \ j \in I,$$

the lagrangian relaxation is

$$\max cx - \lambda(Ax - b)$$
$$Dx \leq d \qquad (1.5)$$
$$x \geq 0, \ x_j \text{ integer}, \ j \in I$$

where λ is a nonnegative vector. Here it is assumed that the constraints $Dx \leq d$ have a special structure which render (1.5) virtually trivial to solve (see Geoffrion 1974 for some illustrative examples). A branch and bound algorithm based on (1.5) would involve some method for choosing values of the multiplier vector λ in order to make (1.5) a tight relaxation.

Another relaxation which has been proposed (see Shapiro 1968) mainly for ILP's is referred to here as the *group-theoretic relaxation*. Let the ILP be represented by

$$\max z = y_{00} - \sum_{j \in R} y_{0j} x_j$$

$$x_i = y_{i0} - \sum_{j \in R} y_{ij} x_j \quad i = 1, \ldots, m \tag{1.6}$$

$$x_i \geq 0, \text{ integer} \quad i = 1, \ldots, m$$

$$x_j \geq 0, \text{ integer}, j \in R$$

where the elements y_{ij} are the coefficients of the simplex tableau resulting from the solution of LPR. Thus the variables x_i are basic and R is the set of nonbasic variables. The relaxation (GTR) results from dropping the non-negativity restriction on the basic variables. It has been shown by Gomory (1965) that the GTR can be solved as a *group knapsack problem*.

1.4 Penalties

The concept of penalties in integer programming was introduced by Driebeek (1966) and Beale and Small (1965). An extensive treatment can be found in Tomlin (1970). The development here closely follows that of Garfinkel and Nemhauser (1972a). Penalties are an integral part of codes 1, 4, 7, 10, and 11.

They are introduced directly after the section on relaxation because they can be thought of as a means for attaining better upper bounds. They can also be used in the branching process as will be seen in Section 1.5

Consider an optimal LP solution at any node which has at least one fractional row given by

$$x_i = y_{i0} - \sum_{j \in R} y_{ij} x_j$$

where $i \in I$, $y_{i0} = [y_{i0}] + f_{i0}$ and $f_{i0} > 0$. Here $[y_{i0}]$ denotes the integer part of y_{i0}. Since x_i is required to be integer, either $x_i \leq [y_{i0}]$ or $x_i \geq [y_{i0}] + 1$. If the constraint $x_i \leq [y_{i0}]$ holds then it can be represented in the simplex tableau as

$$\sum_{j \in R} y_{ij} x_j \geq f_{i0}.$$

The pivot column for the next dual simplex interation will be determined by

$$\frac{y_{0k}}{y_{ik}} = \min_{j \in R} \frac{y_{0j}}{y_{ij}} \quad y_{ij} > 0$$

so that the objective will decrease by the *down penalty*

$$D_i = \min_{j \in R} \frac{f_{i0} y_{0j}}{y_{ij}} \quad y_{ij} > 0. \tag{1.7}$$

Of course D_i is only a lower bound on the decrease in z since additional pivots may be required to obtain feasibility. An *up penalty* U_i corresponding to the condition $x_i \geq [y_{i0}] + 1$ can similarly be defined as

$$U_i = \min_{j \in R} (f_{i0} - 1) \frac{y_{0j}}{y_{ij}} \quad y_{ij} > 0. \tag{1.8}$$

It follows that

$$P_i = \min \{U_i, D_i\}$$

is a valid lower bound on the decrease in z incurred while obtaining feasibility.

In the case of ILP's two other bounds are given below. The first is derived from the simple observation that some nonbasic variable must increase at least to the value one. Thus

$$P = \min_{j \in R} y_{0j} \tag{1.9}$$

is also valid.

The last bound is derived from the cut of the method of integer forms (see Gomory, 1958; Jeroslow, 1977) given by

$$\sum_{j \in R} f_{ij} x_j \geq f_{i0}.$$

The first dual simplex pivot would yield a decrease in z of

$$P'_i = \min_{j \in R} \frac{f_{i0} y_{0j}}{f_{ij}} \quad f_{ij} > 0. \tag{1.10}$$

Finally, a lower bound (penalty) on the decrease in z is

$$P^* = \max \left\{ P, \max_{i, f_{i0} > 0} \max \{P_i, P'_i\} \right\}. \tag{1.11}$$

The penalty calculation (1.11) can be used in a number of ways. The most obvious is to revise the upper bound $\bar{z}_k$ at node k to $\bar{z}_k - P^*$. This may result in immediate fathoming of the node. If not, (1.11) can be used to guide the branching, as we will see in Section 1.5.

A limitation of the usefulness of penalty calculations is the common occurrence of dual degeneracy in many MILP's. For such problems penalties will generally be zero and are therefore not worth calculating. Another more serious drawback of penalties is that they provide information only about the next simplex iteration. In the case where a number of iterations are required to restore primal feasibility, the information contained in the penalties may be very weak. Since penalties are relatively costly to calculate their general usefulness is a matter of question.

1.5 Partitioning

For LP-based branch and bound algorithms the most natural means for separating a problem into subproblems is to introduce a set of contradictory constraints on one of the variables required to be integer. Suppose the relaxed LP has been solved at node k and the solution does not satisfy x_i integer, all $i \in I$. A variable x_r, $r \in I$ is chosen which has fractional value. That is, $y_{r0} = [y_{r0}] + f_{r0}$ where $f_{r0} > 0$. The separation strategy originally proposed by Land and Doig (1960) was to separate S_k into the sets

$$S_k \cap \{x \mid x_r = 0, 1, \ldots, U_r\} \tag{1.12}$$

where U_r is an upper bound on x_r. Clearly (1.8) defines a partition of S_k.

A simplified partitioning rule was introduced by Dakin (1965). In it S_k is partitioned into

$$S_k \cap \{x \mid x_r \leq [y_{r0}]\}$$

and $\tag{1.13}$

$$S_k \cap \{x \mid x_r \geq [y_{r0}] + 1\}.$$

Thus node k has been separated into two successor nodes by appending the contradictory constraints $x_r \leq [y_{r0}]$ and $x_r \geq [y_{r0}] + 1$. Obviously one or other of these constraints must be satisfied by any feasible solution. Rule (1.13) has the obvious advantage over (1.12) of partitioning S_k into only two subsets and is used by all eleven commercial codes, except in the case of special ordered sets (Section 1.11). Code 1 also allows the possibility of branching on a variable which is currently integer.

Note that the penalties derived in (1.11) may provide for limiting the development of a node. For instance, if for some i, $[\bar{z}_k - D_i] \geq z$ but $[\bar{z}_k - U_i] \leq z$, then the node can be developed only in the downward direction without a corresponding increase in the tree size. In other words, this would be a *forced move*. Furthermore, if these conditions hold for more than one variable then these can be simultaneously forced in their appropriate directions without increase in the size of the tree.

If a forced move is not available, and increase in the tree size is therefore necessary, the next important issue is the choice of which fractional variable to choose for (1.13).

This question may be answered in various ways. Here we will discuss the use of 'priorities' and 'quasi-integer' variables, while in Section 1.7 we will

Branch and Bound Methods for Integer Programming

incorporate the concept of 'pseudo costs.' Penalties were at one time highly regarded for guidance in branching, but the experience of Forrest, Hirst, and Tomlin (1974) has indicated that they are not reliable for this purpose.

Priorities

A priority structure is an ordering in importance of the set of variables. Priorities are used to choose a variable for partitioning with higher priority variables chosen first. All commercial codes except Code 4 allow for specification of a priority structure.

A number of methods can be used for establishing priorities. In some cases the user may know (or feel) that certain variables are critical and that others are of secondary importance. When the critical variables have integer values it may be that the secondary variables will naturally assume integer values as well. In this case it certainly makes sense to branch first on the high priority variables.

In the absence of prior information, priorities can be set by ordering the variables by cost, or possibly by use of the down and up 'pseudo costs' of x_i, C_i^D, and C_i^U which are defined in Section 1.7. A rule proposed in Benichou et al. (1971) is to partition on that variable which maximizes

$$\min\{C_i^D f_{i0}, C_i^U(1-f_{i0})\} \qquad (1.14)$$

over $i \in I$. The quantity (1.14) is an estimate of the minimum decrease in $\bar{z}$ over the successor nodes, and the rule is based on the hope of creating two nodes which can both be fathomed.

Quasi-Integer Variables

All commercial codes allow the user to specify a tolerance t such that a variable x_i is considered integer if $\max\{f_{i0}, 1-f_{i0}\} < t$. In general t is very small and the rationale is that computer round-off errors may cause a loss of accuracy. On the other hand, codes 1, 5, 6, 7, and 10 allow the specification of a number $t^* > t$ such that if $\max\{f_{i0}, 1-f_{i0}\} < t^*$, then x_i is called *quasi-integer*. Such variables may be considered computationally less important than those which are more seriously fractional. Branching may be restricted to the other variables in the hope that those which are quasi-integer will become integer-valued in the process.

Finiteness

In general the finiteness of branch and bound algorithms for MILP's based on (1.13) is assured. The only assumption needed to guarantee finiteness is that there exist finite upper bounds U_j on x_j for all $j \in I$. If the problem has a finite optimal solution, such an assumption is clearly justified. The total number of feasible values for the variables $x_j, j \in I$ is then bounded by

$$U(I) = \prod_{j \in I} (1 + U_j).$$

Since the underlying tree structure is nothing more than a successive partitioning of the set

$$X(I) = \{x_j \mid 0 \leq x_j \leq U_j, j \in I\}$$

and since each individual partitioning step yields two nonempty elements of $X(I)$, it follows that the algorithm can do no worse than examine all $U(I)$ elements of $X(I)$. This is not a very comforting bound, but at least it has the merit of being finite.

Finally, to deduce finiteness of the algorithm it must be shown that the calculations at each node are also finite, which follows immediately from the finiteness of the simplex algorithm for linear programming.

1.6 Enumeration Trees

In this section we give two tree representations of the enumeration process described in Sections 1.1–1.5. The first tree, in Figure 1.2, is the tree of

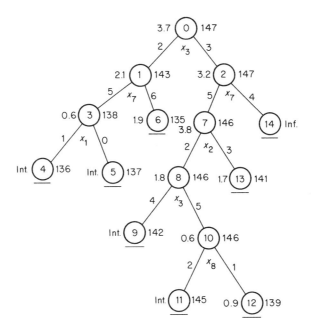

Figure 1.2

Branch and Bound Methods for Integer Programming

Figure 1.1 with additional branching and bounding information illustrated. The particular numbers correspond to the solution of some fictional, unspecified problem.

The numbers of the nodes represent the order in which the subproblems were solved. Thus a total of 14 subproblems were solved. The first integer solution was discovered at node 4 and the optimal solution at node 11. The number to the right of each node k is $\bar{z}_k$, the value of the LP relaxation. The numbers underneath each node and on its descendent edged indicate which was the partitioning variable and what its limits were. Thus node one was partitioned by $x_7 \leq 5$ and $x_7 \geq 6$. Fathomed nodes are underlined, and it is indicated whether integrality, infeasibility, or $\bar{z}_k \leq z$ was the cause of fathoming.

Let us associate one more number with each node k of Figure 1.2, namely its *degree of infeasibility* $d(k)$. By $d(k)$ we mean some measure of how 'far' the solution to the relaxed LP is from being integer. One simple measure which comes to mind is $\sum_{i \in I} \min \{f_{i0}, 1 - f_{i0}\}$ which is the sum of the infeasibilities of the fractional variables in I. This number $d(k)$ is placed to the left of each node k.

The tree of Figure 1.2 yields a summary of the history of the enumeration

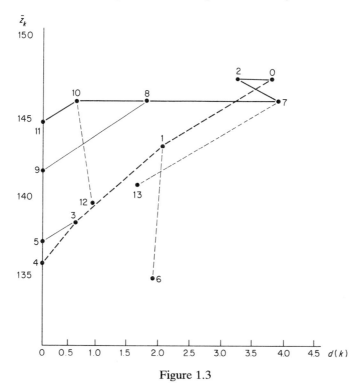

Figure 1.3

process. Another representation of the same tree has been proposed, by Forrest, Hirst, and Tomlin (1974), which seems somewhat more informative. The same problem is illustrated with this tree in Figure 1.3. The heavy solid line shows the path to the optimal solution and the heavy dashed line the path to the first integer solution. Light solid lines show paths to other integer solutions and light dotted lines to noninteger nodes which are fathomed.

1.7 Pseudo Costs

Pseudo costs have often been found to be more accurate indicators of the relative 'importance' of the variables than are their original costs. The purpose of pseudo costs is to estimate the change in objective function value caused by forcing a variable which is currently fractional to be integer. The concept is much the same as that of the penalties introduced in Section 1.4. However, the latter only give 'local' information since they are concerned only with the next dual simplex iteration. Pseudo costs attempt to forecast the change when the variable goes all the way to integrality. Pseudo costs are also not based on local information. They assume that the effect on the objective function associated with changing a variable in a particular direction remains constant throughout the tree. Although there is no theoretical reason to justify this assumption, it has been discovered empirically to be the rule rather than the exception. Pseudo costs are used by codes 5, 6, 7, and 10.

Let the up (down) pseudo costs of x_i be denoted by C_i^U (C_i^D). In general these values are unknown at the beginning of the enumeration process. They can be estimated by the user, or perhaps by the original costs of the variables. They can also be estimated by performing small 'experiments' (Gauthier and Ribière, 1977) previous to the enumeration process. As the enumeration proceeds these estimates can be updated every time a variable is driven to integrality.

The formula used by Gauthier and Ribière (1977) to calculate C_i^U and C_i^D is based on Figure 1.4. Assume that node k has been partitioned based on x_i, $i \in I$ and note that at nodes $k+1$ and $k+2$, x_i will be integer valued. We then take

$$C_i^D = \frac{\bar{z}_k - \bar{z}_{k+1}}{f_{i0}}$$

and (1.15)

$$C_i^U = \frac{\bar{z}_k - \bar{z}_{k+2}}{1 - f_{i0}}.$$

Note that C_i^D and C_i^U are both nonnegative.

Based on (1.15) we may estimate a bound on the objective value of the

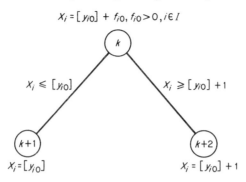

$$x_i = [y_{i0}] + f_{i0}, f_{i0} > 0, i \in I$$

Figure 1.4

best descendent of node k by

$$E_k = \bar{z}_k - \sum_{i \in I} \min \{C_i^D f_{i0}, C_i^U (1 - f_{i0})\}. \tag{1.16}$$

The formula (1.11) assumes 'independence' of the variables x_i and their corresponding penalties. It can be used in the branching process, as we shall see in the next section.

1.8 Branching

By 'branching' we will mean the decision as to what to do next. At any stage the options available are:

(a) Solve the relaxed LP at some node k;
(b) Calculate penalties at some node k;
(c) Select a node k and partition S_k;
(d) Attempt to achieve a better lower bound $\underline{z}$ (see Section 1.9).

At any stage in the enumeration process, the set of nodes which are candidates to be node k in (a–c) above are referred to as the 'live' (waiting) nodes. These are nodes which have been neither fathomed nor partitioned. Termination of the entire enumeration process occurs when this set is empty.

Node selection rules (choice of which unfathomed node to investigate) are less standard than partitioning rules. Sketches of a number of possible rules are given in this section.

The ultimate objective is to solve the problem in as short a time as possible. Solution time is itself a function of the final tree size as well as the amount of computation done at each node of the tree. For the moment, focusing only on the tree size, a rule which in some sense guarantees a smallest tree is 'branch to the greatest upper bound' (BUB). That is, examine next that node k for which $\bar{z}_k$ is greatest over all live nodes.

Disadvantages associated with the BUB rule include the possible problems involved in 'jumping' around the tree from one node to another. Data describing various nodes including the current basis inverse may have to be shifted from internal to external computer storage. Another disadvantage is that there is no initial push to locate a good feasible solution. Thus termination short of optimality because of time limitations may leave the user without a good solution, and the fathoming test (1.3) may not be powerful in the early stages of the enumeration.

A strategy which attempts to overcome these disadvantages is the LIFO strategy in which the next node investigated is one of the successors of the current node if either is unfathomed. The LIFO strategy alleviates some of the storage problems and also strives to find a feasible solution quickly. On the other hand, it may discover a number of inferior solutions which could otherwise have been fathomed by a BUB strategy.

A reasonable compromise may be some mix of the two strategies. That is, use LIFO initially to try to get a good feasible solution and later switch to BUB to try to keep the tree small.

One rule which has great appeal is to always use LIFO in the event that a node has only one successor as previously mentioned in Section 1.4.

In an obvious way we can deduce corresponding rules based on the estimates (1.16) rather than on bounds. It has been claimed by Gauthier and Ribière (1977) that the quantity E_k is generally a superior indicator of the potential 'value' of node k than is $\bar{z}_k$. Thus branching rules based on (1.16) rather than on $\bar{z}_k$ may generally be preferred.

A variation of the same idea is called the 'best projecton method' in Forrest, Hirst, and Tomlin (1974), and is based on the tree of Figure 1.3. Suppose the optimal value of z were known in advance. In Figure 1.3 this is the point $(d, z^*) = (0, 145)$. Then draw a line L between this point and the point $(d(0), \bar{z}_0)$. The empirical observation is that the (heavily lined) optimal path between the two points generally consists of those points closest to the line L. The strategy is then to project point $(d(k), \bar{z}_k)$ onto L by defining

$$P_k = \bar{z}_k - \left(\frac{\bar{z}_0 - z^*}{d(0)}\right) d(k). \tag{1.17}$$

The node which maximizes P_k is examined next.

Of course one drawback of this approach in that z^* is generally unknown. The strategy used by Forrest, Hirst, and Tomlin (1974) is to estimate z^* based on *a priori* information or possibly by a small experiment (solution of a 'pilot' model). The estimate of z^* can then be updated during the enumeration process.

1.9 Fathoming (Lower Bounding)

In the general algorithm given in Section 1.2, fathoming of node k occurs when $\bar{z}_k \leq z$ where z is the value of the best known solution to the MILP. Obviously fathoming will be accelerated if z is high (close to z^*).

There are a number of ways to attempt to get good lower bounds on z^*. It may be the case that a good feasible solution is known *a priori*. This would most likely occur in situations where the MILP is a model of a system which is currently operating and to which improvements are to be made.

In the normal course of enumeration, feasible solutions are discovered at various nodes of the tree. The value of any feasible solution better than z then replaces z. If tree search is relied upon for lower bounding, those nodes which seem to have high probability of containing feasible solutions would be good candidates for early investigation.

Another very appealing method for finding good feasible solutions in S_k is the use of inexact algorithms or heuristics. That is, techniques which are not guaranteed to succeed but which are fast and therefore inexpensive. Very little is lost if a heuristic fails and a great deal is gained if a good feasible solution is found.

Many heuristics have been proposed for ILP's and MILP's. One of the more promising of these is given in Hillier (1969). Others are described by Müller-Merbach (1970), and Senju and Toyoda (1968). To the best of our knowledge none of these has been incorporated in commercial branch and bound codes.

Most heuristics proposed for MILP's are of the 'direct-search' variety. That is, they proceed from one feasible solution to a better one. At each iteration some neighbourhood of the current solution is searched for an improving solution. If none is found, termination can occur with a 'local' (with respect to the neighbourhood) maximum. When a local maximum is found, one has the option of finding another starting point for the search, or of enlarging the neighbourhood around the local maximum. Of course, if the heuristic is to be fast the neighbourhood must be easy to search. For instance, for some x_t

$$N_t(\bar{x}) = \{x \mid x_j = \bar{x}_j, j \neq t\}$$

might be a convenient neighbourhood of a point $\bar{x}$. An improving point in $N_k(\bar{x})$ can easily be found if one exists.

Direct search heuristics are greatly aided by starting at a good feasible point. To this end the heuristic of Hillier (1969) is designed to find a feasible point close to the LP optimal solution.

Heuristics could be incorporated in a branch and bound code at node zero in order to obtain a good initial lower bound. They could also be used later in the tree. It may be convenient to search for a good feasible solution at a node with high $\bar{z}_k$ or E_k and a fairly restricted feasible region.

1.10 Premature Fathoming (ε-Optimality)

By 'premature' fathoming, we mean discarding a node k which does not satisfy (1.3) and may therefore contain an optimal solution. The resulting algorithm then becomes inexact in that an optimal solution is not guaranteed on termination. The motivation is, of course, to save a large amount of computing time and still find what is hopefully a very good, if not optimal, solution.

Premature fathoming can be achieved by changing the definition of the set of live nodes to be those which satisfy

$$\bar{z}_k \geq z_p. \tag{1.18}$$

If $z_p = \underline{z}$ then (1.18) reduces to (1.3).

One common procedure is to let $z_p = \underline{z} - \varepsilon$, where ε is specified by the user. This is sometimes called 'ε-optimality' and is one of the standard and desirable features of branch and bound algorithms. It guarantees that the 'optimal' solution found on termination will be within ε of the true optimum.

One would expect that computation time would decrease as ε increases, and that is generally the case. It is possible, however, for computation time to increase with ε. This can happen if a node containing a good solution is discarded early in the enumeration; this solution, if discovered, would have permitted accelerated fathoming later on.

Another method for premature fathoming is sometimes known as 'truncation', and causes certain 'undesirable' nodes to be fathomed although they do not satisfy any of the previously stated criteria for fathoming. Truncating rules are based on some indication that the node will either not contain a good solution or that exorbitant computations will be required to find one. Codes 1, 5, and 6 contain truncating options which are user controlled. These can be based on $\bar{x}_k$, on E_k, or possibly on the time or number of iterations spent in solving the subproblems at node k.

1.11 Special Ordered Sets

The concept of special ordered sets (SOS) was introduced by Beale and Tomlin (1970). Specially ordered sets of type one (SOS1) are sets of variables in which not more than one member is permitted to be nonzero. These represent commonly occurring constraints in practical problems. A special case of SOS1 is the *set partitioning problem* in which the nonzero variable in each set must be one (see Garfinkel and Nemhauser, 1972b for a survey of theory and applications of these problems). Special ordered sets of type two (SOS2) are sets of variables in which not more than two members are permitted to be nonzero and they must be consecutive. In other words if T

is the index set of an SOS2 then for some k, x_k, $x_{k+1} \geq 0$ and $x_j = 0$, $j \in T - \{k, k+1\}$. Interest in SOS2's is motivated by their use in piecewise linear approximation of nonlinear functions (see Beale, 1977).

For both SOS1's and SOS2's the standard partitioning strategy of Section 1.5 is inappropriate. For instance, consider the constraint $x_1 + \ldots + x_p = 1$ where all x_j are constrained to be integer. Suppose, in some sense, that it is equally probable that any of the p variables will equal one in a best completion (best solution in a descendent node). If a relaxed LP yields $0 < x_j < 1$ and the set is partitioned by the constraints $x_j = 0$ and $x_j = 1$, then the probability is only $1/p$ that the $x_j = 1$ branch leads to a best completion. If p is large, this is clearly undesirable. One obvious remedy is to partition instead with $x_1 + \ldots + x_t = 0$ or $x_{t+1} + \ldots + x_p = 0$ where t is approximately $p/2$. For SOS2's this should be slightly modified to $x_1 + \ldots x_{t-1} = 0$ or $x_{t+1} + \ldots + x_p = 0$.

Codes 1, 2, 3, 6, 7, 9, 10, and 11 contain explicit machinery for handling special ordered sets of both types.

1.12 Options

Some computer codes (see Land and Powell, 1977) contain reasonable options for controlling the enumeration.

One such option is a *stopping rule* for discontinuing the process prior to verification of optimality. A rule based on 'ε-optimality' is discussed in Section 1.10. Another option is to stop when a solution is found that is within a given percentage of the value of the relaxed LP at node zero. It may also be possible to stop after a prespecified number of integer solutions have been located on the tree. This option, although not as intuitively appealing as the others, is based on the supposition that the best solutions tend to be found early and that the latter stages of enumeration are mainly spent in verifying optimality. This has often been found to be the case in practice.

Finally, Codes 1, 2, 5, 6, 7, 8, and 9 allow for part of the tree to be initially specified. The computational advantages of this option, which is often called CASCADE, have been discussed in the context of implicit enumeration by Salkin (1970).

1.13 Computer Codes and Computational Experience

In this section we list the commercial branch and bound codes which are available. We also relate typical published computational experience with some of these codes. Table 1.1, from Land and Powell (1977), gives the presently operating commercial codes.

The first commercial branch and bound code was that of Beale and Small (1965). Computational experience is related in Shaw (1970). Codes of this era were capable of handling a maximum of a few dozen integer variables.

Table 1.1

Code number	Name	Organization	Computer	Date
1	Apex III	Control Data	Cyber 70/72, 73, 74, 76	1975
2	FMPS	Sperry-Univac	1100 Series	1976
3	Haverly-MIP	Haverly Systems	IBM 360 and 370 Series, Univac 90/30	1970
4	LP 400	ICL	System 4–70 and 4–50	1970
5	MIP	IBM	360 and 370 Series	1969
6	MIP/370	IBM	370 Series	1974
7	MPS	Honeywell	Series 60 (Level 66)	1973
8	Ophelie	SIA	CDC 6600	1972
9	Sciconic	Scicon	Univac 1100 Series	1976
10	Tempo	Burroughs	B7000 and B8000 Series	1975
11	XDLA	ICL	1900 Series	1970

In the early 1970's, codes became available which could handle larger problems. Published computational experience distinguished between two generic types of problems. Type one problems are fundamentally large LP's with a few integer variables, and type two problems are highly structured ILP's. Type one problems are more common in practice, and different strategies are generally used for the two types. Much of the published computational experience relates the specific strategies used for handling the different type problems and is too detailed to relate here. We will relate only a few data points for some of these codes, and refer the reader to the individual articles for greater detail or to Geoffrion and Marsten (1972) or Garfinkel and Nemhauser (1972a) for additional detail.

An early version of Code 8 was described by Roy, Benayoun, and Tergny (1970). Their rules were based to a large extent on penalties. They solve, for example, a problem with 1244 constraints and 3908 variables, 24 of which are integer, in about six minutes of CDC-6600 time. Code 5, based on similar concepts, is described by Benichou et al. (1971). A problem with 721 constraints and 1156 variables, 39 of which are integer, is solved in 18 minutes on the IBM 360/75. Tomlin (1970) also describes a similar algorithm and indicates that incorporation of the penalty (1.10) yields reduction in running times of up to 50 percent over an algorithm based on the penalties (1.7)–(1.9).

In the middle 1970's it was discovered that penalty-based codes did not perform well for large problems. This discovery led to experimentation with pseudo costs. It is stressed in Forrest, Hirst, and Tomlin (1974) and Gauthier and Ribière (1977) that the branching strategy is the key element of a branch and bound code and that pseudo costs are good tools for making these decisions. Both of these articles give detailed experience contrasting different strategies.

A good deal of experience is also reported in Breu and Burdet (1974) for binary MILP's.

References

Agin, N. (1966). Optimum Seeking with Branch-and-Bound. *Man. Sci.*, **13**, B176–185.
Balas, E. (1968). A Note on the Branch-and-Bound Principle. *Operations Research*, **16**, 442–445.
Beale, E. M. L. (1977). Branch and Bound Methods for Mathematical Programming Systems. Presented at the Conference on Discrete Optimization, Vancouver, B.C.
Beale, E. M. L. and R. E. Small (1965). Mixed Integer Programming by a Branch and Bound Technique. In W. A. Kalenich (Ed.), *Proc. IFIP Congress*, Vol. 2, Spartan Press. pp. 450–451.
Beale, E. M. L. and J. A. Tomlin (1970). Special Facilities in a General Mathematical Programming System for Nonconvex Problems Using Ordered Sets of Variables. In *Proc. Fifth Int. Conf. on Operational Research*. pp. 447–454.
Benichou, M. et al. (1971). Experiments in Mixed-Integer Linear Programming. *Mathematical Programming*, **1**, 76–94.
Breu, R. and C. A. Burdet (1974). Branch and Bound Experiments in Zero-One Programming. *Mathematical Programming Study*, **2**, 1–50.
Dakin, R. J. (1965). A Tree-Search Algorithm for Mixed Integer Programming Problems. *Computer Journal*, **8**, 250–255.
Driebeek, N. J. (1966). An Algorithm for the Solution of Mixed Integer Programming Problems, *Management Science*, **12**, 576–587.
Forrest, J. J. H., J. P. H. Hirst, and J. A. Tomlin (1974). Practical Solution of Large Mixed Integer Programming Problems with UMPIRE. *Management Science*, **20**, 736–773.
Garfinkel, R. S. and G. L. Nemhauser (1972a). *Integer Programming*, Wiley, New York.
Garfinkel, R. S. and G. L. Nemhauser (1972b). Optimal Set Covering: A Survey. In A. M. Geoffrion (Ed.), *Perspectives in Optimization*, Addison-Wesley. pp. 164–183.
Gauthier, J. M. and F. Ribière (1977). Experiments in Mixed-Integer Linear Programming Using Pseudo-Costs. *Math. Prog*, **12**, 26–47.
Geoffrion, A. M. (1974). Lagrangian Relaxation for Integer Programming. *Mathematical Programming Study*, **2**, 82–114.
Geoffrion, A. M. and R. E. Marsten (1972). Integer Programming Algorithms: A Framework and State-of-the-Art Survey. *Management Science*, **18**, 465–491.
Gomory, R. E. (1958). Outline of an Algorithm for Integer Solutions to Linear Programming. *Bull. Amer. Math. Soc.*, **64**, 175–178.
Gomory, R. E. (1965). On the Relation between Integer and Non-integer Solutions to Linear Programs. *Proc. Nat. Acad. Science*, **53**, 260–265.
Hillier, F. S. (1969). Efficient Heuristic Procedures for Integer Linear Programming with an Interior. *Operations Research*, **17**, 600–637.
Jeroslow, R. (1977). An Introduction to the Theory of Cutting Planes. Presented at the Summer School in Combinatorial Optimization, Urbino, Italy, June 1977.
Land, A. and A. G. Doig (1960). An Automatic Method of Solving Discrete Programming Problems. *Econometrica*, **28**, 497–520.
Land, A. and S. Powell (1977). Computer Codes for Problems of Integer Programming. Presented at the Conference on Discrete Optimization, Vancouver, B.C., 1977.

Lawler, E. L. and D. E. Wood (1966). Branch-and-Bound Methods: A survey. *Operations Research*, **14**, 699–719.

Mitten, L. G. (1970). Branch-and-Bound Methods: General Formulation and Properties. *Operations Research*, **18**, 24–34.

Müller-Merbach, H. (1970). Approximation Methods for Integer Programming, Johannes Gutenberg University, Mainz, Germany.

Rardin, R. L. and B. W. Lin (1977). What Makes Integer Programming Problems Hard to Solve? Report #J-77-13, Industrial Engineering, Georgia Tech. May, 1977.

Roy, B., R. Benayoun, and J. Tergny (1970). From S. E. P. Procedure to the Mixed Ophelie Program. In J. Abadie (Ed.), *Integer and Nonlinear Programming*, American Elsevier. pp. 419–436.

Salkin, H. M. (1970). On the Merit of the Generalized Origin and Restarts in Implicit Enumeration. *Operations Research*, **18**, 549–554.

Senju, S. and Y. Toyoda (1968). An Approach to Linear Programming with 0–1 Variables. *Management Science*, **15**, B196–B207.

Shapiro, J. F. (1968). Group Theoretic Algorithms for the Integer Programming Problem—II: Extension to a general algorithm. *Operations Research*, **16**, 928–947.

Shapiro, J. F. (1971). Generalized Lagrange Multipliers in Integer Programming. *Operations Research*, **19**, 1070–1075.

Shaw, M. (1970). Review of Computational Experience in Solving Large Mixed Integer Programming Problems. In E. H. L. Beale (Ed.), *Applications of Mathematical Programming Techniques*, English Universities Press, London. pp. 406–412.

Tomlin, J. A. (1971). An Improved Branch-and-Bound Method for Integer Programming. *Operations Research*, **19**, 1363–1373.

CHAPTER 2

The Theory of Cutting-Planes

ROBERT JEROSLOW
Georgia Institute of Technology

PART 1
SUBADDITIVE METHODS

A cutting-plane is simply a linear inequality which is valid relative to some set of constraints.

For instance, in a linear program, suppose that the linear constraints

$$\begin{aligned} 2x_1 - x_2 + 3x_3 &\geq 7 \\ 2x_1 + x_2 + 5x_3 &\geq 3 \end{aligned} \quad (2.1)$$

both hold. Then one easily sees that

$$2x_1 + 0x_2 + 4x_3 \geq 5 \quad (2.2)$$

holds, since inequality (2.2) is half the sum of the inequalities in (2.1). In our terminology, (2.2) is a cutting-plane, relative to the constraints (2.1).

Cutting-planes have been used in integer programming, nonlinear programming, and complementarity theory. Cutting-plane algorithms have been successfully employed on set partitioning problems, travelling salesman problems, and certain special structures for Benders' partitioning method.

Now consider the linear constraint in integers

$$2x_1 + 3x_2 + x_3 + 4x_4 = 3 \quad x_1, x_2, x_3, x_4 \text{ nonnegative and integer.} \quad (2.3)$$

We take both sides of (2.3) modulo 2 and obtain

$$x_2 + x_3 = 1 \text{ (modulo 2)} \quad (2.4)$$

and, since x_2 and x_3 are nonnegative integers, from (2.4) we deduce the cutting-plane

$$x_2 + x_3 \geq 1. \quad (2.5)$$

Can we obtain (2.5) from (2.3) in the manner that we obtained (2.2) from (2.1)? That is, does (2.5) derive from taking (an unrestricted in sign) multiplier θ on the equation

$$2x_1 + 3x_2 + x_3 + 4x_4 = 3 \tag{2.6}$$

plus nonnegative multipliers α_j on the nonnegativities

$$x_j \geq 0 \quad j = 1, 2, 3, 4? \tag{2.7}$$

We can even allow the right-hand side to be decreased (e.g. change the right-hand side of (2.2) from 5 to 4), since that still will give a valid inequality.

If we could obtain (2.5) from (2.3) in this way, we would have solutions to

$$\begin{aligned}
3\theta & & &\geq 1 \text{ (constant term)} \\
2\theta + \alpha_1 & & &= 0 \text{ (coeff. of } x_1) \\
3\theta & + \alpha_2 & &= 1 \text{ (coeff. of } x_2) \\
\theta & & + \alpha_3 &= 1 \text{ (coeff, of } x_3) \\
4\theta & & & + \alpha_4 = 0 \text{ (coeff. of } x_4) \\
\alpha_1, \alpha_2, \alpha_3, \alpha_4 & & &\geq 0.
\end{aligned} \tag{2.8}$$

The first line in (2.8) gives $\theta > 0$, then the second line plus $\alpha_1 \geq 0$ gives $\theta = \alpha_1 = 0$, and we have a contradiction.

We see that (2.5) does not arise from (2.3) by taking multiples of the inequalities (2.6), (2.7). The principles of modular arithmetic lead to new cuts, and use the integrality constraints of (2.3) in a basic way.

Note that the vector $(x_1, x_2, x_3, x_4) = (3/2, 0, 0, 0)$ satisfies (2.6) and (2.7), but not (2.5). We say that this vector is 'cut away' by (2.5), hence the name 'cutting-plane'. When cutting-planes are used in algorithms, they are designed to cut away some specific 'current solution' to a linear programming relaxation of an integer program. But for purposes of theory, we use 'cutting-plane' synonymously with 'valid implied linear inequality', whether or not some point is cut away by the inequality. For instance, (2.2) does not cut away any point that satisfies (2.1).

We now have two ways of obtaining cutting-planes for linear constraints on integer variables: taking linear combinations (nonnegative for inequalities, unrestricted in sign for equalities) of the constraints, and using modular arithmetic. A surprising result, due in one form to Chvátal (1973) and in another variant form to Gomory (1963), asserts that these two methods, starting from the linear relaxation, if combined and iterated (perhaps many times), are sufficient to obtain *all* the cutting-planes when an integer program has a bounded feasible region. Gomory's (1963) paper is the beginning of cutting-plane theory in integer programming. A generalization of the result is in Jeroslow (1975b).

The Theory of Cutting-Planes

The result of Chvátal and Gomory refers to *several* applications of modular arithmetic and linear combinations done in sequence, not simply one. For instance, the only solution to

$$2x_1 + 3x_2 = 4 \qquad x_1, x_2 \text{ nonnegative integers} \qquad (2.9)$$

is $(x_1, x_2) = (2, 0)$, hence the inequality

$$-x_2 \geq 0 \qquad (2.10)$$

is valid. Yet (2.10) cannot be obtained by using modular arithmetic followed by linear combinations only once.

To see this, note that the result of taking (2.9) to base m ($m \geq 1$, integer) are the inequalities

$$2 \,(\text{mod } m) x_1 + 3(\text{mod } m) x_2 \geq 4 \,(\text{mod } m) \qquad (2.11)$$

which are

$$2x_1 + 3x_2 \geq 4 \qquad \text{for } m \geq 5 \qquad (2.12\text{a})$$
$$2x_1 + 3x_2 \geq 0 \qquad \text{for } m = 4 \qquad (2.12\text{b})$$
$$2x_1 + 0x_2 \geq 1 \qquad \text{for } m = 3 \qquad (2.12\text{c})$$
$$0x_1 + 1x_2 \geq 0 \qquad \text{for } m = 2 \qquad (2.12\text{d})$$
$$0x_1 + 0x_2 \geq 0 \qquad \text{for } m = 1 \qquad (2.12\text{e})$$

while the remaining linear constraints are

$$2x_1 + 3x_2 = 4 \qquad (2.13\text{a})$$
$$x_1, x_2 \geq 0. \qquad (2.13\text{b})$$

No linear combination (with nonnegative multipliers for inequalities) of (2.12) and (2.13) gives (2.10). Indeed $(x_1, x_2) = (1/2, 1)$ solves (2.12) and (2.13) but not (2.10), and any solution to (2.12) and (2.13) also solves a linear combination of (2.12) and (2.13). (We could have used this technique in place of examining the linear system (2.8) earlier, with $(x_1, x_2, x_3, x_4) = (3/2, 0, 0, 0)$.)

However, as asserted by the result of Chvátal and Gomory, (2.10) can be obtained from (2.9) by several uses of modular arithmetic and linear combinations. To be specific, one use of modular arithmetic gives (2.12c), and with an additional integer variable $y_1 \geq 0$, (2.12c) becomes

$$2x_1 - y_1 = 1. \qquad (2.14)$$

Taking both sides of (2.14) modulo two, we obtain

$$y_1 \geq 1. \qquad (2.15)$$

Adding (2.14) and (2.15) gives

$$2x_1 \geq 2 \qquad (2.16)$$

which can be added to (2.9) multiplied by (−1) to obtain

$$-3x_2 \geq -2. \tag{2.17}$$

Introducing an additional integer variable $y_2 \geq 0$, (2.17) becomes

$$-3x_2 - y_2 = -2. \tag{2.18}$$

Taking the modulus as 3, (2.18) gives

$$2y_2 \geq 1. \tag{2.19}$$

With an integer variable $y_3 \geq 0$, (2.19) becomes

$$2y_2 - y_3 = 1. \tag{2.20}$$

Taking the modulus as 2, (2.20) gives

$$y_3 \geq 1. \tag{2.21}$$

Adding (2.20) and (2.21), we find

$$2y_2 \geq 2 \quad \text{i.e.} \quad y_2 \geq 1. \tag{2.22}$$

Adding (2.18) and (2.22), we have

$$-3x_2 \geq -1 \tag{2.23}$$

and using an additional integer variable $y_4 \geq 0$, we have

$$-3x_2 - y_4 = -1. \tag{2.24}$$

With the modulus of 3, we obtain

$$2y_4 \geq 2 \quad \text{i.e.} \quad y_4 \geq 1. \tag{2.25}$$

Adding (2.24) and (2.25), we have

$$-3x_2 \geq 0 \tag{2.26}$$

and (2.26) gives (2.10).

There is a method to the manipulations above: see Exercise 7 in Appendix 2.1. Exercises 7 and 8 together also serve to relate our use of modular arithmetic here to the operation studied in Chvátal (1973). Essentially (for rational coefficients), Chvátal's operation can be viewed as using repeated modular arithmetic operations implicitly, avoiding the need for explicit additional variables. A different way of introducing additional variables is shown in Gomory (1963).

Instead of allowing the repeated use of some principles of cut construction (as in the result of Chvátal and Gomory), let us consider only a one-time use of a single principle. What is a particularly strong principle in this context? Clearly we must expect such a principle to be significantly more complex than modular arithmetic and linear combinations.

The Theory of Cutting-Planes

A very fruitful principle for 'one-step' cut construction was isolated by Gomory and Johnson (1972), in their studies which depart from and abstract results of Gomory's earlier work on the group problem (Gomory, 1969). This principle has to do with subadditive functions.

A *subadditive function* F is a real-valued function which obeys the restriction

(SUB) $$F(v+w) \leq F(v) + F(w)$$

for all v, w in its domain. Its domain is always a *monoid* M, by which we mean an additive semi-group: (i) $0 \in M$; (ii) if $v, w \in M$ then $v + w \in M$.

We will write the constraints of an integer program as

(IP) $$Ax = b, \ x \geq 0, \text{ integer.}$$

We always assume that A and b are rational in discussions of (IP), and write $A = [a^{(j)}]$ (columns) with $x = (x_1, \ldots, x_r)$. The monoid M for the subadditive functions to be discussed consists of all nonnegative integer combinations of the columns of A.

We next establish the sense in which subadditive functions provide cutting-planes.

Principle IP (Burdet and Johnson, 1975; Gomory and Johnson, 1972; Jeroslow 1974b) If F is a subadditive function on the monoid $M = \{v \mid v = Ax$ for some integer vector $x \geq 0\}$, and $F(0) = 0$, then the inequality

(IPC) $$\sum_{j=1}^{r} F(a^{(j)}) x_j \geq F(b)$$

is valid.

Proof We prove by induction on the sum $\sigma = \sum_{j=1}^{r} x_j$, that for $x \geq 0$ a vector of integers we have

$$\sum_{j=1}^{r} F(a^{(j)}) x_j \geq F(Ax). \qquad (2.27)$$

This will establish (IPC), since $Ax = b$ for solutions to (IP).

For $\sigma = 0$ (2.27) becomes $0 \geq F(0) = 0$, which is true. To go from σ to $\sigma + 1$, let $\sigma + 1 = \sum_{j=1}^{r} x_j$ and, without loss of generality, $x_1 \geq 1$. Using first the inductive hypothesis, and then the subadditivity of F, we have

$$\sum_{j=1}^{r} F(a^{(j)}) x_j = F(a^{(1)}) + F(a^{(1)})(x_1 - 1) + \sum_{j=2}^{r} F(a^{(j)}) x_j$$

$$\geq F(a^{(1)}) + F(a^{(1)}(x_1 - 1) + \sum_{j=2}^{r} a^{(j)} x_j)$$

$$\geq F(a^{(1)} x_1 + \sum_{j=2}^{r} a^{(j)} x_j) = F(Ax). \qquad (2.28)$$

To illustrate Principle IP, we shall derive (2.10) from (2.9) using (IPC) and a suitable subadditive function F. We chose a function F defined on the nonnegative integers b by:

$$F(b) = \begin{cases} -b/3 & \text{if } b \text{ is divisible by 3} \\ -(b-2)/3 & \text{if } (b-2) \text{ is divisible by 3} \\ -(b-4)/3 & \text{if } (b-1) \text{ is divisible by 3}. \end{cases} \quad (2.29)$$

It is easy to check that $F(0)=0$, $F(2)=0$, $F(3)=-1$, and $F(4)=0$, so that (IPC) for this F is indeed (2.10). It remains to prove that F is subadditive. Probably the most efficient way of doing that is by working Exercise 9 in Appendix 2.1.

Linear functions are subadditive (see Exercise 1a) as are functions of modular arithmetic (see Exercise 1e), so Principle IP subsumes our earlier methods for obtaining cutting-planes. It subsumes these methods strictly, as we saw, for the subadditive function F of (2.29) is neither linear nor modular. What is of particular theoretical interest is that Principle IP is, in a certain sense, a 'most general' principle.

To develop this point, let the inequality

(CP) $$\sum_{j=1}^{r} \pi_j x_j \geq \pi_0 \quad x = (x_1, \ldots, x_r)$$

be a cutting plane relative to some set of constraints. Then we say that

(CP)' $$\sum_{j=1}^{r} \pi'_j x_j \geq \pi'_0$$

is a *weakening* of (CP), or that (CP) is a *strengthening* of (CP)', if the following conditions hold: (1) $\pi'_j \geq \pi_j$ for $j = 1, \ldots, r$; (2) $\pi'_0 \leq \pi_0$. Since $x \geq 0$ shall always be required for all constraint sets considered in this chapter, any weakening (CP)' of a valid cut (CP) will also be valid, as $\sum_{j=1}^{r} (\pi'_j - \pi_j) x_j \geq 0$ for $x \geq 0$.

For example, we allow a cut

$$2x_1 - 5x_2 + x_3 \geq 7 \quad (2.30)$$

in nonnegative variables to be weakened to

$$5x_1 - 3x_2 + x_3 \geq 5 \quad (2.31)$$

in which all coefficients are increased or left the same and the constant term is decreased.

We see that a cut already 'contains' all of its weakenings. Certainly, if a cut is added to a linear program in nonnegative variables, the addition of any weakening will not change the optimal value (nor the optimal solution, if it is unique). In this sense, Principle IP is a 'most general' principle, due to the following result.

The Theory of Cutting-Planes

Converse to Principle IP (*Jeroslow 1974b and remarks below*) *If* (IP) *is consistent and the matrix A has only rational quantities, then all valid cuts* (CP) *for* (IP) *are weakenings of a cut* (IPC) *for suitable subadditive function F on* $M = \{v \mid v = Ax \text{ for some integer vector } x \geq 0\}$ *with* $F(0) = 0$.

A rigorous proof for this Converse is not given here, but the idea of the proof is easy to describe and is interesting in its own right. Given (CP), we put $\pi = (\pi_1, \ldots, \pi_r)$ and define F by

$$F(v) = \min\{\pi x \mid Ax = v; x \geq 0, \text{integer}\} \tag{2.32}$$

for $v \in M$, M as described in Principle IP. Incidentally, functions like F in (2.32) are called *value functions*, since they give the optimal value of an integer program as a function of its right-handside v.

Next, with work one shows that $F(0) = 0$ and that $F(v)$ is finite (i.e. not $-\infty$) for all $v \in M$. Clearly, $a^{(j)} \in M$ for $j = 1, \ldots, r$ (use unit vectors) and the fact that (IP) is consistent puts $b \in M$, so that F is defined for all quantities needed in (IPC). Again, unit vectors give $F(a^{(j)}) \leq \pi_j$ for $j = 1, \ldots, r$. Also, the validity of (CP) can be restated as the fact that $\min\{\pi x \mid Ax = b \text{ and } x \geq 0 \text{ is integer}\}$ is at least π_0, i.e. $F(b) \geq \pi_0$. Thus, (CP) is a weakening of (IPC). To see that F is subadditive, use Exercise 9 in Appendix 2.1.

In Principle IP, clearly it would not matter if F were defined on a larger monoid than M. In the Converse to Principle IP, Blair (1975a) has shown that the function F can be defined on R^m, the space of the rows of A in (IP) (we take A to be an m by r matrix). In fact, Blair showed that the function F of (2.32) can be extended to R^m, and his result has since been refined in Blair and Jeroslow (1975b). A related extension-type result, for the mixed-integer group problem, is in Johnson (1974b, Section 7).

Because of Principle IP and its converse, and Blair's result, we know that the problem of obtaining cutting-planes (CP) for the integer program (IP) is 'equivalent' to the problem of constructing all (or, at least, large classes of) subadditive functions on R^m. But this second problem is not so easy, since there are many subadditive functions, and often the ones needed are not simple to compute. For instance, the computation of the function F of (2.32) requires us to solve an integer program parametrically in its right-hand side v, in order to get one cutting-plane (CP) to help solve (IP) for its only right-hand side b.

The main technique for generating subadditive functions to date is to compute extreme points of certain systems of linear inequalities which are intimately connected with these functions. The subadditive functions generated in this way are typically defined on monoids (or groups), and can be extended to all of R^m by different extension techniques (see, for example,

Johnson, 1974b, Section 7). Thus, a subadditive function developed for one monoid, from an (easy) integer program, can often be extended to give a cut for another monoid, from a 'harder' integer program. Many subadditive functions are tabulated in Gomory (1969), Gomory and Johnson (1972), and Johnson (1974b).

This technique of subadditive function generation, via linear inequality systems, is based on Gomory's (1969) work, particularly his Theorem 18, and is extended in Jeroslow (1976) and Johnson (1974a, b).

One aspect of this technique is that the linear inequality systems are often quite large. They can be made smaller, at the cost of using a problem relaxation of (IP) (such as a homomorphic image of a group problem) in place of (IP). Since the subadditive functions which generate cuts for a relaxation of (IP) are often a proper subclass of those for (IP), one loses cutting-planes in return for a more 'tractable' linear inequality system. This remains true even if one obtains facets of the relaxation. The trade-off between the degree of relaxation and the 'strength' of the cutting-planes obtained is not very well understood. In addition to this issue, one needs means for knowing which subadditive function to use at a specific point of computation of some integer programming algorithm. It does not seem realistic to use a 'shotgun' approach, and add on all cuts from the relevant subadditive functions which have been tabulated. Surely different choices of specific cuts will lead to different computational results. Experimentation on the choice of subadditive function to use is given in Johnson and Spielberg (1971); much work remains to be done.

Some newer techniques for generating subadditive functions are given in the papers of Burdet and Johnson (1975) and Blair and Jeroslow (1975). Both techniques yield solutions to integer programs and involve enumerations to build up the functions. Subadditive functions give information one often does not get by other means.

We pointed out in Jeroslow (1974a) that, without cutting-planes, the integer program

$$\min (x_1 + \cdots + x_r)$$
s.t. $\quad 2x_1 + \cdots + 2x_r = r \qquad x_1, \ldots, x_r \in \{0, 1\}$ \hfill (2.33)

requires at least $2^{(r-1)/2}$ nodes to be developed in a branch-and-bound tree, for r an odd integer, before the inconsistency of (2.33) is detected (the left-hand side is even, and the right-hand side is odd).

To see this, note that, at any node in which the variables x_j, $j \in S$ ($S \subseteq \{1, \ldots, r\}$) are set at zero–one values

$$x_j = x_j^0 \qquad j \in S \tag{2.34}$$

with the other variables x_j, $j \notin S$, free, the linear program is consistent if S

The Theory of Cutting-Planes

contains up to (and including) $(r-1)/2$ elements. Indeed, we can solve

$$\sum_{j \notin S} 2x_j = r - \sum_{j \in S} 2x_j^0 \qquad 0 \leq x_j \leq 1, j \in S \tag{2.35}$$

by setting

$$x_j = \left(r - \sum_{j \in S} 2x_j^0\right)/2|S^c| \qquad j \in S \tag{2.36}$$

where $|S^c|$ is the size of the complement S^c of S. The setting (2.36) solves (2.35) since

$$r - \sum_{j \in S} 2x_j^0 \geq r - 2|S| \geq 1 \text{ (using } |S| \leq (r-1)/2) \tag{2.37a}$$

$$r - \sum_{j \in S} 2x_j^0 \leq r \tag{2.37b}$$

and from (2.36) and (2.37) one easily shows that $0 \leq x_j \leq 1$ for $j \notin S$ (using $|S^c| \geq (r+1)/2$).

This means that, in a binary tree-search algorithm, no node can be fathomed if its distance from the root node is $\leq (r-1)/2$, hence there are at least $2^{(r-1)/2}$ nodes in the tree Galil (1974). For $r = 151$, a branch and bound method for (2.33) requires at least 2^{75} nodes, unless the code employs cutting-planes, and on modern computers this means several hundred million years for solution.

In contrast, if $F(b)$ is b modulo 2, upon using (IPC) on the contraints of (2.33), with all $a^{(j)} = 2$ and $b = r$, we obtain

$$0x_1 + \cdots + 0x_r \geq 1 \tag{2.38}$$

for r odd. The addition of the cut (2.38) immediately reveals the inconsistency.

In Appendix 2.2 we give an outline of our extensions in Jeroslow (1976) of Gomory's Theorem 18 (Gomory 1969). The original paper (Gomory 1969) remains recommended reading for those interested in the subadditive approach to cutting-planes.

In Appendix 2.3 we give a characterization, of a subadditive type, for the cutting-planes of a linear constraint set with, in addition to integer variables, both continuous and binary or bivalent (i.e. zero–one) variables. Many, if not most, integer programs in industrial practice are mixed-integer programs in which there are many more continuous variables than integer ones, and the bulk (if not all) of the integer variables are binary. One can accommodate a binary variable z by using a row $z + z' = 1$ in integer variables z and z'. However, this means that a subadditive cut must be based on many constraint rows before one expects it to be effective in general.

Blair pointed out, in a private communication, that there is no class of functions F which both always provide valid cuts (IPC) and gave *all* the valid cuts, for a linear constraint system with binary variables.

For example, given the constraint system

$$2x_1 + 3x_2 + 5x_3 = 8 \qquad x_1, x_2, x_3 \in \{0, 1\} \tag{2.39}$$

the only feasible solution is $(x_1, x_2, x_3) = (0, 1, 1)$, and hence we have the cut

$$-x_1 + x_2 + x_3 \geq 2. \tag{2.40}$$

Presuming that (2.40) is a weakening of a cut (IPC) for some function F in a class of functions that provide valid cuts we have

$$F(2) \leq -1 \quad F(3) \leq 1 \quad F(5) \leq 1 \quad F(8) \geq 2. \tag{2.41a}$$

Actually we have

$$F(5) = 1 \tag{2.41b}$$

for if $F(5) = \alpha < 1$, then

$$-x_1 + x_2 + \alpha x_3 \geq 2 \tag{2.42}$$

would be a valid cut, which it is not (it cuts off $(0, 1, 1)$).

Now consider the constraint system

$$2x_1 + 3x_2 = 5 \qquad x_1, x_2 \in \{0, 1\}. \tag{2.43}$$

Since F provides valid cuts (IPC), using (2.41) we find that

$$-x_1 + x_2 \geq 1 \tag{2.44}$$

is a cut. But this is absurd, since $(x_1, x_2) = (1, 1)$ solves (2.43) but not (2.44). This establishes Blair's point.

Our characterization in Appendix 2.3 uses different functions for the columns of different binary variables, thus avoiding the difficulty with the same function F acting on all columns in (IPC).

For those reader familiar with subadditivity as it was developed for the group problems, we make some remarks concerning the relationship of those developments to the subadditive approach to (IP), as discussed here and in the appendices.

Applying subadditivity directly to (IP) is fairly recent. E. L. Johnson informs us that Araoz (1973) was the first to do so in the context of certain combinatorial programs, under the guidance of Edmonds. Other papers dating from around that time are Jeroslow (1974b) and Johnson (1974), while further work is in Jeroslow (1976) and other papers.

To obtain generalizations of subadditive results from group problems to (IP), mathematical techniques are needed for handling subadditivity in monoids (i.e. semi-groups) rather than groups. There are cases in which a

The Theory of Cutting-Planes

group problem result is easily modified for monoids, but each result must be examined separately to see if this is so. For instance, whenever a result on a group problem has a proof that uses subtraction—and many do—the proof will not generalize to monoids, which are not usually closed under subtraction (as, for example, the nonnegative integers are not closed under subtraction). It then becomes an open problem as to whether the group result is true for monoids, and in what form.

The issue of the *form* of the generalization is quite important, since often the proof, that a generalization does indeed specialize to a known result for a group problem, is not trivial. Very different concepts were developed for group problems (see, for example, the definition of a subadditive function in Johnson, 1974), and it requires a demonstration to show generalized concepts to be equivalent to these in specific contexts.

In addition, there are further difficulties in treating the infinite monoids that arise in connection with a general (IP), as opposed to the finite monoids that are adequate to treat integer covering and packing problems. This matter is touched on in Appendix 2.2 in connection with the representation (2.68).

In some instances, a group result was first found through a proof of its generalization. For instance, our extension result (Blair and Jeroslow, 1975, Theorem 4.6) is easily shown to yield an extension result for subadditive functions on the rational mixed-integer group problem, although previously the group extension result was known only with special hypotheses. Also, the subadditive dual problems were first found in a general setting (see, for example, Jeroslow, 1976).

As regards practical considerations, Araoz (1973) has shown that some integer programs generate semi-group problems which are much *smaller* than their group problem. This occurs even though the semi-group problem exactly represents the integer program, while the group problem is a relaxation of it. Araoz' example serves to emphasize the need for a better understanding of the problem relaxations which are available.

Appendix 2.1 Exercises with Solutions

Exercises

Exercise 1 Show that the following functions are subadditive on their domains:

(a) $F(v) = \lambda v$ for $\lambda \in R^m$ any vector.
(b) $F(v) = \max\{aF_1(v), bF_2(v)\}$ where $a, b \geq 0$ and F_1, F_2 are subadditive on a common domain.
(c) $F(v) = aF_1(v) + bF_2(v)$ where $a, b \geq 0$ and F_1, F_2 are subadditive on a common domain.

(d) $F(v) = \inf_{p \in P} \{G(p) + H(v-p)\}$, where P is a monoid, G a subadditive function defined on P, $v \in K$. K is a group with $P \subseteq K$, and H is subadditive on K. You may also suppose that $F(v)$ is finite for all $v \in K$.

(e) $F_r(v)$, where $r > 0$ is a real number and $F_r(v)$ is defined by the condition (for $v \in R$):

$$F_r(v) = v - nr \text{ provided that } nr \leq v < (n+1)r \text{ and } n \text{ is an integer.} \tag{2.45}$$

(f) $F(v) = \min\{G(v), c\}$ where $c \geq 0$ is a constant and G is a nonnegative subadditive function.

Exercise 2 If F_1 and F_2 are subadditive on a common domain, is $F(v) = \min\{F_1(v), F_2(v)\}$ necessarily subadditive on this domain?

Exercise 3 Suppose that H is subadditive on its domain and that G is subadditive and monotone on the reals R. Show that $F(v) = G(H(v))$ is subadditive on the domain of H. Is F still necessarily subadditive if G is not monotone? Is F still necessarily subadditive if G is not monotone but $H(v) = \lambda v$?

Exercise 4 (Gomory's fractional row cut (Gomory, 1963).) Suppose that x, t_1, t_2, t_3 are nonnegative integer variables constrained by the relation

$$x = 1.7 + 2.3(-t_1) - 0.3(-t_2) + 2(-t_3). \tag{2.46}$$

Show that the cutting-plane

$$0.3t_1 + 0.7t_2 + 0t_3 \geq 0.7 \tag{2.47}$$

is valid. (*Hint:* Use Exercise 1e.)

Exercise 5 (Compare with Dantzig, 1959.) In (2.46) we see that not both variables t_1, t_2 can be at zero in a solution, hence the cutting-plane

$$t_1 + t_2 \geq 1 \tag{2.48}$$

is valid. What subadditive function gives (2.48)? (*Hint:* Use Exercise 3, with H as in the cut for (2.47) and with G suitably chosen via reference to Exercise 1f.) Note that, after multiplying (2.48) by 0.7, (2.48) is a weakening of (2.47).

Exercise 6 Clearly the constraints

$$2x_1 + 5x_2 + 7x_3 + 11x_4 \geq 15 \quad x_1, x_2, x_3, x_4 \text{ all zero or one} \tag{2.49}$$

make it impossible that $x_3 = x_4 = 0$ (since $2 + 5 < 15$), hence the cut

$$x_3 + x_4 \geq 1 \tag{2.50}$$

is valid. Can (2.50) be obtained from a subadditive function F for general integer variables with $F(-1) \leq 0$ (to treat a slack variable for (2.49)),

The Theory of Cutting-Planes

possibly after some linear transformations on certain of the binary variables?

Exercise 7 Suppose that a and b are positive integers and z is an integer variable, and

$$az \geq b \qquad (2.51)$$

holds. Show how one may derive

$$z \geq 1 \qquad (2.52)$$

by repeated use of modular arithmetic plus linear combinations (allowing for a decrease in the right-hand side of a derived inequality). You may use additional integer variables.

Exercise 8 Suppose that $a_0, a_1, \ldots, a_r$ are integer, f is a rational fraction, $0 < f < 1$, and $x_1, \ldots, x_r$ are integer variables. Starting from

$$a_1 x_1 + \cdots + a_r x_r \geq a_0 + f \qquad (2.53)$$

show how to derive

$$a_1 x_1 + \cdots + a_r x_r \geq a_0 + 1 \qquad (2.54)$$

(this is Chvátal's (1973) operation) by repeated use of modular arithmetic plus linear combinations (allowing for a decrease in constant terms). You may use additonal variables and Exercise 7.

Exercise 9 Show that the function F of equation 2.32 is subadditive. You may assume that $F(v)$ is finite for all $v \in M$.

(In regard to the function F of equation (2.29), note that

$$F(b) = \min\{-x_2 \mid 2x_1 + 3x_2 = b, x_1 \text{ and } x_2 \text{ nonnegative integers}\}.$$

Hence this exercise proves F is subadditive.)

Solutions

Exercise 1b We have, for v_1, v_2 arbitrary,

$$aF_1(v_1 + v_2) \leq aF_1(v_1) + aF_1(v_2) \leq F(v_1) + F(v_2)$$
$$bF_2(v_1 + v_2) \leq bF_2(v_1) + bF_2(b_2) \leq F(v_1) + F(v_2).$$

Taking the maximum on the left in both inequalities, we have the desired

$$F(v_1 + v_2) \leq F(v_1) + F(v_2).$$

Exercise 1d Let v_1, v_2 be arbitrary and let $\varepsilon > 0$. Pick $p_1, p_2 \in P$ such that

$$F(v_1) + \varepsilon/2 \geq G(p_1) + H(v_1 - p_1)$$
$$F(v_2) + \varepsilon/2 \geq G(p_2) + H(v_2 - p_2).$$

Then we have

$$F(v_1+v_2) \leq G(p_1+p_2) + H((v_1+v_2)-(p_1+p_2))$$
$$= G(p_1+p_2) + H((v_1-p_1)+(v_2-p_2))$$
$$\leq G(p_1) + G(p_2) + H(v_1-p_1) + H(v_2-p_2)$$
$$\leq F(v_1) + F(v_2) + \varepsilon.$$

Since $\varepsilon > 0$ was arbitrary, $F(v_1+v_2) \leq F(v_1) + F(v_2)$; this is the desired result.

Exercise 1f Let v_1, v_2 be arbitrary. If $G(v_1) \geq c$ then since G (hence F) is nonnegative, we have

$$F(v_1+v_2) \leq c \leq c + F(v_2)$$
$$= F(v_1) + F(v_2).$$

Similarly, we again have $F(v_1+v_2) \leq F(v_1) + F(v_2)$ if $G(v_2) \geq c$. Now suppose that $G(v_1)$ and $G(v_2)$ are both $<c$. Then

$$F(v_1+v_2) \leq G(v_1) + G(v_2)$$
$$= F(v_1) + F(v_2)$$

and all cases have been considered.

Exercise 2 No. Take $F_1(v) = -v$, $F_2(v) = v$, with domain the reals. Then $F(v) = -|v|$, which is not subadditive.

Exercise 3 We have

$$F(v_1+v_2) = G(H(v_1+v_2))$$
$$\leq G(H(v_1)+H(v_2))$$
$$\leq G(H(v_1)) + G(H(v_2)) = F(v_1) + F(v_2).$$

In the above, the first $\leq$ is due to the fact that G is monotone and $H(v_1+v_2) \leq H(v_1) + H(v_2)$.

If G is not monotone, F need not be subadditive. For example, $H(v) = |v|$ is subadditive on the reals, as is $G(v) = -v$, but here $F(v) = -|v|$ is not subadditive.

If $H(v) = \lambda v$ then F is always subadditive, whether or not G is monotone. In the above inequalities, note that the first $\leq$ becomes $=$ since $H(v_1+v_2) = H(v_1) + H(v_2)$.

Exercise 4 By Exercise 1e, $F(v) = v$ modulo 1 is a subadditive function on the reals. Now (2.46) is equivalent to

$$x + 2.3t_1 - 0.3t_2 + 2t_3 = 1.7.$$

Applying F to this equation as in (IPC), one obtains (2.47).

The Theory of Cutting-Planes 35

Exercise 5 Let $H(v) = (v \text{ modulo } 1)/.3$ and let $G(v) = \min\{v, 1\}$. From Exercise 1f, G is subadditive for $v \geq 0$; it is clearly monotone. By Exercise 3, $F(v) = G(H(v)) = \min\{(v \text{ modulo } 1)/.3, 1\}$ is subadditive, and clearly $F(0) = 0$.

We have $F(1.7) = \min\{0.7/0.3, 1\} = 1$, $F(2.3) = \min\{0.3/0.3, 1\} = 1$, $F(-0.3) = \min\{0.7/0.3, 1\} = 1$, $F(2) = \min\{0/0.3, 1\} = 0$, $F(1) = 0$, and (IPC) is (2.48).

Exercise 6 Since x_1 and x_2 are binary variables, so are $x_1' = 1 - x_1$, and $x_2' = 1 - x_2$. Then (2.49) is equivalent to

$$-2x_1' - 5x_2' + 7x_3 + 11x_4 \geq 15 - 7 = 8 \tag{2.49'}$$

and we need only look for a suitable 'truncation', as in Exercise 1f, to turn 7, 11, and 8 into 1, while turning negative numbers (like -2 and -5) into zero.

We note that $G(v) = \max\{0, v\}$ is subadditive by Exercises 1a and 1b, and $G(-2) = G(-5) = 0$, while $G(7) = 7$, $G(11) = 11$, $G(8) = 8$. Then $F(v) = \min\{G(v), 1\}$ is the desired truncation, which gives (2.50) as (IPC).

Exercise 7 If $b \geq a$, (2.51) gives $z \geq b/a \geq 1$, hence (2.52) holds. We may assume then that $0 < b < a$, and use induction on $c = a - b$, with the case $c = 0$ already done.

By introducing an integer variable $y \geq 0$, (2.51) becomes

$$az - y = b. \tag{2.51'}$$

Using a modulus of a in (2.51'), we obtain

$$(a-1)y \geq b.$$

In the above, $c' = (a-1) - b = c - 1 < c$, hence by induction we may derive

$$y \geq 1$$

via repeated uses of modular arithmetic and linear combinations. Adding, we find

$$az \geq b + 1$$

and $c'' = a - (b+1) = c - 1 < c$, hence again the inductive hypothesis holds, and we obtain (2.52).

Exercise 8 Write $f = s/D$, s and D integers, $0 < s < D$. Then by taking multiples and adding an integer variable $y_1 \geq 0$, (2.53) gives

$$Da_1x_1 + \cdots + Da_rx_r - y_1 = Da_0 + s. \tag{2.55}$$

Using modulus D, from (2.55) we obtain the inequality

$$(D-1)y_1 \geq s. \tag{2.56}$$

Then Exercise 7 applies to (2.56) and we may obtain

$$y_1 \geq 1. \tag{2.57}$$

Adding (2.55) and (2.57), we find

$$Da_1 x_1 + \cdots + Da_r x_r \geq Da_0 + (s+1). \tag{2.58}$$

In this manner, we increase s to $(s+1)$, $(s+2)$ etc., until D is reached, and we then have (2.54).

Exercise 9 Let $v, w \in M$ be given. Let $\alpha > F(v)$ and $\beta > F(w)$ be arbitrary, and chose $x^{(1)}$ and $x^{(2)}$ to be vectors of nonnegative integers with $Ax^{(1)} = v$, $\pi x^{(1)} \leq \alpha$ and $Ax^{(2)} = w$, $\pi x^{(2)} \leq \beta$.

Then $x^{(1)} + x^{(2)}$ is a vector of nonnegative integers, $A(x^{(1)} + x^{(2)}) = v + w$, and hence

$$F(v+w) \leq \pi(x^{(1)} + x^{(2)}) \leq \alpha + \beta.$$

Since α resp. β can be chosen as close to $F(v)$ resp. $F(w)$ as desired, (SUB) follows.

Appendix 2.2 Linear Programming Formulations of Subadditivity

Suppose that we wish to obtain all the valid cuts for $S = \{x \geq 0 \mid Ax = b, x \text{ integer}\}$. Throughout this Appendix, we shall assume that A and b are rational, $S \neq \emptyset$, and at first we are willing to try a 'brute force' approach to this problem.

One direct way of characterizing these cuts is to enumerate S by $S = \{x^{(1)}, x^{(2)}, \ldots\}$ and to require that the cut-coefficients $\pi(a^{(j)})$ satisfy the (possibly infinite) linear inequality system

$$\sum_{j=1}^{r} \pi(a^{(j)}) x_j^{(k)} \geq \pi(b) \qquad k = 1, 2, \ldots. \tag{2.59}$$

Indeed, (2.59) simply requires that we obtain a valid cut (CP) if we set $\pi_j = \pi(a^{(j)})$ and $\pi_0 = \pi(b)$. The system (2.59) is one instance of what is termed the 'polar set' for S (see Rockafellar, 1970).

So far, writing (2.59) did not depend on the nature of S, but simply on the fact that $S = \{x^{(1)}, x^{(2)}, \ldots\}$. We now bring the integrality of the variables $x \geq 0$ to bear and simplify (2.59) somewhat.

To be specific, for each $x^{(k)} \in S$, we chose a 'path from 0 to $x^{(k)}$', by which is meant a sequence of vectors $w^{(0)} = 0$, $w^{(1)}, w^{(2)}, \ldots, w^{(q)} = x^{(k)}$ (the sequence and q depending on $x^{(k)}$), such that for each $i = 0, 1, \ldots, q-1$ there is some unit vector $e_j(i)$ (depending on i) with

$$w^{(i)} + e_{j(i)} = w^{(i+1)}. \tag{2.60}$$

The Theory of Cutting-Planes

For instance, if $x^{(k)} = (2, 1, 1)$, we can choose

$$w^{(0)} = (0, 0, 0) \tag{2.61a}$$
$$w^{(1)} = w^{(0)} + e_1 = (1, 0, 0) \tag{2.61b}$$
$$w^{(2)} = w^{(1)} + e_2 = (1, 1, 0) \tag{2.61c}$$
$$w^{(3)} = w^{(2)} + e_1 = (2, 1, 0) \tag{2.61d}$$
$$w^{(4)} = w^{(3)} + e_3 = (2, 1, 1) = x^{(k)} \tag{2.61e}$$

with $q = 4 = 2 + 1 + 1$. Having chosen such a path, we propose to replace

$$\sum_{j=1}^{r} \pi(a^{(j)}) x_j^{(k)} \geq \pi(b) \tag{2.62$_k$}$$

which occurs in (2.59), by the inequalities

$$\pi(Aw^{(i+1)}) \leq \pi(Aw^{(i)}) + \pi(a^{(j(i))}) \quad i = 1, \ldots, q-1 \tag{2.63$_k$}$$

where (2.60) is assumed for each $i = 1, \ldots, q-1$. That is to say, we enlarge the set of variables in (2.62$_k$), which are only $\pi(a^{(1)}), \ldots, \pi(a^{(r)}), \pi(b)$, to include a variable $\pi(Aw^{(i)})$ for each $i = 1, \ldots, q$, and then we replace (2.62$_k$) by the system of inequalities (2.63$_k$).

Any solution to (2.63$_k$) gives a solution to (2.62$_k$) when only the variables $\pi(a^{(1)}), \ldots, \pi(a^{(r)}), \pi(b)$ are considered. In fact, we leave it to the reader to prove by induction on i that (2.63$_k$) implies

$$\pi(Aw^{(i)}) \leq \sum_{j=1}^{r} \pi(a^{(j)}) w_j^{(i)} \quad i = 1, \ldots, q. \tag{2.64}$$

Then from (2.64), by setting $i = q$ and noting that $w^q = x^{(k)}$, $Aw^{(q)} = Ax^{(k)} = b$, we have (2.62$_k$).

Conversely, any solution to (2.62$_k$) can be expanded to a solution to (2.63$_k$), by setting for $i = 1, \ldots q-1$

$$\pi(Aw^{(i)}) = \sum_{j=1}^{r} \pi(a^{(j)}) w_j^{(i)}. \tag{2.65}$$

Therefore the inequality (2.62$_k$) and the system (2.63$_k$) are equivalent.

Now we examine (2.63$_k$) from a different perspective. It has the form of a subadditivity relation

$$\pi(v + v') \leq \pi(v) + \pi(v') \tag{2.66}$$

for $v = Aw^{(i)}, \quad v' = a^{(j(i))}, \quad v + v' = A(w^{(i)} + e_{j(i)}) = Aw^{(i+1)}$.

This is, in elementary form, the *basic insight*, that simply writing down the subadditivity relations on a linear inequality system in variables $\pi(v)$ leads to a characterization of the valid cuts.

One can always write more subadditivity relations (2.66) than the 'bare minimum' required by (2.63_k), since (2.63_k) is enough to imply (2.62_k), and for the converse settings such as

$$\pi'(v) = \inf \left\{ \sum_{j=1}^{r} \pi(a^{(j)}) v_j \,\Big|\, v \geq 0 \text{ is integer and } v = Ax \right\} \quad (2.67)$$

ensure that all the tacked-on subadditivity relations (2.66) hold for π'. A formula like (2.67) is necessary here (in place of the simpler (2.65)) because of multiple paths from 0 to v. Checking that (2.67) works is technically involved, and we omit it here; the details are in Jeroslow (1976).

Our treatment thus far provides a finite linear inequality system characterizing the valid cuts for S finite, via systems (2.63_k) for $k = 1, \ldots, s$ with $S = \{x^{(1)}, \ldots, x^{(s)}\}$. For S infinite, the resulting linear inequality system will not be finite. To make it finite, further work is needed. It follows from a result of Hilbert (1890) that the rationality hypotheses on A, b imply the existence of finitely many integer vectors $x^{(1)}, \ldots, x^{(s)}$ and $d^{(1)}, \ldots, d^{(\sigma)}$, such that solutions $x \in S$ are precisely vectors of the form

$$x = x^{(k)} + \sum_{i=1}^{\sigma} n_i d^{(i)} \quad (2.68)$$

for some $k = 1, \ldots, s$ and some nonnegative integer $n_1, \ldots, n_\sigma$. Furthermore, $Ax^{(k)} = b$, $x^{(k)} \geq 0$ for $k = 1, \ldots, s$, $Ad^{(i)} = 0$, $d^{(i)} \geq 0$, for $i = 1, \ldots, \sigma$, and the $d^{(i)}$ have the property that, if $Au = 0$, $u \geq 0$ then

$$u = \sum_{i=1}^{\sigma} \lambda_i d^{(i)} \quad (2.69)$$

for suitable real scalars $\lambda_i \geq 0$. The proofs of these results are in Jeroslow (1975a).

Since x given by (2.68) is always a member of S, if (CP) is a valid cut then for any integer $n_i \geq 0$ we have

$$\pi_0 \leq \sum_{j=1}^{r} \pi_j x_j = \sum_{j=1}^{r} \pi_j x_j^{(k)} + n_i \sum_{j=1}^{r} \pi_j d_j^{(i)} \quad (2.70)$$

which implies that

$$0 \leq \sum_{j=1}^{r} \pi_j d_j^{(i)} \quad i = 1, \ldots, \sigma \quad (2.71_i)$$

since n_i can be made arbitrarily large.

This suggests the following way of obtaining a finite system from (2.59). We write systems (2.63_k) for only the 'fundamental solutions' $x^{(1)}, \ldots, x^{(s)}$ of (2.68) and, to take care of any remaining solutions, we write systems (2.71_i) for $i = 1, \ldots, \sigma$. This approach does in fact work; details are in

The Theory of Cutting-Planes

Jeroslow (1976). We summarize our discussion in the following result; this result is closely related to results in Gomory (1969), Gomory and Johnson (1972), and Johnson (1974b), and is one of the generalizations of Gomory's (1969) linear inequality formulation of subadditivity for the group problem.

Theorem Suppose that A, b are rational. Then there is a finite set F of pairs of vectors (v, w), with v and w both nonnegative integer sums of columns $a^{(i)}$ of A, and there are finitely many integer vectors $d^{(1)}, \ldots, d^{(\sigma)}$ such that the following holds.

For any valid cut (CP) for $S = \{x \geq 0 \mid Ax = b \text{ and } x \text{ integer}\}$, $S \neq \emptyset$, there is a solution to the finite linear system:

(FS)
$$\pi(0) = 0$$
$$\pi(v + w) \leq \pi(v) + \pi(w) \quad (v, w) \in F$$
$$0 \leq \sum_{j=1}^{r} \pi(a^{(j)}) d_j^{(i)} \quad i = 1, \ldots, \sigma$$

for which

$$\pi_j \geq \pi(a^{(j)}) \quad j = 1, \ldots, r$$
$$\pi_0 \leq \pi(b) \tag{2.72}$$

hold. Conversely, given any solution to (FS), *the inequality*

$$\sum_{j=1}^{r} \pi(a^{(j)}) x_j \geq \pi(b) \tag{2.73}$$

and all of its weakenings are valid cuts.

Furthermore, an arbitrary finite set of subadditivity relations (2.66) *can be added to* (FS), *and our conclusions remain unchanged. (Note: $\pi(a^{(i)})$ and $\pi(b)$ are variables of* (FS).*)*

The theorem can be strengthened in many directions, by making many of the subadditivity relations of (FS) into additivity relations

$$\pi(v + w) = \pi(v) + \pi(w). \tag{2.66'}$$

Some examples of (2.66') occur in the context of 'minimal inequalities'; see Jeroslow (1976) for details. Also, one can deduce various 'subadditive dual problems' from this theorem, as, for example, in Jeroslow (1976).

All the previous analysis depended only on the representation formula (2.68) for $x \in S$, which turns out to be a weaker hypothesis than requiring $S = \{x \geq 0 \mid Ax = b, x \text{ integer}\}$ for some rational A, b. In fact, a representation formula (2.68) holds for Gomory's group problem Gomory (1965, 1969), with the $d^{(i)}$ being unit vectors, so that the constraints $\sum_j \pi(a^{(j)}) d_j^i \geq 0$ for $i = 1, \ldots, \sigma$ simplify to the nonnegativities $\pi(a^{(j)}) \geq 0, j = 1, \ldots, r$. For a proof of the representation formula (2.68) for Gomory's group, see Jeroslow (1975a).

The system (FS) is generally too large for practical use unless both s and σ are not large in (2.68). If (2.68) is literally the solution form to an integer program, usually s is very large; so in typical applications (2.68) arises from a relaxation of an integer program.

One way of obtaining such relaxations from constraints $Ax = b$, $x \geq 0$ integer is via homomorphisms. These are mappings f of the monoid M of nonnegative integer combinations of the columns of $A = [a^{(j)}]$ into another abelian semi-group which satisfy

$$f(v + w) = f(v) + f(w) \quad \text{for all } v, w \in M. \tag{2.74}$$

The solution form (2.68) used then corresponds to the relaxed constraints

$$\sum_{j=1}^{r} f(a^{(j)}) x_j = f(b) \ x \geq 0, \text{ integer} \tag{2.75}$$

and various results ensure the existence of (2.68) in most cases. Gomory's group problem (Gomory, 1965) arises through such a homomorphism.

We need useful problem relaxations which yield reasonable-size problems (FS). So far, it is primarily from the systems (FS) that we have obtained cuts which are quite unrelated to branch and bound; for some other problem relaxations, see Bell and Shapiro (1975).

Appendix 2.3 A Cut Form for Integer Programs with Binary and Continuous Variables

We shall consider constraint sets on a vector (z, x, y) which have the form:

(BMIP)
$$Az + Bx + Cy = d$$
$$z_j \in \{0, 1\} \text{ for } j = 1, \ldots, r$$
$$x_k \geq 0 \text{ and integer for } h = 1, \ldots, s$$
$$y_p \geq 0 \text{ for } p = 1, \ldots, t.$$

In (BMIP), we have $z = (z_1, \ldots, z_r)$, $x = (x_1, \ldots, x_s)$, $y = (y_1, \ldots, y_t)$ and, as always, we assume that A, B, C, and d are rational. We use the notation of column representations $A = [a^{(j)}]$, $B = [b^{(k)}]$, $C = [c^{(p)}]$.

Just as subadditive functions are associated with cuts (IPC) for the integer program (IP), a certain generalization of subadditive functions provides a cut form for the mixed integer program with binary variables (BMIP). This generalization is a *subadditive collection*, which is a set of functions $\{F(S)(\cdot) \mid S \subseteq \{1, \ldots, r\}\}$, one function for each subset S of the index set $\{1, \ldots, r\}$ of the binary variables z.

The Theory of Cutting-Planes

The only stipulation on these functions is that

$$\text{if } S \supseteq S_1 \cup S_2, \ S_1 \cap S_2 = \emptyset,$$
$$v \text{ is in the domain of } F(S_1)(\cdot) \text{ and} \quad (2.76)$$
$$w \text{ is in the domain of } F(S_2)(\cdot), \text{ then}$$
$$F(S)(v+w) \leq F(S_1)(v) + F(S_2)(w).$$

There is also a stipulation on the domain $M(S)$ of these functions $F(S)(\cdot)$ as follows. First, $0 \in M(S)$ for all S, and:

$$\text{if } S \supseteq S_1 \cup S_2 \text{ and } S_1 \cap S_2 = \emptyset, \text{ then } M(S) \supseteq M(S_1) + M(S_2). \quad (2.77)$$

In (2.77), we use the usual notation for the sum of two sets A and B, specifically:

$$A + B = \{a + b \mid a \in A, b \in B\}. \quad (2.78)$$

For one example of a subadditive collection, let G be any subadditive function and put $F(S)(v) = G(v)$ for all v in the domain of G; the domain of each $F(S)(\cdot)$ is that of G (hence a monoid).

Also, every subadditive collection contains a subadditive function $F(\emptyset)(\cdot)$. In fact, applying (2.76) with $S = S_1 = S_2 = \emptyset$, $\emptyset$ the empty set, we have

$$F(\emptyset)(v+w) \leq F(\emptyset)(v) + F(\emptyset)(w) \quad (2.76')$$

which verifies (SUB). Similarly, (2.77) gives $0 \in M(\emptyset)$ and $M(\emptyset) \supseteq M(\emptyset) + M(\emptyset)$, so that $M(\emptyset)$ is indeed a monoid.

Given a subadditive function G, there is another way of constructing a subadditive collection $\{F(S)(\cdot) \mid S \subseteq \{1, \ldots, r\}\}$ from it, as our next result shows.

Lemma *Let G be a subadditive function defined on a monoid M, and let $d^{(j)}$ for $j = 1, \ldots, r$ be any elements of M.*

Put, for $S \subseteq \{1, \ldots, r\}$, and $v \in M$,

$$F(S)(v) = \min \left\{ G\left(v + \sum_{j \in S} d^{(j)} z_j\right) \, \middle| \, z_j \in \{0, 1\} \text{ for } j \in S \right\}. \quad (2.79)$$

Then $\{F(S)(\cdot) \mid S \subseteq \{1, \ldots, r\}\}$ is a subadditive collection with $M(S) = M$ for all $S \subseteq \{1, \ldots, r\}$.

Proof Clearly, $F(S)(v)$ is defined for all $v \in M$, and one easily verifies (2.77). It remains to verify (2.76).

Let $S \supseteq S_1 \cup S_2$, $S_1 \cap S_2 = \emptyset$. Let $v, w \in M$ be given. Choose $z^{(1)}$, $z^{(2)}$ to be binary r-vectors with

$$F(S_1)(v) = G\left(v + \sum_{j \in S_1} d^{(j)} z_j^{(1)}\right); \tag{2.80a}$$

$$F(S_2)(w) = G\left(w + \sum_{j \in S_2} d^{(2)} z_j^{(2)}\right); \tag{2.80b}$$

$$z_j^{(1)} = 0 \quad \text{for} \quad j \notin S_1; \tag{2.80c}$$

$$z_j^{(2)} = 0 \quad \text{for} \quad j \notin S_2. \tag{2.80d}$$

We also have

$$z_j^{(1)} + z_j^{(2)} = 0 \quad j \notin S \tag{2.81a}$$

$$z_j^{(1)} + z_j^{(2)} \in \{0, 1\}, \quad j \in S, \quad \text{as} \quad S_1 \cap S_2 = \emptyset. \tag{2.81b}$$

From (2.80), (2.81), and (2.79), we conclude

$$F(S)(v+w) \leq G\left((v+w) + \sum_{j \in S}(z_j^{(1)} + z_j^{(2)})d^{(j)}\right)$$

$$= G\left((v + \sum_{j \in S_1} z_j^{(1)} d^{(2)}) + (w + \sum_{j \in S_2} z_j^{(2)} d^{(2)})\right)$$

$$\leq G\left(v + \sum_{j \in S_2} z_j^{(1)} d^{(j)}\right) + G\left(w + \sum_{j \in S_2} z_j^{(2)} d^{(j)}\right)$$

$$= F(S_1)(v) + F(S_2)(w).$$

The lemma will provide us with a subadditive collection that is useful for an example of the following principle.

Principle BMIP If $\{F(S)(\cdot) \mid S \subseteq \{1, \ldots, r\}\}$ is a subadditive collection with $a^{(j)} \in M(\{j\})$ for all j, $b^{(k)} \in M(\emptyset)$ for all k, $\bar{F}(\emptyset)(c^{(p)})$ defined for all p, and $F(\emptyset)(0) = 0$, then the inequality

(BMIPC) $\qquad \sum_{j=1}^{r} F(\{j\})(a^{(j)}) z_j + \sum_{k=1}^{s} F(\emptyset)(b^{(k)}) x_k + \sum_{p=1}^{t} \bar{F}(\emptyset)(c^{(p)}) y_p$

$$\geq F(\{1, \ldots, r\})(d)$$

is valid.

In the above, we have used the notation

$$\bar{F}(\emptyset)(v) = \limsup \{F(\emptyset)(\lambda v)/\lambda \mid \lambda \searrow 0^+\} \tag{2.83}$$

where $\bar{F}(\emptyset)(v)$ is defined only if $\{\lambda v \mid \lambda \geq 0\} \subseteq M(\emptyset)$, and the lim sup in (2.83) is finite.

Proof for $t = 0$ Let (z, x) be a solution to (BMIP), and put

$$K = \{j \mid z_j = 1\}. \tag{2.84}$$

The Theory of Cutting-Planes

Applying (2.76) with $S = \{1, \ldots, r\}$, $S_1 = K$, $S_2 = \emptyset$ gives

$$F(\{1, \ldots, r\})(d) = F(\{1, \ldots, r\})(Az + Bx)$$
$$\leq F(K)(Az) + F(\emptyset)(Bx). \tag{2.85}$$

Now from (2.27) in Principle IP (recall that $F(\emptyset)(\cdot)$ is subadditive and $F(\emptyset)(0) = 0$), with B in place of A, we obtain

$$F(\emptyset)(Bx) \leq \sum_{k=1}^{s} F(\emptyset)(b^{(k)}) x_k. \tag{2.86}$$

Thus, to complete verification of (BMIPC), it suffices to show that

$$F(K)(Az) \leq \sum_{j=1}^{r} F(\{j\})(a^{(j)}) z_j. \tag{2.87}$$

For $K = \emptyset$, (2.87) becomes $F(\emptyset)(0) \leq 0$ as $z = 0$, and thus (2.87) follows from the assumption $F(\emptyset)(0) = 0$. For $K \neq \emptyset$, (2.87) is equivalent to

$$F(K)\left(\sum_{j \in K} a^{(j)}\right) \leq \sum_{j \in K} F(\{j\})(a^{(j)}). \tag{2.87'}$$

We now prove (2.87') by induction on the size $|K|$ of K, assuming $|K| \geq 1$. For $|K| = 1$, (2.87') is simply:

$$F(\{j\})(a^{(j)}) \leq F(\{j\})(a^{(j)}) \quad \text{where} \quad K = \{j\}.$$

We henceforth assume $|K| \geq 2$.

Write $K = \{j^*\} \cup K'$, where $j^* \in K$, $K' = K - \{j^*\}$, $|K'| \geq 1$. Then using (2.76) with $S = K$, $S_1 = \{j^*\}$, $S_2 = K'$, we have

$$F(K)\left(\sum_{j \in K} a^{(j)}\right) \leq F(\{j^*\})(a^{(j^*)}) + F(K')\left(\sum_{j \in K'} a^{(j)}\right). \tag{2.88}$$

Since $|K| > |K'| \geq 1$, by the induction hypothesis we have

$$F(K')\left(\sum_{j \in K'} a^{(j)}\right) \leq \sum_{j \in K'} F(\{j\})(a^{(j)}). \tag{2.87''}$$

Now (2.88) and (2.87'') give (2.87).

For $t \geq 0$, the reader may wish to supplement this proof with the discussion in Section 2.7 of Jeroslow (1976).

As an example of Principle BMIP, consider the constraint system

$$z_1 + 3z_2 + 5x_1 + 7x_2 + y_1 = 12.3$$
$$y_1 \geq 0; \ z_1, z_2 \in \{0, 1\}; \ x_1 \text{ and } x_2 \text{ nonnegative integers.} \tag{2.89}$$

Using the subadditive function $G(v) = v \bmod 5$, and putting all $F(S)(\cdot) = G(\cdot)$, we compute from (2.83)

$$\bar{F}(\varnothing)(1) = \limsup \{G(\lambda \cdot 1)/\lambda \mid \lambda \searrow 0^+\}$$
$$= \limsup \{\lambda(\bmod 5)/\lambda \mid \lambda \searrow 0^+\} = 1. \quad (2.90)$$

Therefore, as (BMIPC) we obtain the cut

$$z_1 + 3z_2 + 0x_1 + 2x_2 + y_1 \geq 2.3. \quad (2.91)$$

Now (2.91) is a valid cut even if z_1 and z_2 are general integer variables, as one notes by taking z_1 and z_2 as integer variables in (BMIPC) (hence $r = 0$). Can we do better than (2.91), using the fact that z_2 is a zero–one variable?

We apply the lemma with $G(v) = v \bmod 5$, $d^{(1)} = 0$, $d^{(2)} = -3$. We have

$$F(\{1\})(1) = \min \{(1 + 0z_1) \bmod 5 \mid z_1 \in \{0, 1\}\} \quad (2.92a)$$
$$= 1(\bmod 5) = 1;$$
$$F(\{2\})(3) = \min \{(3 - 3z_2) \bmod 5 \mid z_2 \in \{0, 1\}\} \quad (2.92b)$$
$$= 0 \bmod 5 = 0;$$
$$F(\varnothing)(5) = 5 \bmod 5 = 0; \quad (2.92c)$$
$$F(\varnothing)(7) = 2; \quad (2.92d)$$
$$\bar{F}(\varnothing)(1) = 1. \quad (2.92e)$$

Therefore (BMIPC) gives the following cut, which is a strengthening of (2.91):

$$z_1 + 0z_2 + 0x_1 + 2x_2 + y_1 \geq \min \{(12.3 - 3z_2) \bmod 5 \mid z_2 \in \{0, 1\}\} = 2.3. \quad (2.93)$$

Note that (2.93) is not valid with z_2 as a general integer, since it cuts away the general integer solution $(z_1, z_2, x_1, x_2, y_1) = (0, 4, 0, 0, 0.3)$ to (2.89).

Principle BMIP generalizes both Principle IP and a cut-form for integer and continuous variables given in Jeroslow (1974b), the latter of which becomes principle IP when no continuous variables are present. As with Principle IP, our Principle BMIP is the strongest of its type, in a certain sense that is stated in the next result.

Converse to Principle BMIP *If* (BMIP) *is consistent and A, B, C, d have only rational entries, then any valid cut for* (BMIP) *is a weakening of* (BMIPC) *for a suitable subadditive collection* $\{F(S)(\cdot) \mid S \subseteq \{1, \ldots, r\}\}$ *that satisfies:* $a^{(j)} \in M(\{j\})$, *all j;* $b^{(k)} \in M(\varnothing)$, *all k;* $\bar{F}(\varnothing)(c^{(p)})$ *defined, all p; and* $F(\varnothing)(0) = 0$.

The idea behind a rigorous proof for this Converse is close to the idea involved in a rigorous proof for the Converse to Principle IP. Again one uses

The Theory of Cutting-Planes

a value function idea. Given a valid cut

$$\sum_{j=1}^{r} \pi_j z_j + \sum_{k=1}^{s} \sigma_k x_k + \sum_{p=1}^{t} \theta_p y_p \geq \pi_0 \tag{2.94}$$

for (BMIP), one puts

$$F(S)(v)\left\{\sum_j \pi_j z_j + \sum_k \sigma_k x_k + \sum_p \theta_p y_p \;\middle|\; \right.$$
$$Az + Bx + Cy = d, \; y \geq 0, \; x \geq 0$$
$$\left. \text{and integer, } z \text{ binary, and } z_j = 0 \text{ for } j \notin S \right\}. \tag{2.95}$$

Then one checks (2.76), (2.77), and the domain inclusion requirements of the converse. Finally, one notes that (2.94) is a weakening of (BMIPC).

It is a simple matter to adapt the linear systems of Appendix 2.2 to provide all valid cutting-planes for (BMIP) if $t = 0$. One simply 'expands out' (BMIP) to the following problem in general integer variables:

$$Az + 0z' + Bx + Cy = d$$
$$z_j + z'_j = 1 \; j = 1, \ldots, r \qquad z, z', x \geq 0, \text{ integer.} \tag{2.96}$$

One then utilizes the linear system for (2.96), extracting only those solutions which give nonpositive coefficients for z', and then omitting z' as variables.

Principle BMIP can be generalized somewhat. For instance, if (BMIP) is replaced by a constraint system

$$Az + Bx + Cy \in S$$
$$z_j \in \{0, 1\} \quad \text{for} \quad j = 1, \ldots, r \tag{2.97}$$
$$x_h \geq 0, \text{ integer for } h = 1, \ldots, s$$
$$y_p \geq 0 \quad \text{for} \quad p = 1, \ldots, t$$

where $S \neq \emptyset$ is a set, then the cut form (BMIPC) for a subadditive collection becomes

$$\sum_{j=1}^{r} F(\{j\})(a^{(j)}) z_j + \sum_{k=1}^{s} F(\emptyset)(b^{(k)}) x_k + \sum_{p=1}^{t} \bar{F}(\emptyset)(c^{(p)}) y_p$$
$$\geq \inf\{F(\{1, \ldots, r\})(d) \mid d \in S\}. \tag{2.98}$$

(2.98) can easily be verified by noting that $F(\{1, \ldots, r\})(d)$ can be replaced by $\inf\{F(\{1, \ldots, r\})(d) \mid d \in S\}$ in (2.85), and then continuing the proof as before.

In the case that one wishes only the valid cuts for (BMIP) of the form

$$\sum_{k=1}^{s} \sigma_k x_k + \sum_{p=1}^{t} \theta_p y_p \geq \pi_0 \tag{2.94'}$$

in which $\pi_j = 0$ for $j = 1, \ldots, r$ in (2.94), one can view (BMIP) as

$$Bx + Cy \in S$$
$$x_k \geq 0, \text{ integer for } k = 1, \ldots, s \qquad (2.97')$$
$$y_p \geq 0 \quad \text{for} \quad p = 1, \ldots, t$$

with

$$S = \left\{ d - \sum_{j=1}^{r} a^{(j)} z_j \mid z_j \in \{0, 1\} \quad \text{for} \quad j = 1, \ldots, r \right\}.$$

One obtains from (2.98) the cut form

$$\sum_{k=1}^{s} G(b^{(k)}) x_k + \sum_{p=1}^{t} \bar{G}(c^{(p)}) y_p \geq \inf \left\{ G\left(d - \sum_{j=1}^{r} a^{(j)} z_j \right) \mid z_j \in \{0, 1\}, j = 1, \ldots, r \right\}$$
(2.99)

for a subadditive function $G(\cdot) = F(\emptyset)(\cdot)$ where we see that $r = 0$ in (2.97') so that the subadditive collection reduces to the one function $F(\emptyset)(\cdot) = G$.

Similarly, in place of $S = \{d\}$ as used for (BMIP), one can consider inequality constraints in (BMIP) by using $S = \{v \mid v \geq d\}$. In this connection, monotone subadditive functions are used, and these give the simplification

$$\inf \{ G(v) \mid v \geq d \} = G(d) \qquad (2.100)$$

for G monotone. Many other specializations of the set S are also easily treated.

References

Araoz A. A. (Durand) (1973). *Polyhedral Neopolarities*. Ph.D. Dissertation, *University of Waterloo, Canada*.
Bell, D. E. and J. F. Shapiro (1975). A Finitely-Convergent Duality Theory for Zero–One Integer Programming. *Research Memorandum 75–33, IIASA, July, 1975*.
Blair, C. E. (1975). Topics in Integer Programming. *Ph.D. Dissertation, Carnegie-Mellon University*.
Blair, C. E. (1976). Two Rules for Deducing Valid Inequalities for 0–1 Problems. *SIAM Journal on Applied Mathematics*, **31**, 614–617.
Blair, C. E. and R. G. Jeroslow (1976). The Value Function of a Mixed Integer Program: I and II. *October 1975 and December 1976, GSIA, Carnegie–Mellon University and the University of Illinois at Urbana*.
Burdet, C.-A. and E. L. Johnson (1975). A Subadditive Approach to Solve Linear Integer Programs. *RC5507, IBM Watson Research Center, Yorktown Heights*.
Chvátal, V. (1973). Edmonds Polytopes and a Hierarchy of Combinatorial Problems. *Discrete Mathematics*, **4**, 305–337.
Dantzig, G. B. (1959). Notes on Solving Linear Programs in Integers. *Naval Research Logistics Quarterly*, **6**, 75–76.
Galil, Z. (1974). On the Complexity of Resolution Procedures for Theorem Proving. *IBM Watson Research Center, Yorktown Heights, December 1974*.
Gomory, R. E. (1963). An Algorithm for Integer Solutions to Linear Programs." In Graves and Wolfe (Eds), *Recent Advances in Mathematical Programming*, 269–302.

Gomory, R. E. (1965). On the Relation Between Integer and Non-Integer Solutions to Linear Programs. *Proceedings of the National Academy of Sciences*, **53**, 260–265.

Gomory, R. E. (1969). Some Polyhedra Related to Combinatorial Problems. *Linear Algebra and Its Applications*, **2**, 451–558.

Gomory, R. E. and E. L. Johnson (1972). Some Continuous Functions Related to Corner Polyhedra: I and II *Mathematical Programming*, **3**, 23–85 and 359–389.

Hilbert, D. (1890). Uber die Theorie der algebraischen Formen. *Mathematische Annalen*, **36**, 475–534.

Jeroslow, R. (1974a). Trivial Integer Programs Unsolvable by Branch-and-Bound. *Mathematical Programming*, **6**, 105–109.

Jeroslow, R. (1974b). The Principles of Cutting-Plane Theory: Part I. *Carnegie-Mellon University, February, 1974*.

Jeroslow, R. (1975a). Some Structure and Basis Theorems for Integral Monoids. *Carnegie-Mellon University, July, 1975*.

Jeroslow, R. (1975b). A Generalization of a Theorem of Chvátal and Gomory. In O. L. Mangasarian, R. R. Meyer, and S. M. Robinson (Eds), *Nonlinear Programming*, **2**, Academic Press, New York. pp. 313–332.

Jeroslow, R. (1976). Cutting-Plane Theory: Algebraic Methods. *GSIA, Carnegie-Mellon University, March, 1976*.

Johnson, E. L. (1974a). Integer Programs with Continuous Variables. *IBM Report, Yorktown Heights, Watson Scientific Research Center*.

Johnson, E. L. (1974b). The Group Problem for Mixed Integer Programming. *Mathematical Programming Studies*, Study 2, December 1974, 137–179.

Johnson, E. L. and K. Spielberg (1971). Inequalities in Branch and Bound Programming. *RC3649, IBM Watson Research Center, Yorktown Heights*.

Rockafellar, R. T. (1970). *Convex Analysis*, Princeton University Press, Princeton, New Jersey.

PART 2
DISJUNCTIVE METHODS

The linear inequality

$$3x_1 + x_2 - 7x_3 \geq 5 \qquad (2.101)$$

is a weakening of both

$$x_1 + x_2 - 9x_3 \geq 5 \qquad (2.102)$$

and

$$3x_1 - x_2 - 7x_3 \geq 7. \qquad (2.103)$$

Therefore, (2.101) must hold if *either* (2.102) *or* (2.103) hold, even if one doesn't know which of (2.102) or (2.103) is valid (so long as the variables x_1, x_2, and x_3 are nonnegative). In fact, since $3 = \max\{1, 3\}$, $1 = \max\{1, -1\}$, $-7 = \max\{-9, -7\}$, and $5 = \min\{5, 7\}$, (2.101) is the strongest common weakening of (2.102) and (2.103).

Let us take this simple idea—which first occurs in Owen (1973)—and combine it with some other considerations in order to apply it in different circumstances.

Example 1 (From Owen, 1973) The original application concerns a nonlinear program, with constraints that include those of a type that occur in Kuhn–Tucker and complementarity conditions:

$$x_k, y_k \geq 0 \quad \text{and} \quad x_k y_k = 0 \quad k = 1, \ldots, n. \tag{2.104}$$

These are really purely logical constraints: they say that either $x_k = 0$ or $y_k = 0$.

If the program otherwise has linear constraints, we may solve these, and find that perhaps $x_k y_k > 0$. Taking the rows from the simplex tableau, which we write as

$$x_k = a_{k0} + \sum_{j \in J} a_{kj}(-t_j) \tag{2.105a}$$

$$y_k = b_{k0} + \sum_{j \in J} b_{kj}(-t_j) \tag{2.105b}$$

in the nonbasic variables t_j, $j \in J$, the logical constraints give the disjunctive conditions

$$\text{either} \sum_{j \in J} a_{kj} t_j = a_{k0} \quad \text{or} \quad \sum_{j \in J} b_{kj} t_j = b_{k0}. \tag{2.106}$$

Putting a multiplier λ_1 on the first equation and λ_2 on the other, both unconstrained because of the equality constraints, we obtain the cut

$$\sum_{j \in J} \max\{\lambda_1 a_{kj}, \lambda_2 b_{kj}\} t_j \geq \min\{\lambda_1 a_{k0}, \lambda_2 b_{k0}\} \tag{2.107}$$

as a common weakening. If we put $\lambda_1 = 1/a_{k0}$, $\lambda = 1/b_{k0}$, the right-hand side in (2.107) is unity, and the current solution ($t_j = 0$) is cut away. (We assume, as usual, that nonbasic variables t_j, $j \in J$, are nonnegative.)

For example, suppose that in the current tableau

$$x_1 = 1.2 + 7(-t_1) - 0.3(-t_2) - 2.1(-t_3)$$
$$y_1 = 0.4 - 2(-t_1) - 1.1(-t_2) + 4.2(-t_3). \tag{2.105'}$$

Then, upon taking $\lambda_1 = 1/1.2$, $\lambda_2 = 1/0.4$, (2.107) becomes (rounding up to insure validity)

$$5.84 t_1 - 0.25 t_1 + 1.68 t_3 \geq 1. \tag{2.107'}$$

From our analysis, (2.107') is valid if the condition $x_1 y_1 = 0$ is imposed.

Example 2 From Glover and Klingman, 1973 Glover and Klingman have studied a class of problems which, as one special case, may have constraints of the form:

At least q of the variables x_k are ≥ 1, where $x_k = a_{k0} + \sum_{j \in J} a_{kj}(-t_j)$.

$$\tag{2.108}$$

The Theory of Cutting-Planes

In (2.108), q is a positive integer.

Constraints like (2.108) lead to large disjunctive systems unless $q=1$ (as for (2.105) above) or $q=2$. Specifically, let there be p x_k's, let H be the set of all index subsets h of $\{1,\ldots,p\}$ which have q elements. Then at least one of these $\binom{p}{q}$ systems hold:

$$\sum_{j\in J} a_{kj}(-t_j) \geq 1 - a_{k0} \text{ all } k \in h. \tag{2.109$_h$}$$

One may assign multipliers $\lambda_k^h \geq 0$, $k \in h$, to the hth system, $h \in H$. Within the hth system, one adds to obtain from (2.109$_h$) the inequality

$$\sum_{j\in J}\left(-\sum_{k\in h} \lambda_k^h a_{kj}\right) t_j \geq \sum_{k\in h} \lambda_k^h (1 - a_{k0}). \tag{2.109$'_h$}$$

Indeed, when (2.109$_h$) holds, so does every nonnegative linear combination of the constraints in (2.109$_h$), and (2.109$'_h$) arises by multiplying constraint k in the hth system (2.109$_h$) by $\lambda_k^h \geq 0$ and adding over $k \in h$, then exchanging the order of summation. Since at least one of the (single) linear inequalities in (2.109$'_h$) holds for some $h \in H$, the coefficient of t_j in the cut is

$$\max_{h\in H}\left\{-\sum_{k\in h} \lambda_k^h a_{kj}\right\} \tag{2.110$_j$}$$

and the constant term in the cut is

$$\min_{h\in H}\left\{\sum_{k\in h} \lambda_k^h (1 - a_{k0})\right\}. \tag{2.111}$$

One way of getting an easily-computed cut in (2.110$_j$), (2.111), is to set $\lambda_k^h = 1/(1 - a_{k0})$ for all $k \in h$ and $h \in H$. This setting is possible if all $a_{k0} < 1$, i.e. if all $x_k < 1$. This can often be arranged with constraints such as (2.108), for if (2.108) fails at some tableau, there are $n' < q$ variables x_k with $x_k \geq 1$, and by taking the remaining $q' = q - n'$ variables x_k with $x_k < 1$, one has that at least $q' \geq 1$ of these $x_k \geq 1$. Then one repeats (2.110), (2.111), with q' replacing q.

With the choice $\lambda_k^h = 1/(1 - a_{k0})$, we have $\sum_{k\in h} \lambda_k^h (1 - a_{k0}) \geq |h| = q$, independent of $h \in H$, so the quantity of (2.111) is q. To compute (2.110$_j$), one ranks the quantities $a_{kj}/(1 - a_{k0})$ in increasing order for $k = 1,\ldots,p$. Let $h^*(j) \in H$ be the set corresponding to the q smallest.

Then the jth intercept is:

$$-\sum_{k\in h^*(j)} a_{kj}/(1 - a_{k0}). \tag{2.112$_j$}$$

For example, suppose that
$$x_1 = 0.7 + 2.1(-t_1) - 1.2(-t_2)$$
$$x_2 = 0.5 - 1.5(-t_1) + 1.0(-t_2) \qquad (2.113)$$
$$x_3 = 0.4 + 1.8(-t_1) + 0.6(-t_2)$$

and that at least two of the x_k are to be ≥ 1. Then $H = \{\{1, 2\}, \{1, 3\}, \{2, 3\}\}$, and at least one of the following three systems holds:

$$\begin{aligned} 2.1(-t_1) - 1.2(-t_2) &\geq 0.3 \\ -1.5(-t_1) + 1.0(-t_2) &\geq 0.5 \end{aligned} \qquad (2.109'_{\{1,2\}})$$

or

$$\begin{aligned} 2.1(-t_1) - 1.2(-t_2) &\geq 0.3 \\ 1.8(-t_1) + 0.6(-t_2) &\geq 0.6 \end{aligned} \qquad (2.109'_{\{1,3\}})$$

or

$$\begin{aligned} -1.5(-t_1) + 1.0(-t_2) &\geq 0.5 \\ 1.8(-t_1) + 0.6(-t_2) &\geq 0.6 \end{aligned} \qquad (2.109'_{\{2,3\}})$$

Using the multipliers $\lambda^h = 1/(1 - a_{k0})$, as suggested, we find that at least one of these three inequalities holds:

$$-4t_1 + 2t_2 \geq 2 \qquad (2.114a)$$

or

$$-10t_1 + 3t_2 \geq 2 \qquad (2.114b)$$

or

$$0t_1 - 3t_2 \geq 2. \qquad (2.114c)$$

Therefore, taking a common weakening, we see that

$$0t_1 + 3t_2 \geq 2 \qquad (2.115)$$

holds, i.e. $t_2 \geq 2/3$. One easily checks that the intercept of t_j in (2.115) is that given by (2.112$_j$).

Example 3 (From Balas, 1974a; Glover, 1973; and Gomory, 1960) Suppose that y is a basic integer variable with a tableau row

$$y = a_0 + \sum_{j \in J} a_j(-t_j) \qquad (2.116)$$

such that a_0 is fractional, i.e.

$$a_0 = m + f, \; m \text{ integer}, \; 0 < f < 1. \qquad (2.117)$$

Since either

$$y \leq m \quad \text{or} \quad y \geq m + 1 \qquad (2.118)$$

we know that one of these two inequalities holds:

$$\sum_{j \in J} a_j t_j \geq -m + a_0 = f \qquad (2.119a)$$

The Theory of Cutting-Planes

or
$$\sum_{j \in J} (-a_j) t_j \geq m + 1 - a_0 = 1 - f. \tag{2.119b}$$

Multiplying (2.119a) by $1/f$ and (2.119b) by $1/(1-f)$, and taking a common weakening, we obtain the cut

$$\sum_{j \in J} \max\{a_j/f, -a_j/(1-f)\} t_j \geq 1. \tag{2.120}$$

Other examples of disjunctive cuts are given as exercises in Appendix 2.4.

The disjunctive cuts depend upon logical conditions. Indeed, the term 'disjunctive' was used because some considered an essential aspect of the cut-producing method to be the manner in which it handles the disjunction, i.e. the 'or' connective (either this... or this... or this... holds). As such, once the logical condition is stated from which the cut is to be obtained, the disjunctive cut is valid even if the variables are not integer-constrained. (Of course, as in Example 3, the logical condition itself may depend upon the integrality of some quantity, such as the variable y).

It is a fact that disjunctive cuts are derived from logical conditions, which leads to the surprising versatility of the disjunctive methods. Many of the nonlinearities and nonconvexities of mathematical programming were introduced precisely to model logical conditions. On the other hand, when the variables in the logical condition are integer constrained, there are often ways of strengthening a disjunctive cut by taking this fact into account. To do so, we will state and prove the principle of 'Monoidal Cut Strengthening', and then give an example of it.

Principle MCS (*Compare with Balas and Jeroslow, 1975*) Let $T \neq \emptyset$ *be a set and M a monoid. Suppose that the constraints*

$$\begin{aligned} Ax &\in T + M \\ x &\geq 0, \text{ integer} \end{aligned} \tag{2.121}$$

always imply the cutting-plane

$$\sum_{j=1}^{r} F_j(a^{(j)}) x_j \geq \pi_0 \tag{2.122}$$

for any matrix A, where $F_1, \ldots, F_r$ *are certain (not necessarily subadditive) functions, and* $a^{(j)}$ *is the jth column of* $A = [a^{(j)}]$.

Then the cutting-plane

$$\sum_{j=1}^{r} \left(\inf_{m \in M} F_j(a^{(j)} + m) \right) x_j \geq \pi_0 \tag{2.123}$$

is also valid for the constraints (2.121).

Proof Let $m^{(j)} \in M$ be picked arbitrarily for $j = 1, \ldots, r$. Since M is a monoid, $\sum_{j=1}^{r} m^{(j)} x_j \in M$ for any integer vector $x \geq 0$. Consequently $Ax \in T + M$ implies

$$\sum_{j=1}^{r} (a^{(j)} + m^{(j)}) x_j = \sum_{j=1}^{r} a^{(j)} x_j + \sum_{j=1}^{r} m^{(j)} x_j \in T + M + M = T + M \quad (2.124)$$

(the last equality since M is a monoid), which in turn implies (by the hypothesis)

$$\sum_{j=1}^{r} F_j(a^{(j)} + m^{(j)}) x_j \geq \pi_0. \quad (2.125)$$

In brief, $Ax \in T + M$ implies (2.125) for any arbitrary choice of the $m^{(j)}$, hence (2.123).

Note that the infimum over $m \in M$ in (2.123) need not be computed, since for any $m \in M$, the quantity $F_j(a^{(j)} + m)$ is a valid upper bound on the jth intercept in (2.123).

Example 3 revisited (From Balas, 1974a) If y is an integer variable with tableau row

$$y = 0.5 + 10.5(-t_1) - 5(-t_2) + 3.5(-t_3) \quad (2.116')$$

then from (2.120) we see that

$$21t_1 + 10t_2 + 7t_3 \geq 1 \quad (2.120')$$

is a valid cut. If the variables t_j are continuous in (2.120'), no real strengthening of (2.120') is possible. Indeed, $(y, t_1, t_2, t_3) = (0, 1/28, 0, 1/28)$ solves (2.116'), makes y integral, and makes (2.120') an equality.

Suppose now that the variables t_j are integral in (2.116'). We shall use Principle MCS to strengthen the cut (2.120').

We view (2.116') as

$$\sum_{j \in J} (-a_j) t_j \in T + M \quad (2.126)$$

where $T = \{-0.5\}$, $a_1 = 10.5$, $a_2 = -5$, $a_3 = 3.5$, and M is the set of integers, which is a monoid. Now (2.126) is of the form (2.121) where A has one row, $A = [-a_j]$ (i.e. $a^{(j)} = -a_j$ for $j \in J$), and x is the vector of nonbasic variables t_j, $j \in J$. For T held fixed, our reasoning in Example 3 showed that (2.120) is always valid (i.e. $\pi_0 = 1$ in (2.123)), so in (2.122) we may use

$$F_j(a^{(j)}) = \max\{a_j/f, -a_j/(1-f)\}. \quad (2.122')$$

Therefore the jth intercept of the improved cut (2.123) is

$$\inf\{\max\{(a_j - n)/f, (-a_j + n)/(1-f)\} \mid n \text{ integer}\}. \quad (2.127)$$

The Theory of Cutting-Planes

We need only estimate (2.127). Putting $n = \lfloor a_j \rfloor$, where $\lfloor a_j \rfloor$ is the integer part of a_j, and $a_j = \lfloor a_j \rfloor + f_j$, $0 \leq f_j < 1$, we have

$$\max\{(a_j - n)/f, (-a_j + n)/(1 - f)\} = f_j/f. \tag{2.128a}$$

Next, putting $n = \lfloor a_j \rfloor + 1$, we have:

$$\max\{(a_j - n)/f, (-a_j + n)/(1 - f)\} = (1 - f_j)/(1 - f). \tag{2.128b}$$

To get an upper bound on (2.127) we can use any choice of n, hence we use the most favourable, and conclude that the coefficient of t_j in a valid cut can be taken to be

$$\min\{f_j/f, (1 - f_j)/(1 - f)\}. \tag{2.129}$$

(Actually, (2.129) is equal to (2.127).)

Using (2.129), from (2.116′) we obtain the cut

$$t_1 + 0t_2 + t_3 \geq 1 \tag{2.130}$$

as (2.123). Clearly, (2.130) is stronger than (2.120′). It in fact cuts off the point (0, 1/28, 0, 1/28) that satisfies (2.120′).

The cut (2.130), and cuts desired in this manner with M = the integers, are valid for Gomory's group problem (Gomory, 1965), and are in fact among the cuts given in that paper. They are usually associated with the subadditive methods, but can also be derived as a strengthened disjunctive cut; see Exercise 7 in Appendix 2.4 for a similar derivation of another cut for the group problem. Principle MCS was abstracted from a study of Gomory's addition of unit rows in deriving the cuts in Gomory (1963), and its use goes beyond the setting of the group problem (see, for example, Exercise 6, where a cut with negative coefficient is derived by means of this principle.)

A sense in which Principle MCS strengthens disjunctive cuts is as follows. Since a disjunctive cut is based on a logical condition, any branching scheme which imposes this condition makes the cut redundant on each branch. Thus, the cut (2.120′) would be redundant if either $y \geq 0$ or $y \geq 1$ is imposed on the variable y of (2.116′) in a branch and bound scheme. However a strengthened disjunctive cut, such as (2.130), often is not redundant after branching ((2.130) is not redundant on the branch $i = 0$).

If unstrengthened disjunctive inequalities are used in a branch and bound approach, they provide cuts that allow one to estimate penalties (without the cut actually being added), or the cuts can be added explicitly if the branching is done on a different logical condition than the one from which the cut is derived. For instance, in Example 1, if, in addition to the pair (x_1, y_1) there is a second pair of variables (x_2, y_2) with $x_2 y_2 = 0$ desired, one can branch on $x_2 = 0$ versus $y_2 = 0$ and simultaneously add the cut (2.107). The addition of (2.107) recovers some of the benefit to be obtained by branching on $x_1 = 0$ versus $y_1 = 0$, while one actually branches on a different logical condition.

In order to further discuss the disjunctive methods, we need a precise statement of the main principle involved, and we provide it next. The proof is omitted, since it proceeds by the considerations we have used in working the examples above.

Principle DC *Suppose that at least one of the systems* (S_h) *holds for some* $h \in H$, *where* (S_h) *is*

$$(S_h) \quad \begin{array}{c} A^{(h)}x \geq b^{(h)} \\ x \geq 0 \end{array}$$

and $A^{(h)}$ *is an* $m_h \times r$ *matrix and* $b^{(h)}$ *is an* $m_h \times 1$ *vector.*

Then for any $1 \times m_h$ *vectors* $\lambda^{(h)} \geq 0$ *of nonnegative multipliers, the inequality*

$$(\text{DC}) \quad \left(\sup_{h \in H} \lambda^{(h)} A^{(h)} \right) x \geq \inf_{h \in H} \lambda^{(h)} b^{(h)}$$

holds, as do any of its weakenings. In (DC), $\sup_{h \in H} \lambda^{(h)} A^{(h)}$ *denotes that vector whose* jth *component* $(j = 1, \ldots, r)$ *is* $\sup_{h \in H} v_j^{(h)}$, *and* $v^{(h)} = \lambda^{(h)} A^{(h)}$.

(The sups and the inf of (DC) are assumed to exist. Of course, for H finite, as is usually the case, a 'sup' is a maximum and an 'inf' is a minimum.)

We saw in Part 1 of this chapter that Principle IP can account for all valid cuts (CP) for (IP). Is there an analogous 'adequacy theorem' for Principle DC, relative to a suitable constraint set?

In fact there is, under certain hypotheses. Rather than give the most general statement possible, we cite here one frequently-realized hypothesis from which an adequacy result can be derived.

A Converse to Principle DC (*Blair and Jeroslow* 1976) *Suppose that, for each* $h \in H$, *there is some choice of* $d^{(h)}$ *for which the set*

$$\{x \geq 0 \mid A^{(h)}x \geq d^{(h)}\} \tag{2.131}$$

is both nonempty and bounded, and suppose that H *is finite.*

Then all valid cutting-planes for the set $S = \bigcup_{h \in H} \{x \mid (S_h) \text{ holds}\}$ *occur as weakenings of cuts* (DC).

This converse is useful, since often in the systems (S_h) we have $A^h = [K \mid M^h]^{\text{tr}}$, where the matrix M^h varies with $h \in H$, but K does not, and it is known that $\{x \geq 0 \mid Kx \geq d^*\}$ is bounded and nonempty for some d^*. Then by taking any $x^* \in \{x \geq 0 \mid Kx \geq d^*\}$, it is easy to see that $\{x \geq 0 \mid Kx \geq d^*, M^h x \geq M^h x^*\}$ is bounded and nonempty, thus verifying the hypothesis of the Converse with $d^{(h)} = (d^*, M^h x^*)$. For instance, often the system $Kx \geq d$ contains all the constraints of the linear relaxation of a bounded integer program, while constraints $M^h x \geq c$ are implied by certain values of the integer variables.

The Theory of Cutting-Planes

As a corollary to this Converse, one obtains all valid cuts to a consistent and bounded integer program (IP) as weakenings of (DC), by using as the systems (S_h) the following

$(S_h)'$ $\qquad\qquad\qquad Ax = b \qquad x = h$

as $h \in H$ varies over all integer points which may possibly solve $Ah = b$. By a 'bounded' program (IP), we mean that $\{x \geq 0 \mid Ax = b\}$ is a bounded set, and from bounds for this set the index set H of $(S_h)'$ can be determined. To obtain this corollary, in the Converse one sets

$$A^{(h)} = \begin{bmatrix} A \\ -A \\ I \end{bmatrix} \qquad (2.132)$$

(viewing the equalities $Ax = b$ as $Ax \geq b$ and $(-A)x \geq -b$), and one can use

$$d^{(h)} = \begin{bmatrix} b \\ -b \\ x^* \end{bmatrix} \qquad (2.133)$$

where x^* is any solution to $Ax^* = b$, $x^* \geq 0$.

We remark that a Converse is also valid if each system (S_h) is consistent (Balas, 1975).

When hypotheses are lacking, it is possible for Principle DC to fail to provide all valid cuts. For instance, if either $(-1)x_1 \geq -1$ or $0x_1 \geq 1$ holds, then one obtains only the common weakening $0x_1 \geq 0$, which is worthless. Since $0x_1 \geq 1$ is impossible, we know that $0 \leq x_1 \leq 1$; yet here Principle DC has thrown away information.

A version of Principle DC is also available for H finite, when the variables x are integer-constrained; it is obtained by combining Principle DC with Principle MCS.

Principle SDC (*Balas and Jeroslow, 1975*) Suppose that at least one system (S_h) holds, and that in addition there are 'lower bounds', i.e. all the systems

(*) $\qquad\qquad\qquad A^{(h)}x \geq d^{(h)} \qquad x \geq 0$

hold, where $d^h \leq b^h$ is a suitable vector. Finally, suppose that x is to be an integer vector, and let $a^{(jh)}$ denote the jth column of $A^{(h)}$. Then the inequality

$$\sum_{j=1}^{r} \min\left[\max_{h \in H}\{\lambda^{(h)}(a^{(jh)} + n_h(b^{(h)} - d^{(h)}))\} \;\middle|\; \text{all } n_h \text{ are integer and } \sum_{h=1}^{t} n_h \geq 0\right] x_j$$

$$\geq \min_{h \in H} \lambda^h b^{(h)} \qquad (2.134)$$

is valid.

Proof In Principle MCS, put

$$T = \{(v^{(1)}, \ldots, v^{(t)}) \mid \text{all } v^{(h)} \geq d^{(h)} \text{ and at least one } v^{(h)} \text{ is } \geq b^{(h)}\} \quad (2.135a)$$

$$M = \left\{(n_k(b^{(1)} - d^{(1)}), \ldots, n_t(b^{(t)} - d^{(t)})) \,\bigg|\, \sum_{h=1}^{t} n_h \geq 0 \text{ and all } n_h \text{ are integer}\right\} \quad (2.135b)$$

where $t = |H|$. Also set

$$A = \begin{bmatrix} A^{(1)} \\ \cdot \\ \cdot \\ \cdot \\ A^{(t)} \end{bmatrix}.$$

Then the information we have can be represented as

$$Ax \in T \quad (2.137)$$

which we now relax to

$$Ax \in T + M. \quad (2.138)$$

Now note that (2.138) implies that at least one system (S_h) holds. For if all $n_h = 0$ in (2.135b) then $Ax \in T$ and at least one (S_h) holds; while if some $n_h \neq 0$ then at least one $n_{h*} \geq 1$ and then (S_{h*}) holds. Therefore (2.138) guarantees the validity of the cuts (DC).

Let $a^{(jh)}$ denote the jth column of $A^{(h)}$ for $j = 1, \ldots, r$ and $h = 1, \ldots, t$. Then we write (DC) as

$$\sum_{j=1}^{r} \max_{h \in H} \{\lambda^{(h)} a^{(jh)}\} x_j \geq \min_{h \in H} \lambda^{(h)} b^{(h)} \quad (2.139)$$

and so the strengthened cut (2.123), for (2.139) taken as (2.122) is (2.134). That is, in Principle MCS we take $F_j(a^{(j)}) = \max_{h \in H} \lambda^{(h)} a^{(jh)}$, where $a^{(j)}$ is the jth column of A, so that $a^{(j)\text{tr}} = (a^{(j1)\text{tr}}, a^{(j2)\text{tr}}, \ldots, a^{(jt)\text{tr}})$, $t = |H|$.

We give an application of Principle SDC as Exercise 6 in Appendix 2.4.

The disjunctive cut (DC), (2.139), can also be obtained by subadditive methods. First of all, by the Converse to Principle (BMIP), there is a subadditive collection which yields a strengthening of (2.139). In this case, we can actually give an explicit representation for this subadditive collection, as follows.

In (2.97) of Part 1, we view that the matrices A and B are empty, i.e. there

The Theory of Cutting-Planes

are no binary or integer variables, only continuous ones. We put

$$C = \begin{bmatrix} A^{(1)} \\ \cdot \\ \cdot \\ \cdot \\ A^{(t)} \end{bmatrix} \quad (2.140)$$

with $t = |H|$, and

$$S = \{(v^{(1)}, \ldots, v^{(t)})^{tr} \mid v^{(h)} \geq b^{(h)} \text{ for at least one } h = 1, \ldots, t\} \quad (2.141)$$

where $v^{(h)}$ is of the same dimension as $b^{(h)}$. We also see that the variables x occurring in the systems (S_h) are the variables y of (2.97) of Part 1.

Since there are no binary variables z_j, the subadditive collection of Principle BMIP reduces to one subadditive function $F(\emptyset)(\cdot) = G(\cdot)$ (say), and only quantities $\bar{G}(\cdot)$ enter in (2.98) of Part 1.

For the key step, we choose $\lambda^{(h)} \geq 0$ for $h = 1, \ldots, t$ and set

$$G((v^{(1)}, \ldots, v^{(t)})^T) = \max_{h \in H} \{\lambda^{(h)} v^{(h)}\}. \quad (2.142)$$

By Exercises 1a and 1b of Appendix 2.1, G is subadditive. For any scalar $\rho \geq 0$, $\lambda^{(h)} (\rho v^{(h)})/\rho = \lambda^{(h)} v^h$, and hence using (2.83) of Part 1, we find that $\bar{G}(\cdot) = G(\cdot)$. Finally, we compute

$$\inf \{G((v^{(1)}, \ldots, v^{(t)}) \mid (v^{(1)}, \ldots, v^{(t)}) \in S\} = \inf_{h \in H} \{\lambda^{(h)} b^{(h)}\} \quad (2.143)$$

using $\lambda^{(h)} \geq 0$ and the fact that $v^{(h)} \geq b^{(h)}$ for $h = 1, \ldots, t$.

Combining (2.142), $\bar{G} = G$, and (2.143), the cut (2.98) of Part 1 is (2.139), as desired. This completes our derivation of (DC), (2.139), as a subadditive cut.

The derivation of disjunctive cuts by subadditive methods, or subadditive cuts by disjunctive methods, establishes some theoretical interrelations, as was done in Jeroslow (1974a, b, 1977a). It does not provide a new cut and often, as was the case above, simply parallels the earlier proof in a laborious and more complex fashion.

In an Olympian sense, either the disjunctive or subadditive methods provide a complete theory of valid cuts for the bounded integer program (IP), at least after subadditivity is extended from the group problem to (IP). So it would not be enlightening to claim that a subadditive derivation of a cut discovered by disjunctive methods showed that the cut was simply one illustration of subadditive cuts. Indeed, as we saw, subadditive cuts are a 'special case' of disjunctive cuts. Such claims simply obscure the fact that the broad principles of subadditivity, such as Principle IP, are reductions of the question of cutting-planes to that of subadditive functions. The key issue of the subadditive approach is the construction of suitable subadditive functions. The fact that disjunctive reasoning does provide many useful subadditive functions is a recommendation for disjunctive methods.

Analogous to the linear systems (FS) of Appendix 2.2, Balas (1974b) has associated linear systems with the logical constraints of the disjunctive methods, and provided characterizations of the 'best' cuts (facets, etc.) obtainable from the extreme points of these systems, hence of the 'best' cuts for the relaxations represented by these systems. These linear systems are for the unstrengthened disjunctive cuts (DC) and are closely related to branch and bound. They allow one to compute, for example, facets of the convex span of the union given by

$$\{x \mid Ax = b, x \geq 0, x_1 = 0\} \cup \{x \mid Ax = b, x \geq 0, x_1 = 1\} \qquad (2.144)$$

in the notation of (IP). In contrast, the relaxations associated with the subadditive methods can obtain information not easily obtained from branch and bound, as we saw in Part 1.

As with the subadditive methods, these 'relaxations' can be taken to be the integer program itself, typically at the cost of (an excessively) large linear system. We note that the linear systems we provided in Jeroslow (1974a), equation (108), simplify to Balas' system for the special case of systems (S_h), after certain algebraic manipulations of the type we discuss in Jeroslow (1974a), pp. 77–80 (see Ho, to be published, for details).

In Jeroslow (1974a), we generalized Principle DC by assigning, to every proposition A, a family of cuts $cp(A)$. We used the 'natural mapping' of the lattice of propositions, under the connectives 'or' and 'and', into the lattice of convex polyhedra. If A is the proposition $(S)_1 \vee \cdots \vee (S)_t$, i.e. at least one system (S_h) holds, $cp(A)$ was equal to all weakenings of all disjunctive cuts (DC) for all multipliers $\lambda^h \geq 0$. But the assignment was still defined if A had a more complex form, e.g. $((B \wedge C) \vee D) \wedge E$ for propositions B, C, D, E, although $cp(A)$ depended on the syntactic form of A as much as its truth. We also assigned a linear system to each proposition A. Details of this 'co-proposition' assignment and results about it are in Jeroslow (1974a, 1977a).

We now turn to a theoretical result in the disjunctive methods, due to Balas, which introduces a distinctly new idea that has been fruitful in obtaining other results and providing an alternate proof for an important result (Blair, 1976a). Balas states his result in terms of 'bounded systems with facial constraints', and these systems include the binary integer program:

(BIP) $\qquad Az \geq d, z_j = 0 \quad \text{or} \quad 1 \quad \text{for} \quad j = 1, \ldots, r.$

The features of (BIP) which make it 'facial' are that it can be stated in the form:

$$Az \geq d \quad 0 \leq z \leq e \quad (2.145a)$$

with $e = (1, \ldots, 1)^{tr}$ a vector of ones, plus the constraints

$$(-z_1 \geq 0 \quad \text{or} \quad z_1 \geq 1) \quad (2.145b)$$

and

$$(-z_2 \geq 0 \quad \text{or} \quad z_2 \geq 1)$$

and

$$(-z_r \geq 0 \quad \text{or} \quad z_r \geq 1).$$

Furthermore, in the form (2.145) each inequality $-z_j \geq 0$, or $z_j \geq 1$, appearing in the logical constraints (2.145b), is the 'opposite' of an inequality $z_j \geq 0$, or $z_j \leq 1$, which is implied by the purely linear constraints (2.145a) alone. (In the terminology of Rockafellar (1970), each inequality in (2.145b) provides a *face* of (2.145a)).

Balas abstracts these features of (BIP) to obtain his concept of a 'facial constraint' system, but here we shall content ourselves with stating his result for (BIP) alone. Other examples of facial constraints include linear complementarity problems (Cottle and Dantzig, 1968; Eaves, 1971; Jeroslow, 1978) and generalizations of these (Jeroslow, 1978).

Intersection Theorem for (BIP) (*Following Balas, 1974b, Section 5, Cor. 5.3.1*) *Define inductively the polyhedra*

$$K_0 = \{z \mid Az \geq d, 0 \leq z \leq e\} \quad (2.146a)$$

$$K_{j+1} = \text{convex span of } (K_h \cap \{z_{j+1} \mid z_{j+1} = 0\}) \cup (K_j \cap \{z_{j+1} \mid z_{j+1} = 1\}). \quad (2.146b)$$

Then K_r is the convex span of the solutions to (BIP).

For a proof of this Theorem, see Balas (1974b), Section 5. While it is clear that K_r contains the convex span of solutions to (BIP) (since K_0 does, and, inductively, K_{j+1} does whenever K_j contains this span), it requires a proof using 'faciality' to show that K_r is nothing but the span of feasible (BIP) points.

One result analogous to the Intersection Theorem fails when faciality is not present. For instance, if we begin with the linear inequalities for Figure 2.1 and require that z_1 and z_2 are integer, the only feasible solutions are $(0, 0)$ and $(1, 2)$. The analogue to K_1 is the convex span of

$$\{z_1 \mid z_1 = 0, 0 \leq z_2 \leq 1/2\} \cup \{z \mid z_1 = 1, 3/2 \leq z_2 \leq 2\} \quad (2.147)$$

which gives Figure 2.2.

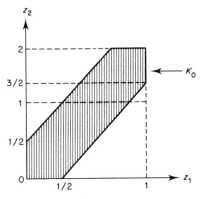

Figure 2.1

Then the analogue to K_2 is the convex span of

$$\{(0, 0)\} \cup \{z \mid z_2 = 1, 1/3 \leq z_1 \leq 2/3\} \cup \{1, 2\} \quad (2.148)$$

which gives Figure 2.3 instead of the convex span of $\{(0, 0)\} \cup \{(1, 2)\}$.

A result seemingly unrelated to the Intersection Theorem is the following inductive characterization of how to generate all the valid cutting-planes for (BIP).

Disjunctive Rule Theorem (Blair, 1976a) *Define sets of inequalities S_0, $S_1, \ldots, S_r$ as follows:*

S_0 *contains all the inequalities in $Az \geq d$, $z \geq 0$, $-z \geq -e$.*

Given S_j, S_{j+1} arises by taking any pair of inequalities

$$\theta_1 z_1 + \cdots + u z_{j+1} + \cdots + \theta_r z_r \geq P \quad (2.149a)$$

and

$$\theta_1 z_1 + \cdots + w z_{j+1} + \cdots + \theta_r z_r \geq T \quad (2.149b)$$

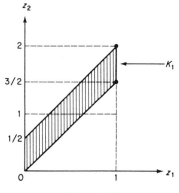

Figure 2.2

The Theory of Cutting-Planes

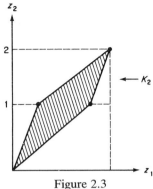

Figure 2.3

which are nonnegative combinations of inequalities in S_j (with their right-hand sides possibly reduced), and which differ only in their coefficient of z_{j+1} and their constant term, and putting into S_{j+1} the inequality

$$\theta_1 z_1 + \cdots + (w + P - T)z_{j+1} + \cdots + \theta_r z_r \geq P. \quad (2.149c)$$

Then S_r consists precisely of all valid inequalities for the convex span of solutions to (BIP).

One direction of Blair's result is easy: specifically, that S_r contains valid inequalities for the span of feasible solutions to (BIP). For S_0 contains such inequalities and, inductively, if S_j contains such inequalites, so does S_{j+1}. Indeed, for a feasible solution z, either $z_{j+1} = 0$, in which case (2.149c) is (2.149a), which was already assumed valid; or $z_{j+1} = 1$, in which case (2.149c) is equivalent to (2.149b), again assumed valid. In either case, (2.149c) is valid. The best proof of Blair's result remains the original one (Blair, 1976).

Blair's result can be sharpened, in that the sets $S_0, S_1, \ldots, S_r$ can be made finite by inductively placing in S_{j+1} only certain of the cuts (2.149c) obtainable from S_j. The conclusion then becomes the assertion, that the valid inequalities are exactly the nonnegative linear combinations of the inequalities in S_r (allowing that a constant term in an inequality may be decreased). See Jeroslow (1978) for details on sharpenings of this type.

It turns out that Blair's result can also be proven from that of Balas (see Jeroslow 1978; full details are given in the technical report preceding the published form). For other closely related results which generalize Blair's theorem, and are derivable using Balas' result, also see Jeroslow (1978).

Another way of looking at the Intersection Theorem is as a kind of 'game', in which there are two players, Player I (the 'indicating player') and Player II (the 'cutting player').

At each round of the game, there is a 'current polyhedron' P consisting of the solutions to (2.145a) plus all cutting-planes which have so far been appended (initially $P = \{z \mid Az \geq d, 0 \leq z \leq e\}$). Player I indicates an extreme point of P which is not a solution to (BIP), i.e. not a zero–one vector. Then Player II must cut off this extreme point with a valid inequality, and, to make his task more difficult, the valid inequality must depend on only one of the facial conditions of (2.145b). That is Player II must pick a single index j^*, and his cutting-plane must be valid for the set $P \cap \{z \mid -z_{j^*} \geq 0 \text{ or } z_{j^*} \geq 1\}$. Then the 'current polyhedron' becomes the previous P intersected with the solutions to the cutting-plane.

In this game, Player II wins only if, after some point of time, Player I cannot indicate an extreme point to cut off, and typically this will be due to the fact that all extreme points of the current polyhedron P are zero–one points. Thus if the game iterates indefinitely, Player I wins.

Balas' result can be cast in this form, for it represents such a game where Player I, the indicating player, allows himself to be restricted in the following way. Initially, he uses index '1' and indicates an extreme point with z_1 fractional; when there is no such point, he changes to index '2' and indicates an extreme point with z_2 fractional, and does not return to z_1 even if later there are extreme points with z_1 fractional; when there are no extreme points with z_2 fractional, he indicates one with z_3 fractional, not again returning to z_2, etc. However, when Player I reaches index r and finds no extreme point with z_r fractional, he then is free to pick any fractional extreme point.

Now the Intersection Theorem states that Player II must win if he plays correctly, since Player I will find no fractional extreme point once he is free (after index r). Indeed, the game starts with $P = K_0$ from (2.146a) and, once K_j is reached, inductively K_{j+1} can be reached by using a defining inequality of K_{j+1} in (2.146b) to cut away the fractional extreme point. (For, clearly, an extreme point of a polyhedron containing K_{j+1} and having z_{j+1} fractional cannot be in K_{j+1}; otherwise it must be an extreme point of K_{j+1}, while (2.146b) shows that all extreme points of K_{j+1} are in either $K_j \cap \{z_{j+1} \mid z_{j+1} = 0\}$ or $K_j \cap \{z_{j+1} \mid z_{j+1} = 1\}$, and so do not have z_{j+1} fractional.) When K_r is reached, by the Intersection Theorem there is no fractional extreme point that Player I can indicate, so he loses.

Since the Intersection Theorem can be understood as the assertion that a certain restricted version of this 'game' has a winning strategy for Player II, it is natural to ask if there is *always* a winning strategy for Player II. In fact there is, as is proven in Jeroslow (1977b), in the generality of bounded facial constraints. The details in that paper show how Player II's strategy consists of an algorithm based on finding cuts by pivoting in certain linear inequality systems until a suitable linear form, determined by the point to be cut off, becomes positive.

What is gained by freeing Player I from restrictions, and considering a

The Theory of Cutting-Planes

general 'game', is a freedom needed for algorithms for bounded systems of facial constraints. For instance, if a facial constraint system is to be solved by a classical dual cutting-plane algorithm, which involves adding a cut and then re-optimizing dually, the next extreme point thus found may have z_2 fractional and z_1 integral, even though there remain extreme points with z_1 fractional. If we view the indicating player (Player I) as showing the new optimal solution to Player II for Player II to cut away, then Player I is not obeying the restrictions placed on him in order to use the Intersection Theorem. Similar considerations for a general 'game' arise from primal algorithms for systems of facial constraints.

For many systems of facial constraints, if often happens that one can arrange for a cutting-plane to both cut off some indicated extreme point and make some known problem constraint redundant in an 'obvious' way. For instance, the cutting-plane may consist of a problem constraint that is strengthened strictly in exactly one coefficient. For another example with (BIP), the problem constraint

$$-8z_1 + 2z_2 + 5z_3 - z_4 \geq 10 \qquad (2.150)$$

is redundant after adding the cutting-plane

$$-8z_1 + 6z_2 + 5z_3 - z_4 \geq 14. \qquad (2.151)$$

To see this, multiply the inequality $-z_2 \geq -1$ by 4, and add it to (2.151) to get (2.150). The constraint (2.151) differs from (2.150) in that one coefficient, that of z_2, has been increased, and no other, while the constraint term is increased by the same amount.

Such redundancies caused by cutting-planes are easy to detect, and then when the cut is added the constraint can be removed. This procedure is then more like a 'constraint tightening' procedure than an ordinary cutting-plane algorithm, since there is no accumulation of new inequalities. Also, problem density remains low. Unfortunately, 'tightening' is not always possible, nor is it known if it produces finite convergence when used as the only technique even if it is always possible. When 'tightening' is possible, the method discussed in Jeroslow (1977b, Appendix II), often produces it. Of course, 'tightening' can be used in alternation with branching, in hybrid algorithms, where the branching feature gurantees convergence.

A more extensive but older survey of the disjunctive methods has appeared in Jeroslow (1977a). For the reader interested in the relationship between the disjunctive cuts and the earlier 'intersection' or 'convexity' cuts, both papers Balas (1975) and Jeroslow (1977a) should be helpful, together with other relevant material in Exercises 3 and 4 of Appendix 2.4. For an alternate formulation of Principle DC, see Glover's (1975) paper; the proof of equivalence is in Jeroslow (1977a, Section 1.2).

Appendix 2.4 Exercises with Solutions

Exercises

Exercise 1 (Balas, 1974a; Zwart, 1972) This exercise is motivated by separable programming, as approximated in its λ- and δ-forms.

Assume that we have continuous nonnegative variables $y_1, \ldots, y_s$ (which, in the intended application, would correspond to certain segments in the separable approximation), which are linked by the logical requirement:

$$\text{if } y_{i+1} > 0 \text{ then } y_i = 1 \text{ for } i = 1, \ldots, s-1. \tag{2.152}$$

Suppose there is a specific index, say $i = k$, for which (2.152) fails. Derive from the two tableau rows

$$y_k = y_{k0} + \sum_{j \in J} y_{kj}(-t_j)$$
$$y_{k+1} = y_{k+1,0} + \sum_{j \in J} y_{k+1,j}(-t_j) \tag{2.153}$$

in the nonbasic variables $t_j \geq 0$, a disjunctive cut that makes the current solution (i.e. all $t_j = 0$) infeasible, using (2.152). (In (2.153), the row for y_k may be a unit row, i.e. possibly y_k is nonbasic).

Exercise 2 Suppose that a set-covering constraint is imposed in zero–one variables x_k, i.e.

$$x_1 + \ldots + x_p \geq 1 \tag{2.154}$$

where the relevant tableau rows (possibly filled out with unit rows) are given by

$$x_k = a_{k0} + \sum_{j \in J} a_{kj}(-t_j). \tag{2.155}$$

Derive a disjunctive cut that makes the current solution infeasible if all $a_{k0} < 1$.

Exercise 3 (The strengthened principle of intersection/convexity cuts, as applied to polyhedra: from Balas, 1974a; Glover 1975) Suppose that, relative to some concept of feasibility, there is no feasible solution which satisfies all the inequalities

$$\sum_{j=1}^{r} a_{ij} x_j \leq a_{i0} \quad i = 1, \ldots, p \tag{2.156}$$

in which every $a_{i0} > 0$. The variables $x = (x_1, \ldots, x_r)$ are nonnegative. Show that the inequality

$$\sum_{j=1}^{r} \max_i \{a_{ij}/a_{i0}\} x_j \geq 1 \tag{2.157}$$

is valid.

The Theory of Cutting-Planes

Exercise 4 (Principle of intersection/convexity cuts for polyhedra: Balas, 1971; Glover, 1973; Young, 1971) With hypotheses as in Exercise 3, let $\bar{\lambda}_j \geq 0$ denote the intersection of the positive x_j-axis with the set of solutions to (2.156), with $\bar{\lambda}_j = \infty$ if there is no intersection.

Show that the inequality

$$\sum_{j=1}^{r} x_j(1/\bar{\lambda}_j) \geq 1 \tag{2.158}$$

is valid, interpreting $1/\infty = 0$.

Exercise 5 (Balas, 1974a) Suppose that the constraint

$$3x_1 + 4x_2 + 8x_3 + 4x_4 + 11x_5 = 15 \tag{2.159}$$

holds, in zero–one variables x_k, and that the tableau rows read as in (2.155). Using the fact that $a_{10} = 0.7$, $a_{20} = 0.3$, and $a_{40} = 0.8$, derive a disjunctive cut which makes the current solution infeasible.

Exercise 6 (Balas and Jeroslow, 1975) Suppose that the constraints

$$x_1 = \frac{1}{6} + \frac{7}{6}(-t_1) - \frac{2}{6}(-t_2) - \frac{3}{6}(-t_3) + \frac{5}{6}(-t_4) + \frac{11}{6}(-t_5)$$

$$x_2 = \frac{2}{6} + \frac{1}{6}(-t_1) + \frac{1}{6}(-t_2) + \frac{1}{6}(-t_3) - \frac{1}{6}(-t_4) + \frac{1}{6}(-t_5) \tag{2.160}$$

$$x_3 = \frac{3}{6} - \frac{2}{6}(-t_1) + \frac{4}{6}(-t_2) - \frac{1}{6}(-t_3) - \frac{1}{6}(-t_4) - \frac{5}{6}(-t_5)$$

are imposed on the binary (i.e. zero or one) variables x_1, x_2, x_3 which are also restrained by the 'set-covering' row

$$x_1 + x_2 + x_3 \geq 1. \tag{2.161}$$

First, note that, by Exercise 2 above, the inequality

$$\frac{2}{3}t_1 + \frac{2}{5}t_2 + \frac{3}{5}t_2 + \frac{1}{3}t_4 + \frac{5}{3}t_5 \geq 1 \tag{2.162}$$

is valid. Then apply Principle SDC, with the $d^{(h)}$ suitably chosen, to obtain the strengthened cut

$$-\frac{1}{5}t_1 + \frac{2}{5}t_2 + \frac{3}{5}t_3 + \frac{1}{4}t_4 - \frac{1}{4}t_5 \geq 1. \tag{2.163}$$

(*Hint*: if properly done, you will find that $b^{(i)} - d^{(i)} = 1$ for $i = 1, 2, 3$.)

Exercise 7 (A cut for Gomory's group problem: Gomory, 1965) Show that the inequality

$$0.83t_1 + 1.16t_2 + 1.23t_3 \geq 1 \tag{2.164}$$

is valid, relative to the constraint set

(GRP)
$$3.7t_1 + 9.3t_3 + 1.5t_3 = 0.5 \pmod{1}$$
$$2.2t_1 + 0.5t_2 + 1.2t_3 = 0.9 \pmod{1}$$
$$t_1, t_2, t_3 \geq 0, \text{ integer.}$$

(*Hint:* use Principle MCS with T and M suitably chosen, and the functions F_i derived from disjunctive considerations. Always chose the easiest multipliers. Example 3 revisited is relevant.)

Exercise 8 Prove Principle SDC by subadditive means. You may use all results from Part 1 and 2 and Appendices.

Solutions

Solutions to the exercises follow. Several exercises have alternate solutions; we give only one.

Exercise 1 We rewrite (2.152) as the disjunctive requirement
either
$$\sum_{j \in J} y_{k+1,j} t_j = y_{k+1,0}$$
or (2.165)
$$\sum_{j \in J} y_{k,j} t_j = y_{k0} - 1.$$

We obtain the cut
$$\sum_{j \in J} \max \{\lambda_1 y_{k+1,j}, \lambda_2 y_{k,j}\} t_j \geq \min \{\lambda_1 y_{k+1,0}, \lambda_2 (y_{k,0} - 1)\}. \quad (2.166)$$

Since (2.152) does not hold for $i = k$, we have $y_{k+1,0} > 0$ and $y_{k,0} < 1$. Hence we may chose multipliers $\lambda_1 = 1/y_{k+1,0}$, $\lambda_2 = 1/(y_{k,0} - 1)$, and the right-hand side in (2.166) becomes unity.

Many other choices of the multipliers make the right-hand side positive, and all these are acceptable.

Exercise 2 We deduce that at least one of the following inequalities hold:
$$\sum_{j \in J} (-a_{kj}) t_j \geq 1 - a_{k0}. \quad (2.167_k)$$

Dividing both sides in (2.167_k) by $(1 - a_{k0}) > 0$, we conclude that the cut
$$\sum_{j \in J} \max_k \{-a_{kj}/\{1 - a_{k0}\}\} t_j \geq 1 \quad (2.168)$$

holds.

The Theory of Cutting-Planes

When (2.154) is a set-partitioning constraint, i.e. equality holds in (2.154), stronger cuts are possible; see Balas (1974a).

Exercise 3 For any feasible solution, at least one of the inequalities

$$\sum_{j=1}^{r} a_{ij} x_j \geq a_{i0} \quad (2.169_i)$$

holds for some $i = 1, \ldots, p$. Divide (2.159_i) by $a_{i0} > 0$, and (2.157) follows, by Principle DC.

Exercise 4 One easily computes

$$\bar{\lambda}_j = \begin{cases} \infty & \text{if } a_{ij} \leq 0 \text{ for all } i \\ \min_i \{a_{i0}/a_{ij} \mid a_{ij} > 0\} & \text{if } a_{ij} > 0 \text{ for some } i \end{cases} \quad (2.170)$$

therefore

$$1/\bar{\lambda}_j = \begin{cases} 0 & \text{if } a_{ij} \leq 0 \text{ for all } i \\ \max_i \{a_{ij}/a_{i0} \mid a_{ij} > 0\} & \text{if } a_{ij} > 0 \text{ for some } i. \end{cases} \quad (2.171)$$

From (2.171), and the fact that

$$\max_i \{a_{ij}/a_{i0} \mid a_{ij} > 0\} = \max_i \{a_{ij}/a_{i0}\} \quad (2.172)$$

if $a_{ij} > 0$ for some i, we see that (2.158) is a weakening of (2.157), hence valid.

Note that, if all $a_{ij} \leq 0$, so that the jth positive coordinate axis lies completely in the set of solutions to (2.156), then the jth coefficient is nonpositive, and usually negative, in (2.157), while it is zero in (2.158). This can provide a strict strengthening of (2.158). However, when the jth positive coordinate axis has a finite intersection value $\bar{\lambda}_j$, the jth intercepts in (2.157) and (2.158) are the same.

This derivation of (2.158) shows that the linear independence assumptions of the earlier arguments in Balas (1971) were not needed. Also, in (2.156), one can delete inequalities which do not intersect the solutions to (2.156) at a point with x positive (i.e. in all coordinates), since that will not change the intersection of the solutions to (2.156) with the $x \geq 0$ (see Jeroslow, 1974b, Section 3). This can further strengthen (2.158).

Exercise 5 Clearly, one cannot have $x_1 = x_2 = x_4 = 0$ in (2.159). Hence $x_1 + x_2 + x_4 \geq 1$, and Exercise 2 applies.

Exercise 6 By logic, either $x_1 \geq 1$ or $x_2 \geq 1$ or $x_3 \geq 1$, i.e. either

$$(S_1) \qquad -\frac{7}{6} t_1 + \frac{2}{6} t_2 + \frac{3}{6} t_3 - \frac{5}{6} t_4 - \frac{11}{6} t_5 \geq \frac{5}{6}$$

or

(S_2) $$-\frac{1}{6}t_1 - \frac{1}{6}t_2 - \frac{1}{6}t_3 + \frac{1}{6}t_4 - \frac{1}{6}t_5 \geq \frac{4}{6}$$

or

(S_3) $$\frac{2}{6}t_1 - \frac{4}{6}t_2 + \frac{1}{6}t_3 + \frac{1}{6}t_4 + \frac{5}{6}t_5 \geq \frac{3}{6}$$

giving $b^{(1)} = 5/6$, $b^{(2)} = 4/6$, $b^{(3)} = 3/6$, and $t = 3$.

Since always $x_1, x_2, x_3 \geq 0$ we similarly obtain in Principle SDC that $d^{(1)} = -1/6$, $d^{(2)} = -2/6$, $d^{(3)} = -3/6$. Therefore $b^{(h)} - d^{(h)} = 1$ for $h = 1, 2, 3$.
From (2.134), a valid cut is

$$\sum_{j=1}^{5} \left(\max_{h=1,2,3} \{\lambda^{(h)}(a^{(jh)} + n_h)\} \right) t_j \geq \min_{h=1,2,3} \lambda^{(h)} b^{(h)} \quad (2.173)$$

whenever the n_h are integer (possibly negative) with $n_1 + n_2 + n_3 \geq 0$, and the $\lambda^{(h)} \geq 0$ are nonnegative scalars. Take, for instance, $\lambda^{(1)} = 6/5$, $\lambda^{(2)} = 6/4$, $\lambda^{(3)} = 6/3$, so that all $\lambda^{(h)} b^{(h)} = 1$. Then with $n_1 = n_2 = n_3 = 0$ we obtain the usual disjunctive cut

$$\frac{2}{3}t_1 + \frac{2}{5}t_2 + \frac{3}{5}t_3 + \frac{1}{3}t_4 + \frac{5}{3}t_5 \geq 1 \quad (2.174)$$

which is (2.162).

Let us now pick (n_1, n_2, n_3) to improve the coefficient 2/3 of t_1 in (2.174), if possible. Using $(n_1, n_2, n_3) = (1, 0, -1)$ will improve the coefficient, and that coefficient becomes

$$\max\left\{\frac{6}{5}\left(-\frac{7}{6}+1\right), \frac{6}{4}\left(\frac{1}{6}+0\right), \frac{6}{3}\left(\frac{2}{6}-1\right)\right\} = -\frac{1}{5}$$

as noted in (2.163).

To improve the coefficient of t_4, which is 1/3 in (2.162), note that $(n_1, n_2, n_3) = (1, 0, -1)$ gives a coefficient of

$$\max\left\{\frac{6}{5}\left(-\frac{5}{6}+1\right), \frac{6}{4}\left(\frac{1}{6}+0\right), \frac{6}{3}\left(\frac{1}{6}-1\right)\right\} = \frac{1}{4}$$

as noted in (2.163).

For an algorithm that finds the best (n_1, n_2, n_3) in problems of this type, see Balas and Jeroslow (1975) from which the above example is drawn.

Exercise 7 View the constraints (GRP) as

$$\binom{a_{11}}{a_{21}} t_1 + \binom{a_{12}}{a_{22}} t_2 + \binom{a_{13}}{a_{23}} t_3 \in T + M \quad (2.175)$$

The Theory of Cutting-Planes

with

$$T = \begin{pmatrix} 0.5 \\ 0.9 \end{pmatrix} \quad M = \left\{ \begin{pmatrix} z_1 \\ z_2 \end{pmatrix} \;\middle|\; z_1 \text{ and } z_2 \text{ integer} \right\} \quad (2.176)$$

and $a_{11} = 3.7$, $a_{12} = 9.3$, etc.
Now

$$\begin{pmatrix} a_{11} \\ a_{21} \end{pmatrix} t_1 + \begin{pmatrix} a_{12} \\ a_{22} \end{pmatrix} t_2 + \begin{pmatrix} a_{13} \\ a_{23} \end{pmatrix} t_3 - \begin{pmatrix} 0.5 \\ 0.9 \end{pmatrix} = \begin{pmatrix} z_1 \\ z_2 \end{pmatrix}$$

is an integer point $(z_1, z_2)^{tr}$, hence one of the following four possibilites must hold:

$$z_1 \geq 0 \quad \text{and} \quad z_2 \geq 0 \quad (2.177a)$$
$$z_1 \geq 0 \quad \text{and} \quad z_2 \leq -1 \quad (2.177b)$$
$$z_1 \leq -1 \quad \text{and} \quad z_2 \geq 0 \quad (2.177c)$$
$$z_1 \leq -1 \quad \text{and} \quad z_2 \leq -1. \quad (2.177d)$$

Hence one of the following four systems holds:

$$a_{11}t_1 + a_{12}t_2 + a_{13}t_3 \geq 0.5 \quad \text{and} \quad a_{21}t_1 + a_{22}t_2 + a_{23}t_3 \geq 0.9 \quad (2.178a)$$

$$a_{11}t_1 + a_{12}t_2 + a_{13}t_3 \geq 0.5 \quad \text{and} \quad (-a_{21})t_1 + (-a_{22})t_2 + (-a_{23})t_3 \geq 0.1 \quad (2.178b)$$

$$(-a_{11})t_1 + (-a_{12})t_2 + (-a_{13})t_3 \geq 0.5 \quad \text{and} \quad a_{21}t_1 + a_{22}t_2 + a_{23}t_3 \geq 0.9 \quad (2.178c)$$

$$(-a_{11})t_1 + (-a_{12})t_2 + (-a_{13})t_3 \geq 0.5 \quad \text{and} \quad (-a_{21})t_1 + (-a_{22})t_2 + (-a_{23})t_3 \geq 0.1. \quad (2.178d)$$

By Principle DC, (2.122) is always valid where

$$F_j((v_1, v_2)^{tr}) = \max \{(v_1/0.5) + (v_2/0.9), (v_1/0.5) - (v_2/0.1),$$
$$-(v_1/0.5) + (v_2/0.9), -(v_1/0.5) - (v_2/0.1)\} \quad (2.179a)$$

and

$$\pi_0 = 1. \quad (2.179b)$$

We then apply Principle MCS to obtain that the jth cut coefficient in (2.123) can be taken to be

$$F_j\left(\begin{pmatrix} a_{1j} \\ a_{2j} \end{pmatrix} + \begin{pmatrix} z_1 \\ z_2 \end{pmatrix} \right) \quad (2.180)$$

for any choice of integers z_1 and z_2, and we therefore pick z_i in an advantageous manner. We chose z_i to be either the negative of the integer part $\lfloor a_{ij} \rfloor$ of a_{ij} or to be $-(\lfloor a_{ij} \rfloor + 1)$, depending on which choice makes (2.180) smallest among these four possibilites.

For each one of the four possibilities, it is easy to identify which one of the four terms in the maximum of (2.179a) gives the value of this maximum. In detail:

$$F_j\left(\begin{pmatrix}v_1\\v_2\end{pmatrix}-\begin{pmatrix}\lfloor v_1\rfloor\\\lfloor v_2\rfloor\end{pmatrix}\right)=\underline{v}_1/0.5+\underline{v}_2/0.9 \qquad (2.181a)$$

$$F_j\left(\begin{pmatrix}v_1\\v_2\end{pmatrix}-\begin{pmatrix}\lfloor v_1\rfloor+1\\\lfloor v_2\rfloor\end{pmatrix}\right)=(1-\underline{v}_1)/0.5+\underline{v}_2/0.9 \qquad (2.181b)$$

$$F_j\left(\begin{pmatrix}v_1\\v_2\end{pmatrix}-\begin{pmatrix}\lfloor v_1\rfloor\\\lfloor v_2\rfloor+1\end{pmatrix}\right)=\underline{v}_1/0.5+(1-\underline{v}_2)/0.1 \qquad (2.181c)$$

$$F_j\left(\begin{pmatrix}v_1\\v_2\end{pmatrix}-\begin{pmatrix}\lfloor v_1\rfloor+1\\\lfloor v_2\rfloor+1\end{pmatrix}\right)=(1-\underline{v}_1)/0.5+(1-\underline{v}_2)/0.1 \qquad (2.181d)$$

where $\underline{x}$ abbreviates the fractional part of the number x, i.e. $\underline{x}=x-\lfloor x\rfloor$.

Therefore, the jth intercept can be taken to be the best of these, i.e. to be

$$\min\{\underline{v}_1/0.5+\underline{v}_2/0.9,\ (1-\underline{v}_1)/0.5+\underline{v}_2/0.9,$$
$$\underline{v}_1/0.5+(1-\underline{v}_2)/0.1,\ (1-\underline{v}_1)/0.5+(1-\underline{v}_2)/0.1\} \qquad (2.182)$$

since that would provide a weakening of (2.123). For example, the coefficient of t_1 will be (rounding up for validity):

$$\min\{0.7/0.5+0.2/0.9,\ 0.3/0.5+0.2/0.9,$$
$$0.7/0.5+0.8/0.1,\ 0.3/0.5+0.8/0.1\}=0.83 \qquad (2.183)$$

In the same manner, the other cut coefficients in (2.164) can be calculated.

Exercise 8 We modify and continue the subadditive derivation of Principle DC in equations (2.140)–(2.143). Again we consider Principle BMIP in Appendix 2.3, using equation (2.97) there, where we see that the matrices A and C of (2.97) are empty, i.e. there are only the (nonnegative) integer variables $x=(x_1,\ldots,x_s)$ and we rename s to r.

We set

$$B=\begin{bmatrix}A^{(1)}\\ \cdot\\ \cdot\\ \cdot\\ A^{(t)}\end{bmatrix} \qquad (2.140')$$

in (2.97), with $t=|H|$, and put S as in (2.141). Again we use the function G of (2.142), which we saw was subadditive. Next we define a new function Q by

$$Q((v^{(1)},\ldots,v^{(t)})^{\text{tr}})=\inf_{m\in M}\ G((v^{(1)},\ldots,v^{(t)})^{\text{tr}}+m) \qquad (2.184)$$

where M is as defined in (2.135b) of the proof of Principle SDC. By Exercise 1d of Appendix (2.1) (with G of 1d being $\equiv 0$, H of 1d set to G of (2.142), K = real space, $P = -M$), we see that Q of (2.184) is subadditive.

Note that $Q(b^{(k)})$, where $b^{(k)}$ is the kth column of B in (2.140'), does indeed provide the kth coefficient of the cut in (2.134). It remains only to verify the constant term of (2.134), and then appeal to the cut-form (2.198) from Appendix 2.3.
We have

$$\inf \{Q((v^{(1)}, \ldots, v^{(t)}) \mid (v^{(1)}, \ldots, v^{(t)}) \in S\}$$
$$= \inf \{G((v^{(1)}, \ldots, v^{(t)})^{tr} + m) \mid (v^{(1)}, \ldots, v^{(t)}) \in S\}$$
$$= \inf \{G((v^{(1)}, \ldots, v^{(t)})^{tr}) \mid (v^{(1)}, \ldots, v^{(t)}) \in S\}$$
$$= \inf_{h \in H} \{\lambda^{(h)} b^{(h)}\} \qquad (2.143')$$

as desired. The crucial third equality in (2.143') is due to the fact that $(v^{(1)}, \ldots, v^{(t)}) \in S$ and $m \in M$ implies $(v^{(1)}, \ldots, v^{(t)}) + m \in S$, by exactly the reasoning as in the proof of Principle SDC.

References

Balas, E. (1971). Intersection Cuts—A New Type of Cutting-Plane for Integer Programming. *Operations Research*, **19**, 19–30.

Balas, E. (1974a). Intersection Cuts from Disjunctive Constraints. *Man. Sci. Res. Rep. No. 330, Carnegie-Mellon University, February, 1974*.

Balas, E. (1974b). Disjunctive Programming: Facets of the Convex Hull of Feasible Points. *MSRR no. 348, Carnegie-Mellon University, July, 1974*.

Balas, E. (1975). Disjunctive Programming: Cutting-Planes from Logical Conditions. Talk given at *SIGMAP-UW Conference, April, 1974*. Published in O. L. Mangasarian, R. R. Meyer, and S. M. Robinson (Eds)., *Nonlinear Programming*, **2**, Academic Press, New York.

Balas, E. and R. Jeroslow (1975). Strengthening Cuts for Mixed Integer Programs. *MSRR no. 359, GSIA, Carnegie-Mellon University, February, 1975*.

Blair, C. E. (1976). Two Rules for Deducing Valid Inequalities for 0–1 Problems. *SIAM Journal on Applied Mathematics*, **31**, 614–617.

Blair, C. E. and R. G. Jeroslow (1976). A Converse for Disjunctive Constraints. *MSRR no. 393, GSIA, Carnegie-Mellon University, June, 1976*.

Cottle, R. W. and G. B. Dantzig (1968). Complementary Pivot Theory of Mathematical Programming. *Linear Algebra and Its Applications*, **1**, 103–125.

Eaves, B. C. (1971). The Linear Complementarity Problem. *Management Science*, **17**, 612–634.

Glover, F. (1973). Convexity Cuts and Cut Search. *Operations Research*, **21**, 123–134.

Glover, F. (1975). Polyhedral Annexation in Mixed Integer and Combinatorial Programming. *Mathematical Programming*, **9**, 161–188.

Glover, F. and D. Klingman, (1973). The Generalized Lattice-Point Problem. *Operations Research*, **21**, 135–141.

Gomory, R. E. (1960). An Algorithm for the Mixed Integer Problem. *RM-2597, RAND Corporation.*
Gomory, R. E. (1963). An Algorithm for Integer Solutions to Linear Programs. In R. Graves and P. Wolfe (Eds), *Recent Advances in Mathematical Programming,* pp. 269–302, McGraw Hill, New York.
Gomory, R. E. (1965). On the Relation Between Integer and Non-Integer Solutions to Linear Programs. *Proceedings of the National Academy of Sciences* **53,** 260–265.
Ho, A. (to be published). Cutting-planes for Disjunctive Programs: Balas' Aggregated Problem. *GSIA, Carnegie-Mellon University.*
Jeroslow, R. (1974a). Cutting-Planes for Relaxations of Integer Programs. *MSRR no. 347, Carnegie-Mellon University, July, 1974.*
Jeroslow, R. (1974b). The Principles of Cutting-Plane Theory: Part I. *GSIA, Carnegie-Mellon University, February, 1974.*
Jeroslow, R. (1977a). Cutting-Plane Theory: Disjunctive Methods. *Annals of Discrete Mathematics,* **1,** 293–339.
Jeroslow, R. (1977b). A Cutting-Plane Game and Its Algorithms. *D.P. 7724, CORE, Belgium, June, 1977.*
Jeroslow, R. (1978). Cutting Planes for Complementarity Constraints. D.P. 7707, CORE, Belgium, April 1977. To appear in *SIAM Journal of Control and Optimization,* January 1978.
Owen, G. (1973). Cutting-Planes for Programs with Disjunctive Constraints. *Journal of Optimization Theory and Its Applications,* **11,** 49–55.
Rockafellar, R. T. (1970). *Convex Analysis,* Princeton University Press, Princeton, New Jersey.
Young, R. D. (1971). Hypercyclindrically-Deduced Cuts in Zero–One Integer Programs. *Operations Research,* **19,** 1393–1405.
Zwart, P. (1972). Intersection Cuts for Separable Programming. *Washington University, St. Louis, January, 1972.*

CHAPTER 3

Subgradient Optimization

CLAUDIO SANDI
IBM Scientific Centre, Pisa

3.1 Introduction

As the term perhaps asserts, the subgradient notion is a generalization of the gradient notion, to be used when dealing with extremum problems for nondifferentiable functions. The definition is however restricted to concave (or likewise to convex) functions, and is simply the following: any vector $a(x_0)$ is a subgradient of the concave function $f(x)$ at x_0 if

$$f(x) - f(x_0) \leq a(x_0)(x - x_0) \quad \forall x \in S$$

where S is the domain of $f(x)$. Geometrically, $f(x_0) + a(x_0)(x - x_0)$ is the equation of a supporting hyperplane to the graph of $f(x)$ at x_0; and, if $f(x)$ is differentiable at x_0, the only such hyperplane is the tangent hyperplane, and $a(x_0) = [\nabla f]_{x_0}$ the gradient of $f(x)$ at x_0. In this sense the subgradient is a generalization of the gradient concept; as far as optimization is concerned, however, gradient methods differ from the subgradient method, since convergence is generally based on different criteria: monotonic increase of $f(x)$ for the former, and monotonic approach to the value of x for which $f(x)$ is maximum for the latter. This will be clear in the next sections, where the subgradient properties and the subgradient method are described in detail; here a survey of main application areas is given, together with some history and general comments. Even if applicable to a wider class of problems, the subgradient method has mainly been used for the solution of the following problem:

$$\begin{cases} \max_x f(x), x \in S \\ f(x) = \min_{i \in I} \{a_i x - b_i\} \end{cases} \quad \text{or} \quad \begin{cases} \min_x f(x), x \in S \\ f(x) = \max_{i \in I} \{a_i x - b_i\} \end{cases} \quad (3.1)$$

where S is a closed convex set of E^n, $a_i x$ is the inner product of the two real n-vectors a_i and x, while b_i is a scalar for all $i \in I = \{1, 2, \ldots, m\}$. As described in Section 3.2, (3.1) is the problem of maximizing (minimizing) a piecewise linear concave (convex) function over S, and may be encountered in a number of situations, the most important of which are illustrated in the following three subsections.

3.1.1 Solution of Linear Inequality Systems

$$\begin{aligned} \text{find} \quad & x = (x_1, \ldots, x_n) \quad \text{(if any)} \\ \text{such that} \quad & Ax \geq b \end{aligned} \quad (3.2)$$

where A is a real $m \times n$ matrix and b is a real m-vector. If, for each row a_i of A $(i = 1, \ldots, m)$, we consider the slack variable as a function of x:

$$s_i(x) = a_i x - b_i \quad (i = 1, \ldots, m)$$

and define:

$$s(x) = \min\{s_i(x) \mid i \in I\}$$

then problem (3.2) becomes:

$$\begin{aligned} \text{find} \quad & x \quad \text{(if any)} \\ \text{such that} \quad & s(x) \geq 0 \end{aligned}$$

which, in turn, recalls the problem of maximizing the feasibility function $s(x) = \min\{a_i x - b_i \mid i \in I\}$, i.e. problem (3.1).

3.1.2 Bounding Functions for Integer Linear Programming Problems

$$P = \begin{cases} \min cx \\ \text{s.t.} \ Ax \geq b \\ x \ \text{integer} \end{cases} \quad (3.3)$$

where cx is the inner product of the real n-vector c with the integer-valued n-vector of unknowns x, A is a real $m \times n$ matrix and b is a real m-vector. Let us introduce the following notation (see, for example, Geoffrion and Marsten, 1972):

$F(P)$ the set of feasible solutions for P

$$\text{opt}(P) = \begin{cases} +\infty & \text{if} \ F(P) = \emptyset \\ \min_x \{cx \mid x \in F(P)\} & \text{otherwise.} \end{cases}$$

P' a relaxation of P, i.e. a minimization problem such that $F(P) \subseteq F(P')$, and the objective function of P' is less than or equal to that of P on $F(P)$.

$P_1 \cdots P_N$ a separation of P, i.e. a list of problems such that $F(P) \subseteq F(P_1) \cup F(P_2) \cup \cdots \cup F(P_N)$.

Subgradient Optimization

$\underline{F}(P)$ the subset of $F(P)$ for which $\min_x \{cx \mid x \in \underline{F}(P)\}$ is known.

$$z^* = \begin{cases} +\infty & \text{if } \underline{F}(P) = \emptyset \\ \min_x \{cx \mid x \in \underline{F}(P)\} & \text{otherwise.} \end{cases}$$

When enumerative methods are used to solve (3.3), the essential point is to avoid direct inspection of (large) subsets of feasible solutions of P. This is accomplished by choosing a problem P_k from the separation list and showing:

$$\{x \mid cx < z^*, x \in F(P'_k)\} = \emptyset \qquad (3.4\text{a})$$

or simply

$$z^* \leq \text{opt}(P'_k) \qquad (3.4\text{b})$$

where P'_k is a 'computationally-convenient' relaxation of P_k. This because (3.4b) implies:

$$z^* \leq \text{opt}(P'_k) \leq \text{opt}(P_k) \leq \min_x \{cx \mid x \in F(P_k) \cap F(P)\}$$

and no improvement can be obtained from the subset $F(P_k) \cap F(P)$ of feasible solutions for P.

Usually $F(P_1), F(P_2), \ldots, F(P_N)$ is a disjoint partition of $F(P)$ and each P_k is obtained from P by assigning fixed integer values to an appropriate subset of variables; if J_k is the index set of such variables, then:

$$P_k = \begin{cases} \min cx \\ \text{s.t. } Ax \geq b \\ x_j = \bar{x}_j \quad j \in J_k \subset J = \{1, \ldots, n\} \\ x \text{ integer} \end{cases} \qquad (3.5)$$

and a relaxation P'_k to be used in (3.4) is generally obtained by dropping the integrality constraints in (3.5):

$$P'_k = \begin{cases} \min cx \\ \text{s.t. } Ax \geq b \\ x_j = \bar{x}_j \quad j \in J_k. \end{cases} \qquad (3.6)$$

Condition (3.4a) under (3.6) is equivalent to the inconsistency of the linear system:

$$\begin{cases} cx < z^* \\ Ax \geq b \\ x_j = \bar{x}_j \quad j \in J_k \end{cases} \qquad (3.7)$$

or, after eliminating the 'fixed' variables x_j, $j \in J_k$:

$$\begin{cases} \tilde{c}\tilde{x} < \tilde{z}^* \\ \tilde{A}\tilde{x} \geq \tilde{b} \end{cases} \quad (3.8)$$

where $\tilde{x}$ is the vector of the 'free' variables and $\tilde{A}, \tilde{b}, \tilde{c}, \tilde{z}^*$ are straightforward. Since a negative value for:

$$\begin{cases} \max_{\tilde{x}} \ f(\tilde{x}) \\ f(\tilde{x}) = \min\left\{\tilde{z}^* - \tilde{c}\tilde{x}; \min_{i \in I}\{\tilde{a}_i\tilde{x} - b_i\}\right\} \end{cases}$$

implies inconsistency for (3.8) (and therefore condition (3.4a) or (3.4b)), here again problem (3.1) is involved.

3.1.3 Bounding Functions for 0–1 Structured Problems

We consider:

$$\begin{cases} \min & cx \\ \text{s.t} & A_1 x \geq b_1 \\ & A_2 x = b_2 \\ & A_3 x \leq b_3 \end{cases} \quad (3.9)$$

where x is an n-vector of 0–1 variables; A_1, A_2, A_3 are $m_1 \times n$, $m_2 \times n$, and $m_3 \times n$ real matrices; and c, b_1, b_2, b_3 are real vectors of conformable dimensions. The last constraint set in (3.9) is supposed to offer such a structure as to make the solution of the problem:

$$\begin{cases} \min & cx + u(b_1 - A_1 x) + v(b_2 - A_2 x) \\ \text{s.t.} & A_3 x \leq b_3; \quad u \geq 0 \end{cases} \quad (3.10)$$

an easy task, for fixed m_1-vector u and m_2-vector v. For any $u \geq 0$ and arbitrary v, (3.10) is a relaxation of (3.9) (see, for example, Geoffrion 1974), since the contribution of $u(b_1 - A_1 x) + v(b_2 - A_2 x)$ to the objective function of (3.10) is nonpositive for all feasible solutions of (3.9) (remember furthermore the usual assumption opt $(P) = +\infty$ when $F(P) = \emptyset$). The best bound we can have from (3.10) is therefore:

$$\begin{cases} \max_{u,v} w(u, v) \quad u \geq 0 \\ w(u, v) = \min_{x}\{cx + u(b_1 - A_1 x) + v(b_2 - A_2 x) \mid A_3 x \leq b_3\}. \end{cases} \quad (3.11)$$

To give (3.11) the form (3.1), let us number the feasible solutions of (3.10)

Subgradient Optimization

as $x^{(1)}, x^{(2)}, \ldots, x^{(N)}$ (even if combinatorially large, they are finite, since x is an n-vector of Boolean variables and therefore $N \leq 2^n$); let further:

π be the $(m_1 + m_2)$-vector (u, v)

c_k be the scalar $cx^{(k)}$, $k \in K = \{1, 2, \ldots, N\}$

v_k be the $(m_1 + m_2)$-vector $(b_1 - A_1 x^{(k)}, b_2 - A_2 x^{(k)})$, $k \in K$,

Then (3.11) becomes

$$\max_{\pi} w(\pi) \quad \pi_1, \pi_2, \ldots, \pi_{m_1} \geq 0 \tag{3.12}$$

$$w(\pi) = \min_{k \in K} \{c_k + v_k \pi\}$$

that is, problem (3.1) again.

As in Sections (3.1.1)–(3.1.3), S of (3.1) may be E^n itself or a polyhedron; in this case (3.1) may be written as the linear programming problem:

$$\begin{aligned} \max \quad & z \\ \text{s.t.} \quad & z \leq a_i x - b_i \quad i \in I = \{1, 2, \ldots, m\} \quad x \in S \end{aligned} \tag{3.13}$$

and may be solved by the simplex method when its dimensions allow the use of existing LP codes. If this is not the case, for instance when m is (combinatorially) large as often happens in (3.1.3) above, (3.1) may be approached by a column generation technique (see, for example, Lasdon, 1970) on the dual of (3.13), provided that the problem of finding an index $h \in I$ such that

$$a_h x - b_h = \min_{i \in I} \{a_i x - b_i\}$$

is an easy task for all $x \in S$. However, the exceedingly slow convergence often encountered in column generation procedures (see, for example, Held and Karp, 1970), has suggested the trial of more effective methods to solve (3.1) for large values of m. The first use of the subgradient method (Held and Karp, 1970, 1971) was in this kind of problem, even though theoretical background can be found earlier in Agmon (1954) and Motzkin and Schoenberg (1954). A number of computerized examples have appeared in Held, Wolfe, and Crowder (1974), while improvements on the method can be found in Crowder (1974) and Sandi (1976).

In Section 3.2 a description of the method is given, together with mathematical contents, numerical examples, and illustrative plots for bi-dimensional problems.

In Section 3.3 some computational results are given, related to situations 3.1.1 and 3.1.2, about the use of the subgradient method for the solution of linear inequality systems, while situation 3.1.3 is illustrated by approaching the travelling salesman problem in terms of subgradient procedures.

Finally in Section 3.4 a few comments are given on positive aspects, difficulties encountered, and possible improvements of the subgradient optimization.

3.2 The Subgradient Method

As pointed out in Section 3.1, the subgradient method is a numerical technique, generally used to solve (3.1), i.e.

$$\begin{cases} \max_x f(x) & x \in S \\ f(x) = \min_{i \in I} \{a_i x - b_i\}. \end{cases}$$

Before describing the method, and for illustrative purposes, a simple example is given in Section 3.2.1; then definitions, properties, and details of the subgradient method are given in Sections 3.2.2 and 3.2.3.

3.2.1 An Example

Consider the two-dimensional problem:

$$\begin{cases} \max_x f(x) & x = (x_1, x_2) \in E^2 \\ f(x) = \min\{(x_1 + x_2 - 12); (9 - x_1); (9 - x_2)\}. \end{cases} \quad (3.14)$$

A plot and some contour lines of the concave and piecewise linear function $f(x)$ are given in Figures 3.1 and 3.2, respectively.

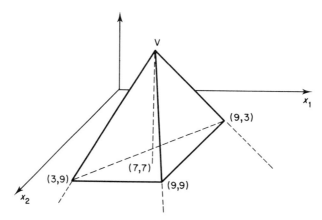

Figure 3.1 Plot of $f(x)$ as defined in (3.14)

Subgradient Optimization

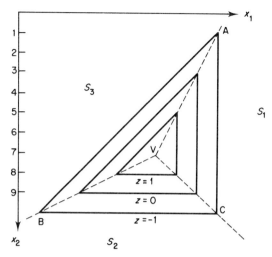

Figure 3.2 Some contour lines of $f(x)$ as defined in (3.14)

From Figure 3.2, E^2 is divided by the rays VA, VB, VC into three disjoint regions S_1, S_2, S_3 such that:

$$f(x) = \begin{cases} (9-x_1) & x \in S_1 \\ (9-x_2) & x \in S_2 \\ (x_1+x_2-12) & x \in S_3 \end{cases}$$

while on the disjoint rays VA, VB, VC, we have:

$$f(x) = \begin{cases} (x_1+x_2-12) & \text{or} \quad (9-x_1) & x \in \text{VA} \\ (x_1+x_2-12) & \text{or} \quad (9-x_2) & x \in \text{VB} \\ (9-x_1) & \text{or} \quad (9-x_2) & x \in \text{VC} \end{cases}$$

and finally:

$$f(x) = (x_1+x_2-12) \quad \text{or} \quad (9-x_1) \quad \text{or} \quad (9-x_2) \qquad x \equiv V$$

which characterizes the max $f(x)$.

Inside S_1, S_2, S_3, $f(x)$ is differentiable, since it coincides with (linear and therefore) differentiable functions; this is not true on the rays VA, VB, VC, which separate different values of ∇f as shown in Figure 3.3.

However, the directional derivative along an arbitrary unit vector d, i.e.

$$f'(x, d) = \lim_{t \to 0^+} (f(x+td) - f(x))/t \qquad (3.15)$$

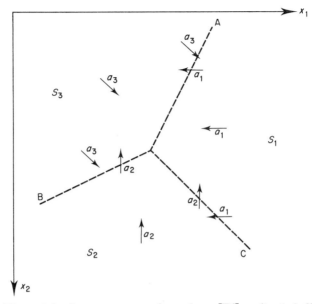

Figure 3.3 Some representations of $a_i = [\nabla f]_{x \in S_i}$ ($i = 1, 2, 3$)

exists for all $x, d \in E^2$, and gives a measure of the rate of change of $f(x)$ along d. Let

$$a_i = [\nabla f]_{x \in S_i} \qquad (i = 1, 2, 3)$$

and let $I(x)$ be the index set of the regions S_i adjacent to or containing the point x, i.e.

$$I(x) = \{i \mid f(x) = (a_i x - b_i), (i = 1, 2, 3)\}$$

and consider the set $A(x)$ of 'active gradients' in x, i.e.

$$A(x) = \{a_i \mid i \in I(x)\}.$$

It is easily shown that:

$$f'(x, d) = \min \{a_i d \mid a_i \in A(x)\} \tag{3.16}$$

and therefore $f'(x, d)$ is simply the (length of the) minimal component along d of the 'active gradients' of $f(x)$ at x. Figure 3.4 tries to visualize $f'(x, d)$ for $x \in S_3$ and $x \in VC$ and indicated d.

Property (3.16) may obviously be generalized as follows:

$$f'(x, d) = \min \{ad \mid a \in \mathbf{A}(x)\} \tag{3.17}$$

Subgradient Optimization

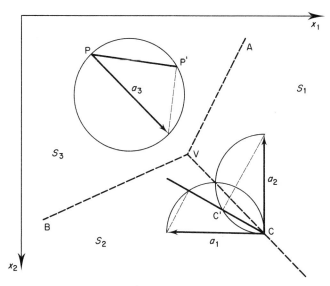

Figure 3.4 $f'(x, d) = |PP'|$, for $x \equiv P \in S_3$ and $d = PP'/|PP'|$.
$f'(x, d) = |CC'|$, for $x \equiv C$ and $d = CC'/|CC'|$

where $\mathbf{A}(x)$ is the set of all the convex combinations of 'active gradients' of $f(x)$ in x, i.e.

$$\mathbf{A}(x) = \left\{ a \;\middle|\; a = \sum_{i \in I(x)} \lambda_i a_i ; \sum_{i \in I(x)} \lambda_i = 1; \lambda_i \geq 0 \right\}.$$

Property (3.17) is illustrated in Figure 3.5: $I(C) = \{1, 2\}$; $A(C) = \{a_1, a_2\}$ is

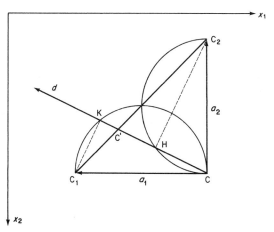

Figure 3.5 Convex combination of 'active gradients' and directional derivatives

the set of the 'active gradients' in C; d is an arbitrary direction; $\mathbf{A}(C) = \{CC' \mid C' \in C_1 C_2\}$; $|CH| \leq ad \leq |CK|$, $a \in \mathbf{A}(C)$; $f'(C, d) = |CH| = a_2 d$.

Peculiar to the element of $\mathbf{A}(x)$ is the property

$$f(y) - f(x) \leq a(y - x)$$

for all $x, y \in E^n$ and $a \in \mathbf{A}(x)$, as will be shown in the following ($\mathbf{A}(x)$ is usually called the subdifferential, while each $a \in \mathbf{A}(x)$ is a subgradient of $f(x)$ at x).

3.2.2 General Properties

After this illustrative example, let us return to the general problem (3.1) and recall some definitions and general properties, referring to Rockafellar (1970) for a complete treatment of the subject. For the sake of simplicity, we suppose $S = E^n$ and $f(x)$ bounded above; the extension to the general case can be found in Held, Wolfe, and Crowder (1974) and Oettli (1972) and does not involve particular difficulties.

Definitions

$$I(x) = \{i \mid f(x) = (a_i x - b_i), i \in I\}$$
$$A(x) = \{a_i \mid i \in I(x)\}$$
$$\mathbf{A}(x) = \left\{ a \mid a = \sum_{i \in I(x)} \lambda_i a_i; \sum_{i \in I(x)} \lambda_i = 1; \lambda_i \geq 0 \right\}$$

Proposition As defined in (3.1), $f(x)$ is concave. In fact:

$$f(z) = a_k z - b_k \qquad k \in I(z) \qquad (3.18a)$$
$$f(x) \leq a_k x - b_k \qquad (3.18b)$$
$$f(y) \leq a_k y - b_k \qquad (3.18c)$$

and, if $z = \lambda x + (1 - \lambda) y$ and $0 \leq \lambda \leq 1$, the sum of (3.18b) and (3.18c), with weights λ and $(1 - \lambda)$ respectively, gives:

$$\lambda f(x) + (1 - \lambda) f(y) \leq f(\lambda x + (1 - \lambda) y) \qquad \forall x, y \in E^n$$

which is a necessary and sufficient condition for the concavity of $f(x)$.

Proposition

$$\{x^* \mid f(x^*) \geq f(x) \, \forall x \in E^n\} \neq \emptyset.$$

In fact the LP problem: $\max \{w \mid w \leq (a_i x - b_i), i \in I\}$ is bounded and has feasible solutions such as $(\bar{x}, \bar{w} = \min \{a_i \bar{x} - b_i\})$; its optimum (x^*, w^*) gives a finite maximum for $f(x)$.

Subgradient Optimization

Definition A vector $a \in E^n$ such that $f(y) - f(x) \leq a(y - x)$, $y \in E^n$, is called a subgradient of $f(x)$ at x.

Definition The set of subgradients of $f(x)$ at x is the subdifferential $\partial f(x)$ of $f(x)$ at x.

Proposition
$$\partial f(x) = \mathbf{A}(x).$$

To prove this last proposition, we show first that if $a \in \mathbf{A}(x)$ then $a \in \partial f(x)$. In fact:

$$f(x) = a_i x - b_i \qquad i \in I(x) \tag{3.19a}$$
$$f(y) \leq a_i y - b_i \qquad i \in I(x). \tag{3.19b}$$

Subtracting (3.19a) from (3.19b), after ordinately multiplying by nonnegative weights λ_i with $\sum \lambda_i = 1$ and $\lambda_i \geq 0$, $i \in I(x)$, we have:

$$f(y) - f(x) \leq \sum_{i \in I(x)} \lambda_i a_i (y - x)$$

hence $\mathbf{A}(x) \subseteq \partial f(x)$. Conversely, suppose $a \in \partial f(x)$; for a small neighbourhood D of x we must have:

$$f(y) = \min_{i \in I(x)} \{a_i y - b_i\} \qquad \forall y \in D.$$

Therefore, if y and $(2x - y)$ are points of D, it follows:

$$f(y) = \min_{i \in I(x)} \{a_i y - b_i\} = a_k y - b_k \qquad k \in I(x)$$

$$f(2x - y) = \min_{i \in I(x)} \{a_i (2x - y) - b_i\} = a_h (2x - y) - b_h \qquad h \in I(x).$$

Also, since $a \in \partial f(x)$:

$$a_k(y - x) = (a_k y - b_k) - (a_k x - b_k) = f(y) - f(x) \leq a(y - x)$$
$$a_h(x - y) = a_h(2x - y) - b_h - (a_h x - b_h) = f(2x - y) - f(x) \leq a(x - y)$$

which gives:

$$a_k(y - x) \leq a(y - x) \leq a_h(y - x) \qquad \forall y \in D \tag{3.20}$$

and therefore $a \in \mathbf{A}(x)$, since otherwise (3.20) does not hold along the direction $a - \bar{a}'$, where $|a - \bar{a}| = \min\{|a - a'| \mid a' \in \mathbf{A}(x)\}$.

3.2.3 The Method

Let us consider (3.1) with $S = E^n$ and $f(x)$ bounded above so that $X^* = \{x^* \mid f(x^*) \geq f(x) \; \forall x \in E^n\} \neq \emptyset$. If t_r are positive quantities, the subgradient method of solution is simply given by the recursive formula:

$$x_0 \text{ arbitrary} \qquad (3.21)$$
$$x_{r+1} = x_r + t_r a_r \qquad a_r \in \mathbf{A}(x_r), t_r > 0$$

and is such that, for any arbitrary $q > 0$,

$$\exists r \Rightarrow f(x^*) - f(x_r) < q \qquad x^* \in X^* \qquad (3.22)$$

provided that:

$$\sum_0^\infty t_r = +\infty \qquad (3.23a)$$

$$t_r \xrightarrow[r \to \infty]{} 0. \qquad (3.23b)$$

This result may be considered as a special case of a more general statement given in Poljak (1967) for the solution of extremum problems. To prove (21)–(23), let us suppose by contradiction that:

$$\exists q > 0 \Rightarrow f(x^*) - f(x_r) \geq q \; \forall r. \qquad (3.24)$$

Since, by definition,

$$a_r x^* - b_r \geq f(x^*) \qquad r \in I(x_r)$$
$$a_r x_r - b_r = f(x_r) \qquad r \in I(x_r)$$

then subtracting and using (3.24) we have

$$a_r(x^* - x_r) \geq q$$

and multiplying by the negative quantity $(-2t_r)$, we have

$$2t_r a_r(x_r - x^*) \leq -2t_r q.$$

It follows that

$$|x_{r+1} - x^*|^2 = |x_r + t_r a_r - x^*|^2$$
$$= |x_r - x^*|^2 + |t_r a_r|^2 + 2 t_r a_r(x_r - x^*)$$
$$\leq |x_r - x^*|^2 + t_r^2 |a_r|^2 - 2 t_r q.$$

Now let $|a^*|^2 = \max\{|a_i|^2 \mid i \in I\}$; under (3.23b) an R can always be found such that:

$$t_r \leq q/|a^*|^2 \qquad \forall r \geq R$$

or equivalently:

$$t_r |a^*|^2 \leq q.$$

Subgradient Optimization

Then we can write:

$$|x_{r+1} - x^*|^2 \leq |x_r - x^*|^2 + t_r(t_r |a^*|^2 - 2q)$$
$$\leq |x_r - x^*|^2 - t_r q \qquad \forall r \geq R.$$

This last inequality, recursively written, gives for any arbitrary integer Q:

$$0 \leq |x_{Q+1} - x^*|^2 \leq |x_R - x^*|^2 - q \sum_R^Q t_r$$

which contradicts (3.23a).

Despite its simplicity, the subgradient method gives rise to a number of problems regarding the rate of convergence in connection with the nature of the 'step size' t_r and with the dimensions of the optimal solution set $\{x^* | f(x^*) \geq f(x), x \in E^n\}$; comments and references on this subject can be found in Held, Wolfe, and Crowder (1974) and Crowder (1974) where preference has been given to the choice:

$$t_r = \lambda_r [\bar{f} - f(x_r)]/|a_r|^2 \qquad 0 < \lambda_r \leq 2 \tag{3.25}$$

with $\bar{f} < f(x^*)$; this ensures for the sequence (3.21) either to contain a point x_r such that $f(x_r) \geq \bar{f}$, or to be such that $f(x_r) \xrightarrow[r \to \infty]{} \bar{f}$. All practical work is based on (3.25) (see, for example, Held and Karp, 1971; Held, Wolfe, and Crowder, 1974; Crowder, 1974; Sandi, 1976 with a 'guess value' for $\bar{f}$ and a number of heuristic criteria for λ_r. Even if convergence is lost in some cases, the method still works in many combinatorial optimization problems; an occasional loss of convergence while computing a bounding function, in fact, may simply involve some more iterations for this kind of problem.

Loss of convergence using (3.25) may also occur when solving linear inequality systems, since in these cases the 'guess value' used in (3.25) is a nonnegative value; when the system is inconsistent, this is an over-estimate of $s(x^*)$, and convergence is lost. The difficulty may be overcome by applying the subgradient method both to the given system $Ax \geq b$ and to its dual:

$$yA = 0; \qquad yb > 0; \qquad y \geq 0$$

since one (and only one) of the two systems is consistent; this, however, calls for double the computational effort required by the method.

Further considerations on the subgradient method can be found after the illustrative examples and the computational experience given in the next section.

3.3 Examples and Computational Experience

In this section a few examples have been chosen to show how the subgradient method behaves when solving (3.1) compared with simplex and column generation methods. The main purpose of these examples is to illustrate basic differences, rather than to give empirical proof of performance for the above methods. However, the extreme simplicity of a subgradient step has to be considered with respect to the relative complexity of a usual pivot operation.

3.3.1 Example 1

Let us consider the following consistent system of $2(n+1)+1$ linear inequalities in E^2 (a plot for $n=6$ is given in Figure 3.6):

$$\begin{aligned}
(n-i)x - iy &\geq -i^2 \quad (i=0, 1, \ldots, n) \\
-ix + (n-i)y &\geq -i^2 \quad (i=0, 1, \ldots, n) \\
x + y &\geq 2n - \tfrac{1}{2}.
\end{aligned} \tag{3.26}$$

For instance, when $n=6$ the system is:

$$\begin{aligned}
6x &\geq 0 \\
5x - y &\geq -1 \\
4x - 2y &\geq -4 \\
&\cdots\cdots\cdots\cdots \\
x + y &\geq 11.5.
\end{aligned} \tag{3.27}$$

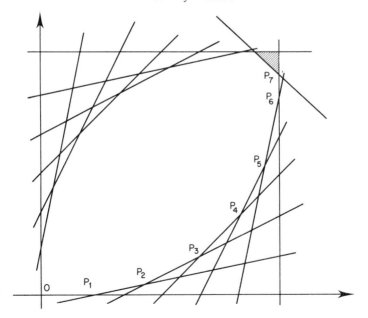

Figure 3.6 Plot of (3.26) for $n=6$; the small shaded area is the solution set

Subgradient Optimization

Table 3.1 Number of iterations required to solve (3.26) for some values of n with the indicated methods

n	Number of iterations	
	Simplex	Subgradient
10	11	8
50	51	16
100	101	19
200	201	22
300	301	24
400	401	25
500	501	25
600	601	27
700	701	27

Clearly, if a simplex technique is used to find a solution for (3.27) starting from the origin, a sequence such as the sequence $O, P_1, P_2, \ldots, P_7$ of Figure 3.6 has to be followed. More precisely, when (3.26) is approached by phase I of the simplex method (using one artificial variable only for the last inequality), $n+1$ iterations are required to obtain a solution.

If a subgradient technique is used to find a solution for (3.26) starting from the origin, the number of steps required is shown in Table 3.1, together with the iteration required by the simplex method. The chosen values of λ_r and $\bar{f}$ in (3.25) are 1.5 and 0, respectively.

3.3.2 Example 2

Let us consider the following linear inequality system (a plot is given in Figure 3.7):

$$\begin{cases} x & \geq 0 \\ y \geq 0 \\ -9x + 10y \geq 90 \\ 10x - 9y \geq 90. \end{cases} \quad (3.28)$$

A solution for (3.28) can be found by phase I of the simplex method, following a sequence such as O, P_1, P_2 of Figure 3.7.

If a subgradient method is used to find a solution for (3.28) starting from the origin and with $\lambda_r = 1.5$ and $\bar{f} = 0$, 1024 steps are required to find the solution (90.00, 90.00). The reason for this large number of steps depends on the strong 'zig-zagging' between the last two inequalities of (3.28). Clearly in this simple case efficient improvements can be given by finding

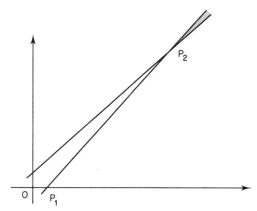

Figure 3.7 Plot of (3.28); the shaded area is the solution set

'persistent subgradient components' or updating empirically the value of r (see Held and Karp, 1974; Crowder, 1974). However, 'zig-zagging' phenomena are peculiar to subgradient techniques, and are hardly removed even when dealing with small problems.

3.3.3 Example 3

Consider the relaxation of the travelling salesman problem in the form given in Held and Karp (1970, 1971). Following these papers, let us introduce the following notation:

K_n undirected graph

$\{1, 2, \ldots, n\}$ vertex set of K_n

c_{ij} $(i, j = 1, \ldots, n)$ edge weights of K_n

$\{1, 2, \ldots, q\}$ index set of the 1-trees of K_n

c_k $(k = 1, 2, \ldots, q)$ weight of the kth 1-tree of K_n.

Let $p = (p_1, p_2, \ldots, p_n)$ be an arbitrary n-vector; with the transformation:

$$c'_{ij} = c_{ij} + p_i + p_j$$

the weights of the 1-trees become

$$c'_k = c_k + \sum_i d_{ki} p_i \qquad (k = 1, 2, \ldots, q)$$

where d_{ki} is the degree of vertex i in the kth 1-tree. A lower bound for a minimum tour of K_n is then given by

$$\max_p \left\{ \min_k \{c_k + v_k p\} \right\} \qquad (3.29)$$

Subgradient Optimization 89

where $v_k = (d_{k1} - 2, d_{k2} - 2, \ldots, d_{kn} - 2)$; or equivalently

$$\begin{cases} \max & w \\ \text{s.t.} & w \leq c_k + v_k p. \end{cases} \quad (3.30)$$

In Held and Karp (1970) this lower bound has been computed both with a column generation technique on the dual of (3.30), i.e.:

$$\begin{cases} \min & \sum_k c_k y_k \\ \text{s.t.} & \sum_k y_k = 1 \\ & \sum_k y_k v_{ki} = 0 \quad (i = 1, \ldots, n) \\ & y_k \geq 0 \quad (k = 1, \ldots, q) \end{cases} \quad (3.31)$$

and with a subgradient maximization of

$$w(p) = \min_k \{c_k + v_k p\}. \quad (3.32)$$

A number of problems with n up to 30 have been approached by the above methods. First a solution w^* to (3.31) has been found using column generation; then a subgradient technique has been used for maximizing (3.32) starting from $p = 0$ and with a step size based on $\bar{w} = w^*$ and $\lambda_r = 1.5$.

Table 3.2 summarizes some results for random generated problems with

Table 3.2 Number of iterations to solve problems (3.31), (3.32); each line refers to samples of size 10, except for $n \geq 20$ (single problems)

	Problem (3.31) Column generation			Problem (3.32) Subgradient		
n	Mean	Min	Max	Mean	Min	Max
8	22.1	8	33	19.4	3	81
9	28.8	17	35	11.3	3	39
10	45.9	18	80	26.6	3	98
11	64.2	35	96	49.8	4	259
12	89.4	36	122	28.3	6	101
13	129.7	88	213	35.7	8	108
14	146.8	86	207	29.2	4	53
15	167.7	87	288	358.3	12	1010
20	502	—	—	367	—	—
20	1610	—	—	266	—	—
20	642	—	—	63	—	—
20	413	—	—	508	—	—
30	4174	—	—	1302	—	—

weights uniformly distributed in (1, 100) and n from 8 to 15; each line refers to samples of size 10; results for single problems of size 20 and 30 are also shown.

3.4 Conclusions

The subgradient method for the solution of (3.1) seems to be very effective for (large) structured problems, while some refinements are needed for general unstructured problems. Often the performance of the method is complementary to that of the (column generation) simplex method; a general explanation of this behaviour may be the following:

(i) The column generation technique uses the basic philosophy of the simplex method, i.e. to iterate along the vertices of a polyhedron towards an optimum. All the constraints defining this polyhedron are always present while determining the iteration step: feasibility has to be maintained during the whole optimization process.

(ii) The subgradient technique uses an iterative procedure, where at each iteration one, and only one, unsatisfied constraint is responsible for the iteration step; all the other constraints are completely ignored.

These are perhaps the reasons for the possible loss of performance for the column generation (slowness) and the subgradient (zig-zagging) techniques respectively. In this context, the subgradient method may be considered as an extremely relaxed method if compared with the simplex method, since only one condition is taken into consideration at each step. Even if the natural effort is towards something in between the two cited methods (see, for example, Sandi, 1976), the major desire is, however, not to lose the peerless simplicity of the subgradient procedures.

References

Agmon, S. (1954). The relaxation method for linear inequalities. *Canadian Journal of Mathematics*, **6**, 382–392.

Crowder, H. (1974). Computational improvement of subgradient optimization. *IBM Technical Report RC* 4907, *Yorktown Heights, New York.*

Geoffrion, A. M. and R. E. Marsten (1972). Integer Programming Algorithms: A Framework and State-of-the-Art Survey. *Management Science*, **18**, 465–491.

Geoffrion, A. M. (1974). Lagrangean Relaxation for Integer Programming. *Mathematical Programming Study*, **2**, 82–114.

Held, M. and R. M. Karp (1970). The travelling-salesman problem and minimum spanning trees. *Operations Research*, **18**, 1138–1162.

Held, M. and R. M. Karp (1971). The travelling-salesman problem and minimum spanning trees: Part II. *Mathematical Programming*, **1**, 6–25.

Held, M., P. Wolfe, and H. P. Crowder (1974). Validation of subgradient optimization. *Mathematical Programming*, **6**, 62–88.

Lasdon, L. S. (1970). *Optimization theory for large systems*, Macmillan, London.

Motzkin, T. and I J. Schoenberg (1954). The relaxation method for linear inequalities. *Canadian Journal of Mathematics*, **6,** 393–404.

Oettli, W. (1972). An iterative method, having linear rate of convergence, for solving a pair of dual linear programs. *Mathematical Programming*, **3,** 302–311.

Poljak, B. T. (1967). A general method of solving extremum problems. *Soviet Mathematics Doklady*, **8,** 593–597. (Translation of *Doklady Akademii Nauk SSSR*, **174.**)

Rockafellar, R. T. (1970). *Convex analysis*, Princeton University Press, Princeton, N.J.

Sandi, C. (1976). A Direct Method for Linear Inequalities. Presented at *Euro II, Stockholm, Sweden, Nov. 29–Dec. 1, 1976*.

CHAPTER 4

A Partial Order in the Solution Space of Bivalent Programs

PETER L. HAMMER
Department of Combinatorics and Optimization, University of Waterloo, Waterloo, Ontario, Canada

SANG NGUYEN
Département d'Informatique, Université de Montreal, Montréal, Québec, Canada

Introduction

Order relations are constructed in the solution space of a linear 0–1 program and they are used to obtain information about the problem (infeasibility, forced values for certain variables, equality or non-equality of certain pairs of variables, etc.). On this basis an algorithm is described for solving 0–1 programs. Good computational experience is reported with the algorithm.

The chapter has three sections. Section 4.1 deals with the extraction of order relations from the constraints, and with the simplifications which can be obtained from the order relations. Section 4.2 gives on this basis an algorithm for the solution of 0–1 programs. The last section summarizes our computational experience, obtained on a CDC-6600.

The ideas of this paper are strongly related to those in Granot and Hammer (1972, 1974).

4.1 Order Relations in Bivalent Programming

4.1.1 Use of Order Relations in Solution Space

Consider a system of linear inequalities

$$\sum_{j=1}^{n} a_{ij}x_j \le b_i \quad (i=1,\ldots,m) \tag{4.1}$$

with bivalent variables

$$x_j \in \{0, 1\} \quad (j=1,\ldots,n). \tag{4.2}$$

Let us put

$$x^\pi = \begin{cases} x & \text{if } \pi = 1 \\ \bar{x}(=1-x) & \text{if } \pi = 0 \end{cases}$$

and denote for the sake of uniformity by x_0 the constant 0 and by $\bar{x}_0$ the constant 1.

Let S denote the set of all n-vectors $X = (x_1, \ldots, x_n)$ satisfying (4.1) and (4.2).

We shall say that the order relation

$$x_j^\alpha \leq x_k^\beta \qquad (j, k = 0, 1, \ldots, n; \alpha, \beta \in \{0, 1\}) \tag{4.3}$$

holds in the system (4.1), (4.2), if it holds for every $X \in S$. Such an order relation can be conveniently represented by the quadruple $(j, \alpha; k, \beta)$. For example $\bar{x}_3 \leq x_7$ will be denoted by $(3, 0; 7, 1)$, while $\bar{x}_4 \leq 0$ (implying $x_4 = 1$) will be denoted by $(4, 0; 0, 1)$.

Let $\tilde{R}$ be the set of all those order relations $(j, \alpha; k, \beta)$ which hold in the system (4.1), (4.2), R an arbitrary subset of $\tilde{R}$, and R^* the subset of $\tilde{R}$ obtained from R by putting

(i) $\forall\, (j, \alpha; k, \beta) \in R \Rightarrow (j, \alpha; k, \beta) \in R^*$
(ii) $\forall\, (j, \alpha; k, \beta) \in R^* \Rightarrow (k, \bar{\beta}; j, \bar{\alpha}) \in R^*$.

The knowledge of R^* allows us to obtain valuable information about the structure of the solution space S.

Some of the most important conclusions which can be obtained from R^* are the following:

(1) $(0, 0; 0, 1) \in R^* \Rightarrow S = \emptyset$
(2-1) $(j, \alpha; 0, 1) \in R^* \Rightarrow \forall X \in S, x_j = \bar{\alpha}$
(2-2) $(j, \alpha; j, \bar{\alpha}) \in R^* \Rightarrow \forall X \in S, x_j = \bar{\alpha}$
(2-3) $(j, \alpha; k, \beta) \in R^*, (j, \alpha; k, \bar{\beta}) \in R^* \Rightarrow \forall X \in S, x_j = \bar{\alpha}$
(3) $(j, \alpha; k, \beta) \in R^*, (k, \beta; j, \alpha) \in R^* \Rightarrow \forall X \in S, x_j^\alpha = x_k^\beta$.

Their proof is obvious and is omitted.

In the first case we see that the system (4.1), (4.2) (implying $1 \leq 0$) is infeasible; in the three variants of the second case we see that (at least) a variable has a forced value; in the third case we see that the system can be condensed.

We shall sometimes say that R^* is of type $t = 1, 2, 3$ if it is in case t but not in $t-1$. It will be said to be of type 4 if it is not of type 1, 2, 3.

Obviously, the larger the subset R of $\tilde{R}$, the greater the chances of obtaining direct conclusions about the system (4.1), (4.2). Similarly, the larger R^* the more conclusions we can get from it about our system.

4.1.2 Construction and Transitive Closure of R*

For the construction of R we shall first rewrite (4.1) in the form

$$\sum_{j=1}^{n} |a_{ij}| x_j^{\alpha_{ij}} \leq b_i^* \quad (i=1,\ldots,m) \tag{4.4}$$

where

$$\alpha_{ij} = \begin{cases} 1 & \text{if } a_{ij} \geq 0 \\ 0 & \text{if } a_{ij} < 0 \end{cases}$$

and

$$b_i^* = b_i - \sum_{j=1}^{n} \min(a_{ij}, 0).$$

Before constructing R we may notice that any inequality of (4.4) with

$$b_i^* \geq \sum_{j=1}^{n} |a_{ij}|$$

is redundant and can be eliminated from the system.

Construction of R:

I. $\exists i, b_i^* < 0 \Rightarrow (0, 0; 0, 1) \in R$

II. $\exists i, j, |a_{ij}| > b_i^* \Rightarrow \{j, \alpha_{ij}; 0, 1\} \in R$

III. $\exists i, j_1, j_2, |a_{ij_1}| + |a_{ij_2}| > b_i^* \Rightarrow \{j_1, \alpha_{ij_1}, j_2, \bar{\alpha}_{ij_2}\} \in R.$

As soon as steps I or II can be applied, the process stops either because the system is infeasible, or because it is forcing a variable and hence we can first perform this operation and then repeat the process for the new system.

Once R is constructed we pass to the construction of R^* using the rules (i. and (ii) of Section 4.1.1.

Finally, R^* can be enlarged to its transitive closure $\mathrm{TC}(R^*)$ by the repeated application of the following rules

(ζ) $(j, \alpha; k, \beta) \in R^* \Rightarrow (j, \alpha; k, \beta) \in \mathrm{TC}(R^*)$

(τ) $(j, \alpha; k, \beta) \in \mathrm{TC}(R^*), (k, \beta; h, \gamma) \in \mathrm{TC}(R^*) \Rightarrow (j, \alpha; h, \gamma) \in \mathrm{TC}(R^*).$

Obviously,

$$R \subseteq R^* \subseteq \mathrm{TC}(R^*) \subseteq \tilde{R}.$$

It is also clear that the process of constructing $\mathrm{TC}(R^*)$ is to be stopped, even if not yet completed, as soon as the already constructed subset of $\mathrm{TC}(R^*)$ is of type 1, 2, or 3.

4.1.3 First Example

Consider the system

$$-6x_1 + 8x_2 + 6x_3 - x_4 - 5x_5 + 3x_6 \leq -1$$
$$-2x_1 - 7x_2 + 5x_3 - 5x_4 + 5x_5 + 4x_6 \leq -4$$
$$-8x_1 - x_2 + 4x_3 + 4x_4 + 4x_5 - 3x_6 \leq -1$$

where
$$x_j \in \{0, 1\} \quad (j = 1, \ldots, 6).$$

Rewriting the system as
$$6\bar{x}_1 + 8x_2 + 6x_3 + \bar{x}_4 + 5\bar{x}_5 + 3x_6 \leq 11$$
$$2\bar{x}_1 + 7\bar{x}_2 + 5x_3 + 5\bar{x}_4 + 5x_5 + 4x_6 \leq 10$$
$$8\bar{x}_1 + x_2 + 4x_3 + 4x_4 + 4x_5 + 3\bar{x}_6 \leq 11$$
$$x_j \in \{0, 1\} \quad (j = 1, \ldots, 6)$$

(in order to make the relations clearer we shall write them here in the ordinary way, say $\bar{x}_1 \leq \bar{x}_2$, instead of $(1, 0; 2, 0)$).

The first inequality yields:
$$\bar{x}_1 \leq \bar{x}_2 \quad \bar{x}_1 \leq \bar{x}_3 \quad x_2 \leq \bar{x}_3 \quad x_2 \leq x_5$$

the second inequality yields
$$\bar{x}_2 \leq \bar{x}_3 \quad \bar{x}_2 \leq x_4 \quad \bar{x}_2 \leq \bar{x}_5 \quad \bar{x}_2 \leq \bar{x}_6$$

the third yields
$$\bar{x}_1 \leq \bar{x}_3 \quad \bar{x}_1 \leq \bar{x}_4 \quad \bar{x}_1 \leq \bar{x}_5.$$

Thus, R^* will consist of the inequalities

$$\bar{x}_1 \leq \bar{x}_2 \quad \bar{x}_1 \leq \bar{x}_3 \quad \bar{x}_1 \leq \bar{x}_4 \quad \bar{x}_1 \leq \bar{x}_5$$
$$x_2 \leq x_3 \quad x_2 \leq x_5 \quad \bar{x}_2 \leq \bar{x}_3 \quad \bar{x}_2 \leq x_4$$
$$\bar{x}_2 \leq \bar{x}_5 \quad \bar{x}_2 \leq \bar{x}_6$$

and of the relations

$$x_2 \leq x_1 \quad x_3 \leq x_1 \quad x_4 \leq x_1 \quad x_5 \leq x_1$$
$$x_3 \leq \bar{x}_2 \quad \bar{x}_5 \leq \bar{x}_2 \quad x_3 \leq x_2 \quad \bar{x}_4 \leq x_2$$
$$x_5 \leq x_2 \quad x_6 \leq x_2$$

obtained from the first set of relations by applying rule (ii).

We see that R^* is of type 2 (because $x_2 \leq x_3$ and $\bar{x}_2 \leq \bar{x}_3$), implying
$$x_3 = 0.$$

Introducing $x_3 = 0$ in R^*, we get R_1^* consisting of

$$\bar{x}_1 \leq \bar{x}_2 \quad \bar{x}_1 \leq \bar{x}_4 \quad \bar{x}_1 \leq \bar{x}_5 \quad x_2 \leq x_5$$
$$\bar{x}_2 \leq x_4 \quad \bar{x}_2 \leq \bar{x}_5 \quad \bar{x}_2 \leq \bar{x}_6 \quad x_2 \leq x_1$$
$$x_4 \leq x_1 \quad x_5 \leq x_1 \quad \bar{x}_5 \leq \bar{x}_2 \quad \bar{x}_4 \leq x_2$$
$$x_5 \leq x_2 \quad x_6 \leq x_2$$

(where the trivial relations $\bar{x}_1 \leq \bar{x}_3 = 1$ etc. were omitted).

A Partial Order in the Solution Space of Bivalent Programs

We can construct now $\text{TC}(R_1^*)$, which will contain all the relations in R_1^*, as well as

$$\bar{x}_1 \leqslant x_4 \quad (\text{for, } \bar{x}_1 \leqslant \bar{x}_2 \text{ and } \bar{x}_2 \leqslant x_4)$$

and

$$\bar{x}_4 \leqslant x_1 \quad (\text{for, } \bar{x}_1 \leqslant x_4).$$

Before continuing the construction of $\text{TC}(R_1^*)$ we remark that $\bar{x}_1 \leqslant \bar{x}_4$ and $\bar{x}_1 \leqslant x_4$ imply $\bar{x}_1 = 0$, or, equivalently,

$$x_1 = 1.$$

Replacing x_1 by 1 leads to R_2^*:

$$x_2 \leqslant x_5 \quad \bar{x}_2 \leqslant x_4 \quad \bar{x}_2 \leqslant \bar{x}_5 \quad \bar{x}_2 \leqslant \bar{x}_6$$
$$\bar{x}_5 \leqslant \bar{x}_2 \quad \bar{x}_4 \leqslant x_2 \quad x_5 \leqslant x_2 \quad \bar{x}_6 \leqslant x_2.$$

We see that R_2^* is of type 3 (for, $x_2 \leqslant x_5$ and $x_5 \leqslant x_2$), and we get

$$x_2 = x_5.$$

Replacing x_5 by x_2 in R_2^*, we get R_3^*:

$$\bar{x}_2 \leqslant x_4 \quad \bar{x}_2 \leqslant \bar{x}_6 \quad \bar{x}_4 \leqslant x_2 \quad \bar{x}_6 \leqslant x_2.$$

It is easy to see that $\text{TC}(R_3^*) = R_3^*$ is of type 4, and no further conclusions can be obtained from it.

Thus, returning to our original system, introducing there $x_3 = 0$, $x_1 = 1$ and replacing x_5 by x_2, we get:

$$3x_2 - x_4 + 3x_6 \leqslant 5$$
$$-2x_2 - 5x_4 + 4x_6 \leqslant -2$$
$$3x_2 + 4x_4 - 3x_6 \leqslant 7.$$

Obviously, the last inequality is redundant, and hence the system reduces to

$$3x_2 + \bar{x}_4 + 3x_6 \leqslant 6$$
$$2\bar{x}_2 + 5\bar{x}_4 + 4x_6 \leqslant 5$$

implying that $R_4^* = R_3^* \cup \{(2, 0; 4, 1), (2, 0; 6, 0), (4, 0; 6, 0)\}$.

Thus, our original system was considerably simplified by the use of its implicit order relations. We shall see that even more information can be obtained by using, beside R^* and $\text{TC}(R^*)$, relations between triplets of variables.

4.1.4 Ternary Relations

There is a possibility of obtaining a usually richer set R of order relations, if besides the obviously forced variables (unary relations $x_i^\alpha \leqslant 0$) and directly readable order relations (binary relations $x_j^\alpha \leq x_k^\beta$) we take into consideration also the existence of ternary relations in S. However, this option requires substantial computing time and the available experience does not justify a definite recommendation for or against its utilization. (The possibility of using general n-ary relations was discussed in Granot and Hammer, 1972, 1974.)

A ternary relation can be detected in an inequality

$$\sum_{j=1}^{n} |a_{ij}| x_j^{\alpha_{ij}} \leq b_i^*$$

if there are three distinct indices, j, k, l in $\{1, \ldots, n\}$, such that

$$|a_{ij}| + |a_{ik}| + |a_{il}| > b_i^*$$

but

$$|a_{ij}| + |a_{ik}| \leq b_i^* \qquad |a_{ij}| + |a_{il}| \leq b_i^* \qquad |a_{ik}| + |a_{il}| \leq b_i^*.$$

In this case every $X \in S$ must satisfy

$$x_j^{\alpha_{ij}} x_k^{\alpha_{ik}} x_l^{\alpha_{il}} = 0. \qquad (4.5)$$

If $j < k < l$, we shall denote such a relation by

$$(j, \alpha_{ij}; k, \alpha_{ik}; l, \alpha_{il}) \qquad (4.6)$$

and the set of all ternary relations obtained in this way will be denoted by T.

The following remarks are immediate:

R1. If $x_j = \bar{\alpha}_{ij}$ (i.e. $(j, \alpha_{ij}; 0, 1) \in R^*$) then (4.6) can be omitted.
R2. If $x_j = \alpha_{ij}$ (i.e. $(j, \bar{\alpha}_{ij}; 0, 1) \in R^*$) then (4.6) reduces to the order relation $(k, \alpha_{ik}; l, \bar{\alpha}_{il})$.
R3. If $(j, \alpha_{ij}; k, \bar{\alpha}_{ik}) \in R^*$ then (4.6) can be omitted.
R4. If $(j, \alpha_{ij}; k, \alpha_{ik}) \in R^*$ holds, then (4.6) reduces to the order relation $(j, \alpha_{ij}; l, \bar{\alpha}_{il})$.
R5. If $(j, \alpha_{ij}; k, \alpha_{ik}; l, \alpha_{il})$ and $(j, \alpha_{ij}; k, \alpha_{ik}; l, \bar{\alpha}_{il})$ hold, then both can be replaced by the order relation $(j, \alpha_{ij}; k, \bar{\alpha}_{ik})$. Similar rules apply if the role of j is played by k or by l, etc.

Hence the knowledge of R and of T leads to the construction of R' (with $R \subseteq R' \subseteq \tilde{R}$) and of $T' (\subseteq T)$; the repeated application of this idea may be very efficient in constructing a large subset of $\tilde{R}$.

4.1.5 Second Example

Consider the system

$$-6x_1 + 8x_2 + 6x_3 - x_4 - 4x_5 + 3x_6 \leq 1$$
$$-2x_1 - 6x_2 - 5x_3 + 2x_4 + 5x_5 + 4x_6 \leq -2$$
$$-7x_1 - x_2 \qquad + 4x_4 \qquad - 3x_6 \leq 0$$
$$2x_2 \qquad\qquad + 2x_5 + x_6 \leq 4$$

A Partial Order in the Solution Space of Bivalent Programs

with
$$x_j \in \{0, 1\} \quad (j = 1, \ldots, 6).$$

Rewriting the system as

$$6\bar{x}_1 + 8x_2 + 6x_3 + \bar{x}_4 + 4\bar{x}_5 + 3x_6 \leq 12$$
$$2\bar{x}_1 + 6\bar{x}_2 + 5\bar{x}_3 + 2x_4 + 5x_5 + 4x_6 \leq 11$$
$$7\bar{x}_1 + \bar{x}_2 + 4x_4 \qquad\qquad + 3\bar{x}_6 \leq 11$$
$$2x_2 \qquad\qquad\qquad + 2x_5 + x_6 \leq 4$$

we get for R^* the following set of order relations

$$\bar{x}_1 \leq \bar{x}_2 \qquad x_2 \leq x_1 \qquad x_2 \leq \bar{x}_3 \qquad x_3 \leq \bar{x}_2.$$

$R^*(= \text{TC}(R^*))$ is of type 4, and no conclusion can be drawn.
Let us construct T:

obtained from the first constraint
$$\begin{cases} (2, 1; 4, 0; 5, 0) \\ (2, 1; 5, 0; 6, 1) \\ (1, 0; 3, 1; 4, 0) \\ (1, 0; 3, 1; 5, 0) \\ (1, 0; 3, 1; 6, 1) \\ (1, 0; 5, 0; 6, 1) \\ (3, 1; 5, 0; 6, 1) \end{cases}$$

obtained from the second constraint
$$\begin{cases} (1, 0; 2, 0; 3, 0) \\ (1, 0; 2, 0; 5, 1) \\ (1, 0; 2, 0; 6, 1) \\ (1, 0; 3, 0; 5, 1) \\ (2, 0; 3, 0; 4, 1) \\ (2, 0; 3, 0; 5, 1) \\ (2, 0; 3, 0; 6, 1) \\ (2, 0; 4, 1; 5, 1) \\ (2, 0; 4, 1; 6, 1) \\ (2, 0; 5, 1; 6, 1) \\ (3, 0; 4, 1; 5, 1) \\ (3, 0; 5, 1; 6, 1) \end{cases}$$

obtained from the third constraint
$$\begin{cases} (1, 0; 2, 0; 4, 1) \\ (1, 0; 4, 1; 6, 0) \end{cases}$$

obtained from the fourth constraint
$$\begin{cases} (2, 1; 5, 1; 6, 1). \end{cases}$$

Now
$(1, 0, 2, 0)$ and $(1, 0; 2, 0; 3, 0)$ give $(1, 0; 3, 1)$ (by R4)
$(1, 0; 2, 0)$ and $(1, 0; 2, 0; 5, 1)$ give $(1, 0; 5, 0)$ (by R4)
$(1, 0; 2, 0)$ and $(1, 0; 2, 0; 6, 1)$ give $(1, 0; 6, 0)$ (by R4)
$(1, 0; 2, 0)$ and $(1, 0; 2, 0; 4, 1)$ give $(1, 0; 4, 0)$ (by R4)
$(1, 0; 3, 1)$ and $(1, 0; 3, 1; 4, 0)$ give $(1, 0; 4, 1)$ (by R4)

but
$(1, 0; 4, 0)$ and $(1, 0; 4, 1)$ give $(1, 0; 0, 1)$

or
$$x_1 = 1.$$

Hence R^* becomes now
$(2, 1; 3, 0)$ $(3, 1; 2, 0)$

and T becomes
$(2, 1; 4, 0; 5, 0)$ $(2, 1; 5, 0; 6, 1)$ $(3, 1; 5, 0; 6, 1)$
$(2, 0; 3, 0; 4, 1)$ $(2, 0; 3, 0; 5, 1)$ $(2, 0; 3, 0; 6, 1)$
$(2, 0; 4, 1; 5, 1)$ $(2, 0; 4, 1; 6, 1)$ $(2, 0; 5, 1; 6, 1)$
$(3, 0; 4, 1; 5, 1)$ $(3, 0; 5, 1; 6, 1)$ $(2, 1; 5, 1; 6, 1).$

Now
$(2, 1; 5, 0; 6, 1)$ and $(2, 1; 5, 1; 6, 1)$ give $(2, 1; 6, 0)$ (by R5)
$(2, 0; 5, 1; 6, 1)$ and $(2, 1; 5, 1; 6, 1)$ give $(5, 1; 6, 0)$ (by R5)
$(2, 1; 6, 0)$ and $(2, 0; 3, 0; 6, 1)$ give $(3, 0; 6, 0)$ (by R4)
$(2, 1; 6, 0)$ and $(2, 0; 4, 1; 6, 1)$ give $(4, 1; 6, 0)$ (by R4)
$(2, 1; 6, 0)$ and $(2, 0; 5, 1; 6, 1)$ give $(5, 1; 6, 0)$ (by R4)
$(5, 1; 6, 0)$ and $(3, 1; 5, 0; 6, 1)$ give $(3, 1; 6, 0)$ (by R4).

But now $(3, 0; 6, 0)$ and $(3, 1; 6, 0)$ give respectively $(6, 1; 3, 1)$ and $(6, 1; 3, 0)$, implying $(6, 1; 0, 1)$, or simply
$$x_6 = 0.$$

The introduction of $x_6 = 0$ into R leaves it unchanged, while T reduces to
$(2, 1; 4, 0; 5, 0)$ $(2, 0; 3, 0; 4, 1)$ $(2, 0; 3, 0; 5, 1)$
$(2, 0; 4, 1; 5, 1)$ $(3, 0; 4, 1; 5, 1).$

No further conclusions can be drawn from here, and the system reduces to
$$8x_2 + 6x_3 + \bar{x}_4 + 4\bar{x}_5 \leq 12$$
$$6\bar{x}_2 + 5\bar{x}_3 + 2x_4 + 5x_5 \leq 11$$
$$x_2, x_3, x_4 \in \{0, 1\} \qquad x_1 = 1 \qquad x_6 = 0$$

A Partial Order in the Solution Space of Bivalent Programs

(the other two constraints are redundant).

4.2 Solving Bivalent Programs

The ideas of Section 4.1 can be used for the solution of 0–1 programs. They are not sufficient in themselves to solve such problems, but—if combined with other procedures—can lead with remarkable efficiency to the solution.

Given a linear bivalent program:

$$\text{maximize} \sum_{j=1}^{n} a_{0j} x_j \qquad (4.7)$$

subject to

$$\sum_{j=1}^{n} a_{ij} x_j \leq b_i \quad (i = 1, \ldots, m) \qquad (4.8)$$

$$x_j \in \{0, 1\} \quad (j = 1, \ldots, n) \qquad (4.9)$$

(where, without any loss of generality, we may assume that all $a_{0j} \geq 0$, and with almost no loss of generality that all a_{0j} are integers), the order relations discussed in Section 4.1 can lead to: (A) recognition of infeasibility, (B) determination of values for certain variables, (C) reduction of the size of the problem due to condensations $x_j^\alpha = x_k^\beta$, etc. The list is unfortunately to be completed by: (D) no obvious conclusion.

In this last situation it seems advisable to solve the linear program P': max (4.7) subject to (4.8) and to

$$0 \leq x_j \leq 1 \quad (j = 1, \ldots, n). \qquad (4.10)$$

Let an optimal solution to this problem be $(\xi_1, \ldots, \xi_n)$ and let the optimal value of (4.7) be ζ. If the integer part $[\zeta]$ of ζ is denoted by z, then we may add the constraint

$$\sum_{j=1}^{n} a_{0j} x_j \leq z \qquad (4.11)$$

to (4.8). The resulting order relations and ternary relations can be added to R^* and to T respectively, having as possible effect the reduction to one of the situations A, B, or C.

Another possibility is to add to the optimal simplex tableau of P' those order relations $x_j^\alpha \leq x_k^\beta$ from R^* which are violated by its optimal solution (i.e. for which $\xi_j^\alpha > \xi_k^\beta$), as 'cuts' and to solve the corresponding post-optimization problem. This idea seems to us very promising.

A third possibility is to construct the surrogate constraint of Balas (1969), Geoffrion (1969), or Glover (1968) in order to derive from it new elements of R^* and T.

Finally, if none of the above approaches brings us to one of the situations

A, B, or C, we have to branch, i.e. to select a variable x_{i_0}, and examine separately the two subproblems resulting for $x_{i_0}=1$ and $x_{i_0}=0$.

It is to be mentioned that if $(x_1^*, \ldots, x_n^*)$ is a solution of P, with $\sum_{j=1}^{n} a_{0j}x_j^* = w^*$, the inequality

$$\sum_{j=1}^{n}(-a_{0_j})x_j^* \leq w^* - 1 \qquad (4.12)$$

to be satisfied by all those solutions of P which are better than $(x_1^*, \ldots, x_n^*)$ (if any) produces usually many useful new elements of R^* and T.

All the above can be summarized in the following algorithm:

Step I Produce R and R^*; if R^* is of type 1, stop; if it is of type 2 or 3, introduce the results in R^*, until arriving at a new R^* of type 1 or 4. If during this process there was an R^* of type 2 or 3, introduce the results in the original system and return to I. Otherwise go to:

Step II Produce T and enlarge R^*; if the new R^* is strictly larger than the original one, return to I; otherwise go to:

Step III Solve P'. Choose an ε ($0<\varepsilon<1$), put

$$x_j' = \begin{cases} 1 & \text{if } \xi_j \geq \varepsilon \\ 0 & \text{if } \xi_j < \varepsilon. \end{cases}$$

If $(x_1', \ldots, x_n')$ is a solution of P and

$$\sum_{j=1}^{n} a_{0j}x_j' = z \qquad (4.13)$$

the optimum has been found. If $(x_1', \ldots, x_n')$ is a solution of P but (4.13) does not hold, introduce the new constraint

$$\sum_{j=1}^{n} a_{0j}x_j \leq z \qquad (4.14)$$

and enlarge R^* and T by the new relations resulting from (4.14); if this is possible return to I. Otherwise go to:

Step IV Add to P' those inequalities (cuts) of R^* which are not fulfilled by $(\xi_1, \ldots, \xi_n)$ and return to III. If none, go to:

Step V Add to P' a surrogate constraint, and enlarge R^* and T by the resulting binary and ternary relations. If this is possible, return to I, otherwise go to:

Step VI Choose a variable x_j and examine separately the subproblems where $x_j=1$ and $x_j=0$. In the branching process, add the constraint (4.12) if w^* is the best known value of the objective function.

Example Maximize:
$$2x_1+3x_2+3x_3+5x_4+5x_5+6x_6 \tag{4.15}$$
subject to the constraints of the second Example (Section 4.1.5).

Steps I and II were examined in (4.1.5), and gave $x_1 = 1$, $x_6 = 0$, reducing the system of constraints to
$$8x_2+6x_3 - x_4-4x_5 \leqslant 12 \tag{4.16}$$
$$-6x_2-5x_3+2x_4+5x_5 \leqslant 11 \tag{4.17}$$
$$x_2, \quad x_3, \quad x_4 \in \{0, 1\} \tag{4.18}$$

and producing
$$R^* = \{(2, 1; 3, 0); (3, 1; 2, 0)\}$$
$$T = \{(2, 1; 4, 0; 5, 0); (2, 0; 3, 0; 4, 1); (2, 0; 3, 0; 5, 1);$$
$$(2, 0; 4, 1; 5, 1); (3, 0; 4, 1; 5, 1)\}.$$

Step III gives now
$$\xi_2 = \tfrac{3}{4} \quad \xi_3 = 1 \quad \xi_4 = 1 \quad \xi_5 = 1.$$
Choosing $\varepsilon = \tfrac{1}{2}$ we get $X' = (1, 1, 1, 1)$ which does not solve the constraints of P. The corresponding constraint $3x_2+3x_3+5x_4+5x_5 \leqslant 15$ does not enlarge either R^* or T.

Step IV requires the solution of the linear programming problem: maximize (4.15) subject to (4.16), (4.17), and to $x_2 \leqslant \bar{x}_3$, or
$$x_2+x_3 \leqslant 1 \tag{4.19}$$
with
$$0 \leqslant x_j \leqslant 1 \quad (j=2, 3, 4, 5). \tag{4.20}$$

The optimum of this problem is $(1, 0, 1, 4/5)$, the value of the objective function being 12. The corresponding 'rounded' integer point $(1, 0, 1, 1)$ does not satisfy our constraints. The new constraint
$$3x_2+3x_3+5x_4+5x_5 \leqslant 12$$
adds to T the elements $(2, 1; 4, 1; 5, 1)$ and $(3, 1; 4, 1; 5, 1)$. But $(2, 0; 4, 1; 5, 1)$ and $(2, 1; 4, 1; 5, 1)$ produce a new element $(4, 1; 5, 0)$ of R^*.

We return to Step IV adding the constraint given by $(4, 1; 5, 0)$:
$$x_4+x_5 \leqslant 1. \tag{4.21}$$

Solving (by post-optimization techniques) the resulting linear program (maximize (4.15) subject to (4.16), (4.17), (4.19), (4.20), (4.21)) we find that its optimum
$$x_2 = 0 \quad x_3 = 1 \quad x_4 = 0 \quad x_5 = 1$$

is integer, and hence the problem is solved. Thus, the optimal solution of P is $(1, 0, 1, 0, 1, 0)$.

4.3 Computational Experience

A variant of the above described algorithm was programmed in Fortran and test problems were run on a CDC-6600; the version of our algorithm which was coded differs from the one described in the paper in two points.

(P1) Ternary relations were not taken into account;
(P2) The binary relations were only used in order to fix certain variables, or as cuts; however conclusions of the form $x_i = x_j$ or $x_i = \bar{x}_j$ which can be derived from them were not taken into account.

The authors believe that by implementing these two additions (especially the second one), the algorithm could be much improved.

Under its present form the program requires approximately $2(m+5)^2 + 2mn + 12(m+n) + 10\,000$ words. The number of binary relations taken into account was limited to 500, and that of 'binary cuts' to 5.

For certain test problems of Petersen (1967) and of Bouvier and Messoumian (1965), the performance was compared to that of Geoffrion's algorithm; the results are given in Table 4.1. A sequence of test problems was generated randomly. All of them involve 15 constraints and the coefficients matrix has a density of 0.25, the coefficients being uniformly distributed between -100 and $+100$. The durations and the numbers of iterations reported in Table 4.2 are the averages for five test problems of the given dimensions.

Table 4.1

	Problem m, n	Aposs Duration (s)	Iterations	Geoffrion Duration (s)	Iterations
Petersen					
	10, 28	1.74	74	3.00	167
	5, 39	2.70	191	9.17	237
	5, 50	6.62	399	16.17	405
Bouvier and Messoumian					
	20, 20	0.90	6	1.05	27
	20, 20	12.88	156	12.27	163
	20, 23	34.70	572	69.93	1057
	20, 25	2.14	48	2.00	55
	20, 27	43.75	703	66.07	879
	20, 28	60.05	869	22.82	337

A Partial Order in the Solution Space of Bivalent Programs

Table 4.2

Number of variables	Duration (s)	Number of iterations
30	0.35	20
40	0.90	45
50	7.86	343
60	5.59	198
70	10.70	472
80	14.14	486
90	10.44	370
100	20.19	653
110	5.00	212
120	32.71	1088
130	17.96	513
140	85.41	2261
150	21.82	542
160	15.87	406
170	65.21	1578
180	50.21	968
190	23.65	632

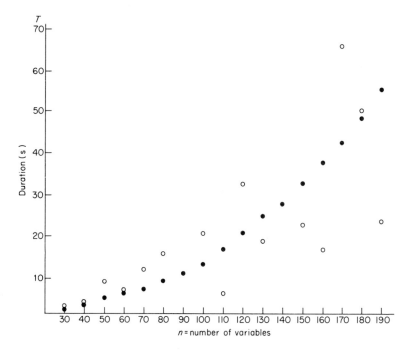

Figure 4.1   ○ Experimental time (average for 5 problems). ● Value of the function $t = A \times n^p$ ($A = 4.75 \times 10^{-4}$; $p = 2.22$)

By least square fitting we arrive at the theoretical curve of Figure 4.1, showing that the computational time grows proportionally with $n^{2.22}$.

The largest solved problem involved 22 constraints and 578 variables; the optimal solution was found in 1.12 seconds, and 300 additional seconds were insufficient to prove optimality.

Conclusions

(1) The performance of the presently coded form of the algorithm compares favourably with that of other known algorithms.

(2) The use of order relations in the presently coded form of the algorithm does not seem to influence substantially the total computing time, but reduces the number of iterations.

(3) The introduction of the improvement (P.1) leading to the detection of more order relations), and especially that of (P.2) (allowing a reduction of the number n of variables to be considered), could improve (perhaps substantially) the performance of the presently coded version of the algorithm.

(4) The algorithm can easily be generalized for quadratic (or other types of nonlinear) 0–1 programming, offering in that case (perhaps major) advantages over other methods.

References

Balas, E. (1967). Discrete Programming by the Filter Method. *Operations Research*, **15,** 915–957.

Bouvier, B. and G. Messoumian (1965). Programmes linéaires en variables bivalentes—Algorithme de Balas. Université de Grenoble, France, juin, 1965.

Geoffrion, A. M. (1969). An Improved Implicit Enumeration Approach for Integer Programming. *Operations Research*, **17,** 437–454.

Glover, F. (1968). Surrogate Constraints. *Operations Research*, **16,** 741–749.

Granot, F. and P. L. Hammer (1972). On the Use of Boolean Functions in 0–1 Programming. *Methods of Operations Research*, **12,** 154–184.

Granot, F. and P. L. Hammer (1974). On the Role of Generalized Covering Problems. *Cahiers du Centre d'Etudes de Recherche Opérationnelle*, **16,** 277–289.

Petersen, C. C. (1967). Computational Experience with Variants of the Balas Algorithm Applied to the Selection of R & D Projects. *Management Science*, **13,** 736–750.

CHAPTER 5

The Complexity of Combinatorial Optimization Algorithms and the Challenge of Heuristics

FRANCESCO MAFFIOLI
Istituto di Elettrotecnica ed Elettronica, Politecnico di Milano

5.1 Introduction

In recent years there has been a growing interest in the subject of this chapter and as a consequence a very large body of literature has appeared. This survey is neither meant to be exhaustive, nor meant to present the subject in a new light. It consists essentially in reviewing, in as elementary a manner as possible, what the author believes are the fundamentals of the subject bearing in mind its practical rather than theoretical implications: this implies often the use of nonspecialist language.

The following section will introduce the criteria for measuring the efficiency of algorithms and the consequent classification of existing methods. Section 5.3 will present examples of efficient algorithms, while Section 5.4 will deal with problems for which it is very unlikely that efficient methods will ever be found and which are therefore classified as 'hard' problems, presenting briefly the rationale upon which this conjecture is based. At this point the reader should already be sufficiently motivated towards the subject of the second part of this work, that is heuristic methods, their design, and their evaluation both from a deterministic (Section 5.6) and probabilistic (Section 5.7) point of view.

5.2 Measuring Complexity and Classes of Problems

Efficiency results always from an attempt to optimize some combination of resources. In the case of computer algorithms the two principal resources are: (a) *time* i.e. number of elementary computational steps; (b) *space* i.e. magnitude of internal memory required.

We need also a norm by which two algorithms for the same problem may be compared. The two most frequently used norms are the *worst case* norm and the *expected case* norm. Algorithm A' is better than A'' for a given problem P of size n according to the worst case (expected case) norm if the longest (average) time that A' takes for any single problem (for all these problems) is smaller than the corresponding norm for A''. Similar comparisons can be made in terms of memory space rather than time.

Let $c_A(n)$ be the norm measuring the complexity (in time or space, worst or average case) of algorithm A for a problem P of size n. This function may contain in general logarithmic, polynomial, exponential, and factorial terms.

A fairly common and motivated assumption says that algorithm A is *efficient* if $c_A(n)$ contains terms of the first two types only. In fact let $c_{A'}(n) = k'n^h$ $(h>0)$, while $c_{A''}(n) = k''a^n$ $(a>1)$ where k', k'', h, a are arbitrary constants. Then there exists an $\bar{n}$ sufficiently large for which $c_{A''}(\bar{n}) > c_{A'}(\bar{n})$. It is of course possible that $k' \gg k''$ and a is very near to 1 so that $\bar{n}$ would be extremely large. This is however a very rare occurrence, while practical combinatorial optimization problem sizes are very often quite large.

Algorithms which are efficient in the above mentioned sense w.r.t. the worst case norm will be called *polynomial bounded* (in time and/or space).

Of great theoretical relevance is also the problem of evaluating lower bounds to the complexity of algorithms, i.e. the question of how near to the maximum possible theoretical efficiency a given algorithm is. This aspect of complexity theory will be considered only very superficially in this work.

Finding the maximum (minimum) among n given numbers, finding the maximum *and* the minimum among n given numbers (Pohl, 1970), constructing an ordered list from two given ordered lists of numbers (Knuth, 1973), maximizing a unimodal function (by Fibonacci method) (Wilde, 1964) are all problems for which the corresponding algorithms have been shown to be optimal w.r.t. the worst case time norm. Dijkstra (1959) and Prim (1957) methods for computing shortest paths and shortest spanning trees of weighted complete graphs appear to be optimal within a multiplicative factor. Both are $O(n^2)$, where by this notation we mean that the highest order term of the polynomial giving their worst case time norm varies as a quadratic function of the number of vertices of the graph.

A similar situation applies to Hopcroft and Tarjan's (1971) method for determining whether a graph is planar and to Blum's algorithm for finding the median of n numbers (Blum et al., 1972).

Several sorting methods exceed only by a little the lower bound of $O(n \log n)$ for their worst case time norm and may be therefore classified as quasi-optimal.

A large class is then constituted by those problems for which polynomial bounded methods exist. The recently published book of Lawler (1976) is probably the most exhaustive and authoritative reference on this matter, which includes problems such as bipartite matching, shortest path, network flow, and two matroids intersection, many of which are studied in other contributions to this volume.

(Apparently) nonpolynomial bounded problems will be considered in later sections where a brief account of the works of Cook, Karp, Lawler, and many others will be summarized.

5.3 Good Combinatorial Optimization Algorithms

This section does not have the purpose of reviewing all efficient algorithms: even briefly this task would be enormous (the reader can refer to Lawler (1976) as well as to other parts of this volume). Our intention is rather to clarify by some examples how the complexity of an algorithm is evaluated with respect to the previously defined norms.

As a first example let us consider a rectangular $m \times n$ matrix A and let $\mathcal{I}$ denote the family of all linearly independent sets of columns of A (the elements of A are considered in any field). Let any column be weighted by a real number w_j, $j = 1, 2, \ldots, n$, and assume without loss of generality that $w_1 \geq w_2 \geq w_3 \geq \cdots \geq w_n$. It is required to find a set I of columns of A which belongs to $\mathcal{I}$ (i.e. such that all columns of I are linearly independent) and such that the sum of their weights is maximum.

A 'greedy' algorithm is proved to yield the optimum solution in this case and may be summarized by the following rule: put into the optimum set of columns I the column of largest weight disregarding a column whenever it would destroy the independence of I if adjoined to it. A more detailed implementation of this algorithm follows (Lawler, 1976).

Step 0 $k \leftarrow 1$

Step 1 If column k is zero, go to Step 2. Otherwise, choose any nonzero entry in the column, say a_{ik}. Subtract a_{ij}/a_{ik} times column k from each column $j > k$.

Step 2 If $k < n$, set $k \leftarrow k+1$ and go to Step 1. Otherwise stop. The nonzero columns identify the set I.

Note that in this implementation the testing for linear independence is carried out systematically by Gaussian elimination.

It is easy to see that the main loop of the algorithm is executed n times and that at worst m comparisons will be needed to test if a given column is zero. The process of modifying the remaining part of matrix A once a nonzero element is found requires as many elementary operations as there are elements in this part of the matrix, hence again a number of steps which is of the order of nm. The whole algorithm is therefore $O(n^2m)$ w.r.t. the worst case time norm. The memory space required is obviously of the order of magnitude of the number of elements of the matrix i.e. $O(mn)$.

The task of identifying the broader possible class of combinatorial optimization problems for which good algorithms exist has been attempted with considerable success in Lawler (1976). Roughly speaking this class corresponds to those problems for which the concept of *augmenting sequence* may be defined and exploited as a tool for reaching the optimum by successive augmentations of an initial suboptimum solution.

Most problems of this class may be approached as network flow problems. Network flow problems have seen in recent years the development of successively better algorithms (Edmonds and Karp, 1972; Dinic, 1970; Karzanov, 1974) so that now an $O(n^3)$ method exists for their solution, while Edmonds and Karp (1972) were the first to prove that a careful implementation of the Ford and Fulkerson method would yield a polynomial bounded computation (of $O(n^5)$).

Two generalizations of network flow problems have been successfully solved by polynomial bounded algorithms: nonbipartite matching problems (Edmonds, 1965) and matroid intersection problems (Lawler, 1975). Both problems may be viewed as particular cases of 2-parity matroid problems, or of bitroid problems (Sakarowitch, 1976). If solved by a polynomial bounded computation, any one of these problems would constitute nowadays the most general class of problems solvable in polynomial time. Unfortunately we do not yet know if this is possible, although augmenting sequences do exist for the 2-parity problem (see Lawler, 1976). The generality of such problems arises from the fact that their formulation requires a very simple structure. Let in fact E be any (finite) set and $P(E)$ the family of all its subsets. Let $\mathscr{I} \subset P(E)$ be such that for all $I \in \mathscr{I}$, $I' \subset I \to I' \in \mathscr{I}$. Then $H = (E, \mathscr{I})$ is an independence system and the members of $\mathscr{I}$ are called its independent sets. Maximal independent sets are called the bases of H, $\mathscr{D} = P(E) - \mathscr{I}$ is the set of all dependent sets, and minimal dependent sets C are called the circuits of H. If any $I \in \mathscr{I}$ is such that, for all elements $e \in E - I$, $I \cup \{e\}$ contains at most two circuits, H is a *bitroid*, if $I \cup \{e\}$ contains at most one circuit, H is a *matroid*. Once a weighting function $w : E \to Z$ is given, the bitroid problem requires to find an independent set of maximum total weight. The fact that a greedy algorithm gives an optimum solution to the previously presented matrix problem can be viewed in a new light considering that the set of the columns of a matrix and the family of its subsets containing only linearly independent columns constitute a matroid. In fact not only does the greedy algorithm give an optimum solution to any (one-) matroid problem, but also the converse is true, that is if the greedy algorithm reaches always an optimum the structure is a matroid.

On the other hand, as far as matching problems are concerned a fairly complete set of efficient algorithms is available; their time performances are summarized in Table 5.1. The first polynomial bounded method for unweighted nonbipartite matching of $O(n^4)$ was due to Edmonds (1965).

Table 5.1 Matching problems m is the number of arcs and n the number of vertices of the given graph

	unweighted	weighted
Bipartite	Hopcroft–Karp (1973) $O(mn)$	$O(n^3)$
Nonbipartite	Even–Kariv (1975) $O(n^{2.5})$	Gabow–Lawler (Gabow 1976) $O(n^3)$

A very interesting and active field has also been concerned in recent years with particular cases of problems of relevant practical importance for which the exploitation of their structure could yield better methods of solution. It would be impossible to list here all such cases (see Maffioli, 1973); nevertheless we cannot avoid mentioning the recently developed field of *computational geometry* (Shamos, 1975; Preparata and Hong, 1977). It may well be said that these studies are the first which show a considerable advantage in working with the euclidean metric in comparison with a completely general metric, for several graph optimization problems.

Consider for instance the minimum spanning tree problem, for which it is well known that Prim's algorithm (Prim, 1957) is $O(n^2)$ and is to be preferred when the original graph is complete, while Kruskaal's algorithm (Kruskaal, 1956) having undergone several improvements (Cheriton and Tarjan, 1976; Kerschenbaum and Van Slyke, 1972; Yao, 1975) is $O(m \log \log n)$ and has to be preferred for (even moderately) sparse graphs.

In Shamos (1975) it is shown that this problem, together with many others, is solvable in $O(n \log n)$ when the euclidean metric is assumed.

The key step is the construction of the so-called *Voronoi diagram* of a set of n points (vertices) of the plane, i.e. the subdivision of the plane in n regions such that each region contains all points of the plane which are nearer to one of the given points than to every one of the others. This construction may be obtained by a careful divide-and-conquer scheme in $O(n \log n)$ steps.

The geometric dual of the Voronoi diagram has only $O(n)$ edges and it may be shown that the minimum spanning tree of this geometric dual is also the minimum spanning tree of the complete graph having as vertices the given n points. Thus the minimum spanning tree can be constructed in $O(n \log \log n)$ steps once the Voronoi diagram is known and therefore the construction of the diagram itself, being $O(n \log n)$, requires a time which is of a greater order of magnitude than that for obtaining the tree, hence dominating the overall computation time.

5.4 Hard Problems and Reducibility

A problem P1 is said to be *reducible* to another problem P2 if there exists a way of coding by a polynomial method any instance of P1 into an instance of P2, so that a hypothetical polynomial bounded algorithm for the latter could solve the first as well. When this happens we will indicate as a short-hand notation that $P1 \propto P2$.

The results of Cook (1971) imply that roughly speaking all problems solvable by a polynomial depth backtrack search are reducible to the following problem.

Satisfiability Problem (SAT) A boolean expression in the form of product of sums, e.g. $(A+\bar{B})(\bar{A}+\bar{B})(A+\bar{B}+C)$, is said to be satisfiable if there is some assignment of 0's and 1's to its variables such that the expression evaluates to 1. For example for the above expression it is sufficient to put $B=0$ and the expression is seen to be satisfiable irrespective of the values of A and C. Given any boolean expression in this form the problem is to determine if it is satisfiable or not.

A key notion in the development of a formal theory of complexity of problems is the notion of deterministic and nondeterministic Turing machines (see Aho, Hopcraft, and Ullman, 1974). Loosely speaking, for our purposes we can think of these as suitable simplified models of computing machines which, given an instance of a problem, are able to *recognize* whether or not it satisfies a given property.

However, while the deterministic machine models an ordinary computer, a nondeterministic machine has the rather unusual feature of being able to duplicate its current state in zero time whenever convenient: this amounts to having at disposal an infinite degree of parallelism, so that a polynomial depth backtrack search would be a polynomial algorithm for such a machine.

$\mathcal{P}$ is the class of (recognition) problems which may be solved by a polynomial method on a deterministic machine; $\mathcal{NP}$ is similarly defined for nondeterministic machines.

A problem P is called *NP-hard* if $P' \propto P$ for every $P \in \mathcal{NP}$. If moreover $P \in \mathcal{NP}$, P is called *NP-complete* (Karp, 1972).

Cook (1971) proved that every problem in $\mathcal{NP}$ is reducible to SAT: the rather complicated reduction he used is beyond the scope of this chapter. Thus SAT was the first *NP*-complete problem. Given this result, if a problem $P \in \mathcal{NP}$ is such that $P' \propto P$, with P' *NP*-complete, then P is also proved to be *NP*-complete.

In recent years the list of *NP*-complete problems has grown quite a bit; the following are only a few examples.

Clique Problem (CLI) Given a graph and an integer k, does the graph contain a complete subgraph (clique) with k vertices?

The Complexity of Combinatorial Optimization Algorithms

Vertex Covering (VCO) Given a graph and an integer l, does there exist a selection of no more than l vertices which will cover all the arcs?

Steiner Network Problem (STE) Given an integrally arc-weighted graph, a specified subset of its vertices and an integer k, is it possible to connect together all the vertices of the subset by means of a tree whose total weight does not exceed k?

Sequencing Problem (SEQ) Given a set of jobs, each with a known processing time, deadline, and penalty, and an integer k is it possible to sequence the jobs on a single machine in such a way that the sum of the penalties for the late jobs does not exceed k?

Network Design Problem (NDP) Given an undirected graph $G = (V, E)$, a weight function $L: E \to \mathbb{N}$, a budged B, and a threshold $C(B, C \in \mathbb{N})$, does there exist a subgraph $G' = (V, E')$ of G with weight $\leq B$ and such that the sum of the lengths of the shortest paths in G' between all vertex pairs is $\leq C$?

Set packing (STP) Given a finite set S, a finite family $\mathscr{S}$ of subsets of S and an integer l, does $\mathscr{S}$ contain a subfamily $\mathscr{S}'$ of at least l pairwise disjoint sets?

Set cover (STC) Given a finite set S, a finite family $\mathscr{S}$ of subsets of S and an integer l, does $\mathscr{S}$ contain a subfamily $\mathscr{S}'$ of at most l sets such that $U_{S' \in \mathscr{S}'} S' = S$?

Exact cover (XCO) Given a finite set S and a finite family $\mathscr{S}$ of subsets of S, does $\mathscr{S}$ contain a subfamily $\mathscr{S}'$ of pairwise disjoint sets such that $U_{s' \in \mathscr{S}'} S' = S$?

Directed Hamiltonian circuit (DHA) Given a directed graph, does it contain a directed cycle passing through each vertex exactly once?

Undirected Hamiltonian circuit (UHA) Given an undirected graph, does it contain a cycle passing through each vertex exactly once?

Knapsack (KNA) Given positive integers $a_1, \ldots, a_t, b$, does there exist a subset $S \subset \{1, \ldots, t\}$ such that $\sum_{j \in S} a_j = b$?

Partition problem (PAR) Given positive integers $a_1, a_2, \ldots, a_t$, does there exist a subset S of $\{1, 2, \ldots, t\}$ such that $\sum_{j \in S} a_j = \sum_{j \notin S} a_j$?

3-dimensional matching (3DM) Given 3 sets $S_i = \{\sigma_{i1}, \sigma_{i2}, \ldots, \sigma_{is}\}$, $i = 1, 2, 3$, and t triples of the form $\{\sigma_{1h}, \sigma_{2l}, \sigma_{3m}\}$, $1 \leq h, l, m \leq s$, do there exist s pairwise disjoint triples?

Constrained spanning tree (CST) Given an undirected graph $G = (N, A)$, a specified vertex $\rho \in N$, and an integer k, does there exist a spanning tree of G such that each subtree incident (i.e. connected by an edge) to ρ contains no more than k vertices?

Constrained path-tree (CPT) Given a undirected graph $G = (N, A)$ integrally weighted on A, a specified vertex $\rho \in N$, an integer r, and an integer k, does there exist a spanning tree of G such that the number of edges in any path from ρ to every other vertex is not greater than k and of total weight $\leq r$?

Chromatic number (CHR) Given a graph G and a positive integer k is there a function $\phi : N \to \{1, 2, \ldots, k\}$ such that if u, v are adjacent vertices then $\phi(u) \neq \phi(v)$?

The reductions connecting these problems together are given in Table 5.2.

All the preceding problems are 'recognition' rather than 'optimization' problems. NP-completeness is always proved with respect to *recognition* problems. The corresponding optimization problem can often not formally be shown to belong to $\mathcal{NP}$, but it might be called NP-hard in the sense that the existence of a good algorithm for its solution would imply that $\mathcal{P} = \mathcal{NP}$. This is very unlikely since the possibility of solving efficiently all the above mentioned problems and many others, for which in spite of considerable research effort no good algorithms have been found so far, would be implied. As a corollary it is very unlikely that any one of these problems belongs to $\mathcal{P}$.

Two simple reductions of Table 5.2 are proved in the following as examples.

Theorem 5.1 SAT $\propto$ CLI

Proof Let $G = (N, A)$ be the graph to be constructed for CLI. Then for every literal occurring in every clause let there be a node of N. The set of arcs of G is then formed by pairing all vertices of N provided that they do not belong to the same clause and/or they do not correspond to a literal and its complement. Finally k is set equal to the number of clauses. An example is shown in Figure 5.1.

Table 5.2 Reductions of problems

SAT (with >2 literals per clause) $\propto$

	CLI $\propto$			CHR $\propto$				
STP	VCO $\propto$		XCO $\propto$					
	DHA $\propto$	STC	3DM $\propto$			KNA $\propto$		STE
	UHA		CPT	NDP	CST	SEQ	PAR	

The Complexity of Combinatorial Optimization Algorithms

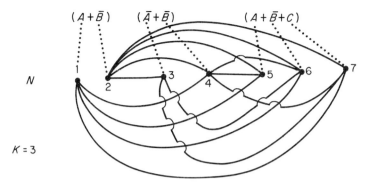

Figure 5.1

The proof follows easily realizing that from a purely combinatorial point of view, SAT is equivalent to finding at least one letter for each clause, without selecting a complementary pair.

Theorem 5.2   CLI ∝ VCO

Proof Let G' be the graph of VCO and G the graph for CLI. It is then sufficient to let G' be equal to the complement of G and $l = |N| - k$ (see Figure 5.2 where the graph of Figure 5.1 is redrawn in a different way).

The interested reader can see Karp (1972) for the proofs of NP-completeness of 3DM, DHA, UHA, SEQ, STE, KNA, PAR, CHR, XCO, STC, and STP,

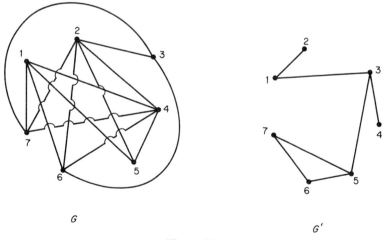

Figure 5.2

while NDP has been proved to be *NP*-complete in Johnson, Lenstra, and Rinnooy Kan (1977). CST has been proved recently to be *NP*-complete by Lenstra reducing the Exact 3-covering problem to it and by Camerini directly from the 3DM. Similarly, CPT can be proved *NP*-complete by a reduction from 3DM (Camerini, 1978, private communication).

Since the early works in this field it has become more and more commonplace to justify the effort for solving a given problem by implicit enumeration methods such as branch and bound, heuristically guided search, etc., by first showing some evidence that the given problem (often in a simpler formulation) belongs to the class of 'hard' problems of Cook and Karp. Many new, interesting results have appeared recently, among them the successful search for the simplest instances of *NP*-complete problems (Garey, Johnson, and Stockmeyer, 1974), for geometric *NP*-complete problems (Garey, Graham, and Johnson, 1976), the classification of scheduling problems (Lenstra, Rinnooy Kan, and Brucker, 1976), the finding of space- (i.e. memory-) complete (and therefore also time-complete) problems (Tarjan and Even, 1975).

For instance, not only is the VCO *NP*-complete for general graphs, but also for graphs with degrees of the vertices not greater than three, while DHA is *NP*-complete even in the euclidean case.

Before we leave this topic, it is perhaps worthwhile to give a few words of warning in order to prevent possible misunderstandings.

Consider PAR. An algorithm exists for its solution which is $O(t \sum a_i)$ (Dreyfus and Bellman, 1962). This means that if we take as dimension of the problem the encoding of the data as integers or binary numbers, this algorithm runs in exponential time, while it would be linearly running in terms of a *unary* encoding of the data. However finding an algorithm polynomial in terms of the binary encoding of the data for PAR would still amount to showing that $\mathcal{P} = \mathcal{NP}$.

The algorithms of this kind are called *pseudo-polynomial*.

Strong NP-complete problems are those for which no algorithm is known which runs in polynomial time even if the data are encoded in unary form (Garey and Johnson, 1976c).

Another example of an *NP*-complete problem which is not strong is the knapsack problem.

5.5 The Need for Heuristic Methods

Real world combinatorial optimization problems have unfortunately quite often in common the following unpleasant features: (a) they are of large dimensionality, (b) they are much more cumbersome than the problems mentioned so far, (c) when they can be decomposed into simpler subproblems, these are quite often of the hard kind. On the other hand real life problems only seldom require that an optimum solution be found, a suboptimum solution being usually what is really needed, provided a certain evidence that it is not too bad may be provided.

These facts together with the results of the previous section, which are highly discouraging (to say the least) for any one willing to try to find a polynomial bounded method for solving any one of the *NP*-complete problems, make the need for good heuristic methods one of the most important challenges for any researcher in the field of combinatorial optimization.

The first thing to be done is of course to clarify what 'good' means as far as a heuristic algorithm is concerned. We have here more or less the same situation we have found for exact optimization algorithms: for these algorithms many norms, i.e. efficiency criteria, have been shown to be available, which apply to the running time or/and the memory requirements. Here running times and memory requirements must be assumed polynomial in the size of the problem, otherwise we would have a heuristic having an asymptotic behaviour as bad as an exact implicit enumeration method.

Instead many norms may be proposed to estimate how close a given heuristic is to the optimum. We will consider separately the case of provably good heuristics with respect to deterministic and probabilistic norms in the next sections.

However, before going into this let us remark that an *a posteriori* evaluation of any heuristic is possible in many cases with a high degree of accuracy.

In fact for many hard problems (in the sense of Section 5.4) implicit enumeration methods exist of the kind usually called branch and bound, heuristically guided search, etc.

All these methods use some bounding algorithm as a subroutine in order to reduce the number of explicitly analysed solutions of the problem. When the bounding method is good, even the bound obtained at the beginning of the algorithm constitutes a rather close estimate of the value of the optimum and therefore, together with the value of the heuristic solution obtained, makes it possible to establish a valid upper bound to the error by which the heuristically obtained solution is affected.

One of the first hard problems to enjoy this possiblity has been the travelling salesman problem (Held and Karp, 1971), and the technique used in that case has been extended to a very large class of hard problems in Camerini and Maffioli (1975, 1977) and Maffioli (1975). The reader is referred to the literature since even a brief survey of the subgradient techniques (Wolfe, Held, and Crowder, 1974) used to obtain the bounds is beyond the scope of this work. A polynomial bounding technique is also available for the travelling salesman problem (Christofides, 1972) and some attempts to do the same for other problems are currently under investigation (Camerini and Maffioli, 1976).

In dealing in general with heuristic methods, two factors emerge (Weiner 1975): (A) there is a large number of fairly standard techniques to guide the designs of heuristic methods for a given problem, but (B) at the present time such a design is still an experimental art, rather than a precise science.

Among the more common heuristic techniques we may list the following (Weiner 1975):

constructive methods, that build solutions using heuristic rules often in a sequential deterministic manner;

local transformation methods, that aim to improve existing solutions by the local search of a well defined neighbourhood of the solution space;

decomposition methods, that break the problem into smaller parts;

feature extraction methods, both statistical and combinatorial, aiming to reduce the size of the problem (for instance removing from the problem features common to several solutions obtained by different methods or different applications of the same method);

inductive methods, that solve first small embedded problems and then add inductively to these solutions;

approximation methods, transforming an intractable problem into a tractable one.

In the following two sections we shall present several problems for which heuristic methods are guaranteed to yield a solution which is not worse than a certain function of the optimum value either in the worst possible case (Section 5.6) or in the average case (Section 5.7).

Three types of bounds are considered, which are normally increasingly closer to the optimum:

(a) a single bound for all instances of a certain problem,
(b) a data-dependent bound,
(c) an *a posteriori* bound.

We have already mentioned a technique for bounds of the type (c). Most of the following will belong to the (b) class.

5.6 Deterministic Heuristic Evaluation

Given a combinatorial optimization problem P, let w_0 be the value of an optimum solution, and consider a heuristic method which yields a solution to P. Let w_h be the value obtained by this method. In this section we shall survey the results concerning heuristic methods for which it is known that $w_h \geq f(w_0)$ or $w_h \leq f(w_0)$ for respectively maximization or minimization problems. An annotated bibliography is available (Garey and Johnson,

1976b) which subdivides the combinatorial optimization problems for which results of the above mentioned type are available into four classes: packing, scheduling, colouring, and routing and placement problems.

Johnson (1974c) presents results relevant to the following NP-complete problem.

Bin Packing Problem (BIN) Given a list of n real numbers in $(0, 1]$, $L = \{a_1, a_2, \ldots, a_n\}$, place the elements of L into a minimum number l_0 of 'bins' so that no bin contains numbers whose sum exceeds one.

BIN is relevant to several practical problems (stock cutting, table formatting, prepackaging, file allocation, see Brown, 1971; Johnson, 1974c) and may be solved efficiently for instance by the following simple heuristic: arrange L in nonincreasing order and, to place a_i, find the least j such that bin B_j is filled to level $\beta \leq 1 - a_i$, starting with $j = i = 1$, and all bins B_j empty. Let l_h be the number of bins employed by this method. Then Johnson (1974c) proves that

$$\max_{\mathscr{L}_0}\left(\frac{l_h}{l_0}\right) = \frac{11}{9}$$

where $\mathscr{L}_0$ is the set of all lists having l_0 as optimum result.

In Garey et al. (1975) the results of Johnson (1974c) for BIN are generalized to multidimensional problems for which vectors instead of numbers are to be packed.

The second interesting result on packing problems is relevant to the knapsack problem (see Section 5.4). In Ibarra and Kim (1975), linear time algorithms for this problem are presented using rounding techniques and dynamic programming and with performance bounds which are arbitrarily close to optimal.

To guarantee a shortfall no more than $1/k$ times optimal, the time required is $O(k^2 n)$. The same work proves also that a 'greedy' algorithm for the same problem may give a result as bad as $\frac{1}{2}$ the optimum.

Scheduling problems are approached similarly and the reader is referred to Lenstra, Rinnooy Kan, and Brucker (1976).

Graph colouring with less than twice the optimal number of colours is shown in Garey and Johnson (1976a) to be just as hard as finding an optimal colouring for general graphs. However there are good heuristics for particular classes of graphs. For instance, Matula, Marble, and Isaacson (1972) present a heuristic which guarantees never to use more than 5 colours on planar graphs. Let us note here that as far as planar graphs are concerned colouring with no more than 4 colours is NP-complete (Karp, 1975) although the four colour conjecture has been finally proved in Appel and Haken (1978).

The somehow negative results for general colouring problems should have prepared the reader for other negative results. One of the more important concerns the travelling salesman problem.

We know that DHA is NP-complete (see Section 5.4). Then any algorithm which for a given arc-weighted (complete) graph would ensure finding a hamiltonian cycle of weight less than or equal to a given constant times a minimum weight cycle would enable one to solve DHA by, for instance, assigning a weight of 1 to the arcs of the given graph and a very large weight to the arcs of the complement. It indeed follows therefore that finding any approximation to the optimum solution is just as hard as finding an optimum solution to the general travelling salesman problem.

A warning is however necessary at this point. From a practical point of view in many cases it may well happen that not an optimal hamiltonian cycle is sought, but rather what could be called an *optimal eulerian subgraph*. In other words if going from node n_i to node n_j costs more by the direct link than by a path which uses node n_k, for several problems there is no reason for not going through n_k.

The problem reduces therefore easily to a travelling salesman problem in a complete graph having the same set of vertices and where arcs (i, j) are weighted by the cost of the shortest path connecting n_i and n_j in the original graph. The new graph obeys obviously the triangular inequality. For this kind of travelling salesman problem the simple heuristic described in Figure 5.3 (Christofides, 1976) guarantees a solution never worse than 3/2 the optimum solution. No better worst case performance is known for other heuristics, but practical computational experience (Rosenkrantz, Stearns, and Lewis, 1974) shows that some of the commonly used methods perform quite well.

For the euclidean metric case (that is when not only is the triangular inequality respected, but the given points are effectively points on a two-dimensional surface), taking into account the results of Shamos (1975) and Rosenkrantz, Stearns, and Lewis (1974), a possibly effective heuristic could be the following.

Step 1 Construct the convex hull of the given points ($O(n \log n)$ steps). We know that points on the convex hull will never appear in a different order on the shortest hamiltonian cycle.

Step 2 Insert one at a time the other points taking always into account the farthest point from the already formed cycle among the remaining set of points.

This is illustrated in Figure 5.4. A possible variation suggests performing Step 2 first for all points belonging to the next convex hull in order to balance the effect of taking the farthest or the nearest next point.

Another NP-complete problem with similar properties in STE. In the euclidean case where auxiliary vertices may be placed everywhere in the

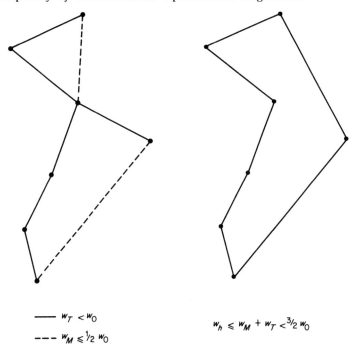

Figure 5.3 T, shortest spanning tree; M matching of odd degree vertices in T; w_0, weight of optimum tour; w_h, weight of heuristically obtained tour

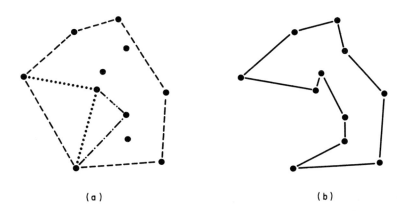

(a)   (b)

Figure 5.4

plane, a strongly supported conjecture (Gilbert and Pollack, 1968) states that the ratio between the optimum tree and a spanning tree without auxiliary vertices is never less than $\sqrt{3}/2$, while a lower bound of $1/\sqrt{3}$ has been proved in Graham and Huang (1976). In Hwang (1976) the rectilinear metric Steiner problem is considered and the spanning tree without auxiliary vertices is shown to be no more than 50% longer than the minimum length rectilinear Steiner tree.

Algorithms for placement or allocation problems have been analysed in Chandra and Wong (1975), Karp, McKellar, and Wong (1975), and Cornuejols, Fisher, and Nemhauser (1975). Some of these heuristics have very good worst case behaviour. For instance, the two-dimensional placement problem of Karp, McKellar, and Wong (1975) is shown to be always solvable within an *additive* constant of optimum (most of other results mentioned so far are within multiplicative constant optimum).

5.7 Probabilistic Heuristic Evaluation

The fact that computational experience shows many heuristics to work quite well even if their worst case behaviour is quite bad has recently encouraged researchers to pay increasing attention to methods which can be proved to work well on the average. Unfortunately it is very difficult, if not impossible, in most cases to know the real probability distribution of the various problem instances.

What is normally done therefore is to assume a certain probability distribution and derive a result which we hope will depend very little from this assumption.

Some definitions from probability theory (Feller, 1968) concerning problem distributions are given here.

Let S_i be the sample space of problems of size i and x a random sample from S_i, $\pi(x)$ a property of x. Then

$$p_i = \text{Prob}\{\pi(x) \text{ holds}\} = 1 - q_i$$

Definition 5.1 We say that π holds *almost everywhere* if

$$\sum_{i=1}^{\infty} q_i < \infty.$$

Definition 5.2 We say that π holds with probability which tends to one if

$$\lim_{i \to \infty} p_i = 1.$$

Let us remark that the second definition establishes a weaker requirement than the first and that we are looking for asymptotic properties in both cases.

The Complexity of Combinatorial Optimization Algorithms

The main reference for this section is the work of Karp (1976) and again some of the more interesting results are obtained for the euclidean travelling salesman problem.

Let in fact T_n be a random variable equal to the length of a shortest tour through n points given completely at random inside a region of the plane of area A. The following results hold (Beardwood, Halton, and Hammersley, 1959):

(A) $T_n \leq \gamma \sqrt{(An)}$ (γ a constant),

(B) there exists a constant β such that, for all $\varepsilon > 0$

$$\beta - \varepsilon < \frac{T_n}{\sqrt{(nA)}} < \beta + \varepsilon \tag{5.1}$$

almost everywhere.

Result (A) establishes a worst case behaviour, while result (B) establishes an average behaviour. Experiments tend to indicate a value of β of about 0.75.

For convenience let now $A = 1$ be the area of a unit square of the plane and consider the following heuristic, assuming that problems of size $\leq t$ may be solved optimally by our computer sufficiently quickly.

Step 1 Subdivide A into smaller regions A_i, each one containing n_i of the given points, until $n_i \leq t$ for all i, by the technique of Figure 5.5, i.e. dividing A into rectangles obtained always by a segment parallel to the smaller side.

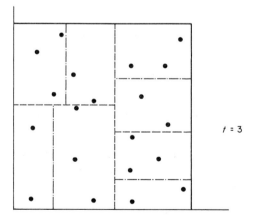

Figure 5.5

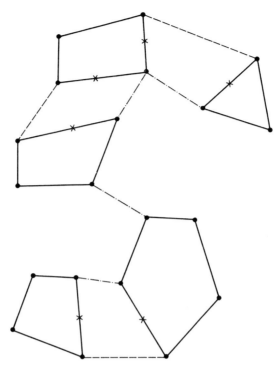

Figure 5.6

Step 2 Construct shortest tours for each region A_i.

Step 3 Treat each subtour as a point and construct a minimum spanning tree connecting them all.

Step 4 Obtain an eulerian tour taking the small tours and twice each tree edge.

Step 5 Eliminate pairs of edges from the eulerian tour in order to get a hamiltonian tour of smaller length (see Figure 5.6).

Let now c_i be the cost of the ith subtour, c_0 the cost of the optimum tour, and c_h the cost of the tour obtained by the algorithm. The following results are reported in Karp (1976).

Theorem 5.3 (Worst case) There exist constants α_1 and α_2 such that

$$\sum_i c_i - \alpha_1 \sqrt{\left(\frac{n}{t}\right)} \leq c_0 \leq c_h \leq \sum_i c_i + \alpha_2 \sqrt{\left(\frac{n}{t}\right)}$$

Corollary 5.1 $c_h - c_0 \leq (\alpha_1 + \alpha_2)\sqrt{\binom{n}{t}}$

Theorem 5.4 (Average case) *There exists a δ such that*

$$c_h - c_0 \leq \delta \sqrt{\binom{n}{t}}$$

almost everywhere.

Algorithms based on similar partitioning principles and with similar properties apply to Steiner tree problems as well as to the same problems in other metric or in spaces of higher dimensionality.

For a different density function $f(x, y)$ governing the distribution of the points into the unit square, if

$$I = \iint_A \sqrt{(f(x, y))} \, dx \, dy$$

a similar result applies since now instead of (5.1) we have (Beardwood, Halton, and Hammersley, 1959)

$$\beta I - \varepsilon \leq \frac{T_n}{\sqrt{(nA)}} \leq \beta I + \varepsilon.$$

A completely different kind of approach, due to Posa (1976), may be applied to the so-called Bottleneck Travelling Salesman Problem, i.e. the problem of finding a hamiltonian tour in a given arc-weighted graph such that the length of its longest arc is minimized. The approach is based upon the theory of random graphs.

Theorem 5.5 (Posa, 1976) *Consider a graph G with n vertices. If for any subset S of the vertices of G there exists at least $\min(2|S|, n-|S|)$ vertices $\notin S$ joined to vertices of S, G possesses a hamiltonian tour.*

Theorem 5.6 *For any $\alpha > 1$ if a graph contains $m > \alpha n \ln(n)$ arcs, the condition of Theorem 5.5 holds with probability approaching 1 as $n \to \infty$.*

Posa gives also a constructive algorithm for finding such a tour. To solve by this heuristic the bottleneck travelling salesman problem it is then sufficient to put into G enough arcs to make it connected, with no pendant vertices and with at least $n \ln(n)$ arcs starting with the smallest arcs, and then apply Posa's method.

Finally let us consider the case of a travelling salesman problem in a directed graph with a general metric (problems of this kind arise for instance in connection with scheduling problems). Assume the d_{ij}, the distances from vertex i to vertex j, are drawn independently from a common distribution uniform in $[0, 1]$. Then all $n!$ permutations of the vertices are equally likely to be optimum.

Let us consider the solution of the corresponding assignment problem. The probability that this gives immediately a single tour is small ($\simeq e/n$). However the following is claimed to be true (Karp, 1976):

> The number of cycles that solving the assignment problem will create tends to $\log n$ with probability one. Then the longest of such cycles will have a length $m \geq n/\log n$ and the cost of patching the smaller tours into the largest one can be shown to be bounded by $(\log n)^{2+\varepsilon}/\sqrt{n}$ with probability one.

Although this section has been devoted almost entirely to the travelling salesman problem, we believe that it outlines sufficiently the various kinds of approach one can take for evaluating the average behaviour of approximate algorithms. The reader interested in this field is referred to Karp (1976) for further information.

5.8 Concluding Remarks

In this survey we have attempted to give a rapid look at some of the more active and promising areas for research in combinatorial optimization: complexity of algorithms (mainly with respect to worst case behaviour) and heuristic methods both from the point of view of their worst case and average case behaviour. Concerning complexity of algorithms w.r.t. their average time norm, we cannot finish without mentioning the efforts being made towards establishing the expected performance of tree-search methods (branch and bound, etc.). In Bellmore and Malone (1971), polynomial-bounded expected performance has been claimed. However, some serious doubts exist about the adequacy of the proofs and have been pointed out in Lenstra and Rinnooy Kan (1976). Following a much more careful approach, Karp (1976) proposes a probabilistic analysis of tree-search methods, exemplifying its application for STC.

Roughly speaking the idea is to examine the nodes of the search tree to determine their likelihood of yielding a near optimal solution and to avoid branching from unpromising nodes. In this way the number of nodes actually generated grows linearly, rather than exponentially, with problem size. Despite this, nearly optimal solutions are obtained almost everywhere, even if the probabilistic model assumed shows that any tree-search method having constant positive probability of finding the optimum must almost always explores an exponential number of nodes of the search tree.

References

Aho, A. V., J. E. Hopcroft, and J. D. Ullman (1974). *The Design and Analysis of Computer Algorithms*, Addison-Wesley, Reading, Massachusetts.

Appel, K., W. Hanken and K. Koch (1977). Every Planar Map is four Colorable. *Illinois J. Math.*, **21,** 429–567.

Beardwood, J., J. H. Halton, and J. M. Hammersley (1959). The shortest path through many points. *Proc. Camb. Phil. Soc.*, **55,** 299–327.

Bellmore, M. and J. C. Malone (1971). Pathology of Travelling-Salesman Subtour-Elimination Algorithms. *Opns. Res.*, **19,** 278–367.

Blum, M., R. Floyd, V. Pratt, R. Rivest, and R. Tarjan (1972). Linear time bounds for media computation. *4th Annual ACM Symp. on Theory of Computing, 1972.*

Brown, A. R. (1971). *Optimum packing and depletion.* American Elsevier, N.Y.

Camerini, P. M. and F. Maffioli (1975). Bounds for 3-matroid intersection problems. *Inf. Proc. Letters*, **3,** 81–83.

Camerini, P. M. and F. Maffioli (1976). Polynomial bounding for NP-hard problems. *Proc. 9th Int. Symp. on Math. Programm.*, Budapest, August 1976.

Camerini, P. M. and F. Maffioli (1978). Heuristically guided algorithm for k-parity matroid problems. *Discrete Math.* **21,** 103–116.

Chandra, A. K. and C. K. Wong (1975). Worst-case analysis of a placement algorithm related to storage allocation. *SIAM J. Computing*, **4,** 249–263.

Cheriton D. and R. Tarjan (1976). Finding minimum spanning trees. *Siam J. Computing*, **5,** 724–742.

Christofides, N. (1972). Bounds for the travelling salesman problem. *Opns. Res.*, **20,** 1044–1056.

Christofides, N. (1976). Worst case analysis of a new heuristic for the travelling salesman problem. Abstract in J. F. Traub (Ed.), *Proceedings of the Symposium on New Directions and Recent Results in Algorithms and Complexity*, Academic Press, N.Y. (Also: Carnegie-Mellon U., Report WP 62-75-76, Feb. 1976).

Cook, S. A. (1971). The complexity of theorem proving procedures. *Proc. 3rd Annual ACM Symposium on Theory of Computing, 1971.*

Cornuejols, G., M. L. Fisher, and G. L. Nemhauser (1975). An analysis of heuristic and relaxations for the uncapacitated location problem. *Cornell Univ. Oper. Res. Dept. Techn. Rep. 271.*

Dijkstra, E. (1959). A note on two problems in connection with graphs. *Numer. Math.*, **1,** 269–271.

Dinic, E. A. (1970). Algorithm for Solution of a problem of maximum flow in network with power estimation. *Sov. Math. Dokl.*, **11,** 1277–1280.

Dreyfus S. E. and R. E. Bellman (1962). *Applied Dynamic Programming*, Princeton University Press, Princeton, N.J.

Edmonds, J. (1965). Paths, trees and flowers. *Canad. J. Math.*, **17,** 449–467.

Edmonds, J. and R. M. Karp (1972). Theoretical improvements in algorithmic efficiency for network flow problems. *J. ACM*, **19,** 248–264.

Even, S. and O. Kariv (1975). On $O(n^{2.5})$ algorithm for maximum matching in general graphs. *Proc. 16th Ann. Symp. on Foundations of Comp. Sc.*, IEEE, N.Y. pp. 100–112.

Feller, L. (1968). *An introduction to probability theory and its applications*, Wiley, New York.

Gabow H. (1976). An efficient implementation of Edmonds algorithm for maximum matching on graphs. *J. ACM*, **23,** 221–234.

Garey, M. R., R. L. Graham, and D. S. Johnson (1976). Some NP-complete geometric problems. *Proc. 8th Ann. ACM Symp. on Theory of Computing, 1976.*

Garey M. R., R. L. Graham, D. S. Johnson, and A. C. C. Yao Resource constrained scheduling as generalized bin packing. *J. Comb. Theory (Series A)* **21,** 257–298.

Garey, M. R. and D. S. Johnson (1976a). The complexity of near-optimal graph coloring *J. ACM,* **23,** 43–49.

Garey, M. R. and D. S. Johnson (1976b). Approximation algorithms for combinatorial problems. In J. F. Traub (Ed.), *Proc. Symp on New Directions and Recent Results in Algorithms and Complexity,* Academic Press, N.Y.

Garey, M. R., D. S. Johnson, and L. Stockmeyer (1974). Some simplified NP-complete problems. *Proc. 6th ACM Ann. Symp. on Theory and Computing, 1974.*

Gilbert, E. N. and H. O. Pollack (1968). Steiner Minimal trees. *SIAM J. Appl. Math.,* **16,** 1–29.

Graham R. L. and F. K. Hwang (1976). A remark on Steiner minimal trees. *Bull. Inst. Math. Acad. Sinica,* **4,** 177–182.

Held, M. and R. M. Karp (1971). The travelling salesman problem and minimum spanning trees: part II. *Math. Programm.,* **1,** 6–25.

Hopcroft, J. and R. Tarjan (1971). Planarity testing in $V \log V$ steps: extended abstract. *Proc. IFIP Congress,* North-Holland, Amsterdam.

Hopcroft, J. and R. M. Karp (1973). A $n^{5/2}$ Algorithm for Maximum Matchings in Bipartite Graphs. *SIAM J. Computing,* **2,** 225–231.

Hwang, F. K. (1976). On Steiner minimal tree with rectilinear distance. *SIAM J. Appl. Math.,* **30,** 104–114.

Ibarra, O. and C. Kim, (1975). Fast approximation algorithms for the knapsack and sum of subsets problems. *J. ACM,* **22,** 463–468.

Johnson, D. S. (1974a). Approximation algorithms for combinatorial problems. *Proc. 5th Ann. ACM Symp. on Theory of Computing, 1973,* pp. 38–49; also in *J. Computer and System Sciences,* **9,** 256–278.

Johnson, D. S. (1974b). Fast algorithms for bin packing. *J. Computer and System Sciences,* **8,** 272–314.

Johnson, D. S., A. Demers, J. D. Ullmann, M. R. Garey, and R. L. Graham (1974). Worst-case performance bounds for simple one-dimensional packing algorithms. *SIAM J. Computing,* **3,** 299–326.

Johnson, D. S., J. K. Lenstra, and A. H. G. Rinnooy Kan (1977). The complexity of the network design problem. Math. Centrum, Dept. of Opns. Res. Int. Rep., Amsterdam.

Karp, R. M. (1972). Reducibility among combinatorial problems. In: R. E. Miller and J. W. Thatcher (Eds.), *Complexity of Computer Computations,* Plenum Press, New York. pp. 85–103.

Karp, R. M. (1975). On the computational complexity of combinatorial problems. *Networks,* **5,** 45–68.

Karp, R. M. (1976). The probabilistic analysis of some combinatorial search algorithms. In J. F. Traub (Ed.), *Proc. Symp. on New Directions and Recent Results in Algorithms and Complexity,* Academic Press, N.Y.

Karp, R. M., A. C. McKellar, and C. K. Wong (1975). Near-optimal solutions to a 2-dimensional placement problem. *SIAM J. Computing,* **4,** 271–286.

Karzanov A. V. (1974). Determining the maximal flow in a network by the method of preflows. *Soviet Math. Dokl.,* **15,** 434–437.

Kerschenbaum, A. and R. Van Slyke (1972). Computing Minimum Spanning Trees Efficiently. *Proc. ACM 25th Ann. Conf., Aug. 1972.*

Knuth, D. E. (1973). *Sorting and Searching, the Art of Computer Programming,* Vol. 3, Addison Wesley, N.Y.

Kruskaal, J. B. Jr. (1956). On the shortest spanning subtree of a graph and the travelling salesman problem. *Proc. Amer. Math. Soc.*, **7,** 48–50.

Lawler, E. L. (1975). Matroid intersection algorithms. *Math. Programm.*, **9,** 31–56.

Lawler, E. L. (1976). *Combinatorial optimization: networks and matroids*, Holt, Rinehardt and Winston, N.Y.

Lenstra, J. K. and A. H. G. Rinnooy Kan (1976). A note on the expected performance of branch-and-bound algorithms. Int. Rep. 63/76, Mathematish Centrum, Amsterdam.

Lenstra, J. K., A. H. G. Rinnooy Kan, and P. Brucker (1977). Complexity of machine scheduling problems. *Ann. Discrete Math.* **1,** 343–362.

Maffioli F. (1973). The travelling salesman problem and its implications. Summer School in Combinatorial Optimization, Udine, 1973. (In S. Rinaldi (Ed.), *Combinatorial Optimization*, Springer Verlag, Berlin.)

Maffioli F. (1975). Subgradient optimization, matroid problems and heuristic evaluation. *IFIP Conf. on Optimization Techniques, Nice, 1975.*

Matula, D. M., G. Marble, and J. D. Isaacson (1972). Graph coloring algorithms. In R. C. Read (Ed.), *Graph Theory and Computing*, Academic Press, N.Y.

Pohl, I. (1970). A sorting problem and its complexity. U. of California at S. Cruz internal report.

Posa, L. (1976). Hamilton circuits in random graphs. *Discrete Math.*, **14,** 359–364.

Preparata, F. P., and S. J. Hong (1977). Convex hulls of finite sets of points in two and three dimensions. *Comm. ACM*, **20,** 87–93.

Prim, R. C. (1957). Shortest interconnection networks and some generalizations. *B.S.T.J.*, **36,** 1389–1401.

Rosenkrantz, D. J., R. E. Stearns, and P. M. Lewis (1974). Approximate algorithms for the travelling salesperson problem. *Proc. 15th Annual IEEE Symp. on Switching and Automata Th.* pp. 33–42.

Sakarowitch, M. (1976). Deux ou trois choses que je sais des bitroides. Univ. Sc. et Medic. de Grenoble, Math. Appl. et Informatique R. R. 55, Nov. 1976.

Shamos, M. I. (1975). Geometric complexity. *Proc. 7th Annual ACM Symposium on Theory of Computing.* pp. 224–233.

Tarjan, R. E. and S. Even (1975). A combinatorial problem which is complete in polynomial space. *Proc. ACM Symp. on Theory of Computing.*

Weiner, P. (1975). Heuristics. *Networks*, **5,** 101–103.

Wilde, D. J. (1964). *Optimum seeking methods*, Prentice Hall, N.J.

Wolfe, P., M. Held, and H. Crowder (1974). Validation of subgradient optimization. *Math. Programming*, **6,** 62–88.

Yao, C. C. (1975). An $O(|E|\log\log|V|)$ Algorithm for Finding Minimum Spanning Trees. *Inform. Proc. Lett.*, **4,** 21–25.

CHAPTER 6

The Travelling Salesman Problem

NICOS CHRISTOFIDES
Department of Management Science, Imperial College, London

6.1. Introduction

Consider a complete directed or nondirected graph $G = (N, A)$, where N is a set of n vertices and A is a set of m arcs. Let c_{ij} be the cost of arc (i, j). A *hamiltonian circuit* of G is a circuit passing through every vertex of G once and once only. The cost of such a circuit is the sum of the costs of its arcs. The *travelling salesman problem* (TSP) is the problem of finding in a graph G with a given matrix $[c_{ij}]$ a hamiltonian circuit having the least cost. The TSP is an archetypical combinatorial optimization problem with some real life applications, but, more importantly, one which forms the substructure of a large number of other problems arising in practical situations, including vehicle routing (see Chapter 11 of this volume) and in sequencing and scheduling.

The purpose of the present chapter is to review some of the exact and approximate procedures that have been suggested for the solution of TSP's with particular emphasis on their computational aspects. (See also Burkard (1977))

6.1.1 Formulations of the TSP

Let

$x_{ij} = 1$ if arc (i, j) is in the optimal TSP tour (Hamiltonian circuit)

$\phantom{x_{ij}} = 0$ otherwise.

The problem is then:

$$P_A \begin{cases} \min & \sum_{i=1}^{n} \sum_{j=1}^{n} c_{ij} x_{ij} & (6.1) \\ \text{s.t.} & \sum_{i=1}^{n} x_{ij} = 1 & \forall j \in N & (6.2) \\ & \sum_{j=1}^{n} x_{ij} = 1 & \forall i \in N & (6.3) \\ & x_{ij} \in \{0, 1\} & \forall i, j \in N & (6.4) \end{cases}$$

and

$$x_{ij} \text{ must form a tour.} \qquad (6.5)$$

This last constraint can be written in a number of different ways, three of which are:

$$\sum_{i \in S_t} \sum_{j \in \bar{S}_t} x_{ij} \geq 1 \qquad \forall S_t \subset N \qquad (6.5a)$$

$$\sum_{i \in S_t} \sum_{j \in S_t} x_{ij} \leq |S_t| - 1 \qquad \forall S_t \subset N \qquad (6.5b)$$

$$\sum_{(i,j) \in \Phi} x_{ij} \leq |S_t| - 1 \qquad \forall S_t \subset N \text{ and all } \Phi \in \chi(S_t) \quad (6.5c)$$

where $\bar{S}_t = N - S_t$, $|S_t|$ is the cardinality of S_t and $\chi(S_t)$ is the family of all hamiltonian circuits of the induced subgraph $\langle S_t \rangle$ defined by the vertex set S_t. We use Φ to represent both the hamiltonian circuit and the set of arcs forming that circuit.

Let $K_t \equiv (S_t, \bar{S}_t)$ be the set of arcs (i, j) with $i \in S_t$ and $j \in \bar{S}_t$. K_t will be referred to as an *arc cut-set* of G.

Constraints (6.5a) express the fact that at least one arc in the TSP tour must belong to any arc cut-set K_t of G. Constraints (6.5b) and (6.5c) are two expressions of the fact that no subtour through the subset of vertices defined by S_t can exist as part of the TSP solution. It can be easily shown that for any feasible solution (6.5a), (6.5b), and (6.5c) are equivalent.

In the special case when G is a nondirected graph, let l be the index of the arc set A and c_l the cost of arc l. Let $x_l = 1$ if arc l is in the solution; $x_l = 0$ otherwise. The symmetric TSP can then be formulated as:

$$\text{TSP}_s \begin{cases} \min & \sum_{l=1}^{m} c_l x_l & (6.6) \\ \text{s.t.} & \sum_{l=1}^{m} x_l = n & (6.7) \\ & \sum_{l \in K_t} x_l \geq 2 \quad \forall K_t \equiv (S_t, \bar{S}_t), S_t \subset N & (6.8) \\ & x_l \in \{0, 1\} \quad l = 1, \ldots, m. & (6.9) \end{cases}$$

It is worthwhile to note here that constraints (6.7), (6.8) are equivalent to the two sets of constraints:

$$\sum_{l \in K_t} x_l \geq 1 \qquad \forall K_t \equiv (S_t, \bar{S}_t), S_t \subset N \qquad (6.10a)$$

and

$$\sum_{l \in A_i} x_l = 2 \qquad i = 1, \ldots, n \qquad (6.10b)$$

where A_i is the set of arcs incident at vertex i.

The Travelling Salesman Problem

6.2 Exact Solution Procedures with Branch and Bound

Most of the exact methods used for solving the TSP are of the branch and bound variety where, at each node of the branch and bound tree, lower bounds are computed by solving related problems which are relaxations of the original TSP. Camerini *et al.* (1975), Gabovič (1970), Held and Karp (1971), Padberg and Hong (1977), Rubinshtein (1971). As for all branch and bound methods, the quality of the computed bounds has a much greater influence on the effectiveness of the algorithm than any branching rules that may be used to generate the subproblems during the search. In view of this, we will concentrate our attention on the problems of calculating bounds and assume that a depth-first search, and branching schemes similar to those introduced by Bellmore and Malone (1971) are used for the tree search procedure. For further discussion of lower bounds for the TSP see Christofides (1972), D'Atri (1977).

6.2.1 Bounds from the Shortest Spanning Tree (SST)

Symmetric TSP

Consider the case when G is a nondirected graph and let $\hat{\Phi}$ be a hamiltonian circuit which is the solution to the TSP. Let v be a distinguished vertex on $\hat{\Phi}$. If the two arcs incident at v together with vertex v are removed, the remaining graph is a path, P, through vertices $X - \{v\}$. Clearly, if T_v is the SST of graph $\langle X - \{v\} \rangle$, that is, the subgraph induced by the set of vertices $X - \{v\}$, then the cost of T_v is a lower bound on the cost of P.

The two arcs which have been removed from $\hat{\Phi}$ to obtain P have a total cost at least as great as the sum of the two shortest arcs—say (i_1, v) and (i_2, v)—incident at v. Let the graph composed of the arcs in T_v and arcs (i_1, v) and (i_2, v) be called G_v. It is clear that the total cost of the arcs in G_v is a lower bound on the cost of $\hat{\Phi}$. The mathematical formulation of the problem whose solution is the graph G_v is problem P_T below.

$$P_T \begin{cases} \min & \sum_{l=1}^{m} c_l x_l & (6.11) \\ \text{s.t.} & \sum_{l=1}^{m} x_l = n & (6.12) \\ & \sum_{l \in K_t} x_l \geq 1 \quad \forall K_t \equiv (S_t, \bar{S}_t), S_t \subset N & (6.13) \\ & \sum_{l \in A_v} x_l = 2 & (6.14) \\ & x_l \in \{0, 1\}, \quad l = 1, \ldots, m & (6.15) \end{cases}$$

Problem P_T is clearly the same as problem TSP_s with constraints (6.8) replaced by constraints (6.10a) and constraints (6.10b) ignored.

A better bound on the value of problem TSP_s can be derived if constraints (6.10b) are included in the objective function of P_T by means of lagrange multipliers λ, that is, by solving the problem:

$$\min_x \left(\sum_{l=1}^m c_l x_l \right) + \sum_{i \in N} \lambda_i \left(\sum_{l \in A_i} x_l - 2 \right)$$

subject to constraints (6.12)–(6.15); that is, by minimizing:

$$P_T(\lambda) \begin{cases} \min_x \left[\sum_{l=1}^m (c_l + \lambda_{i_l} + \lambda_{j_l}) - 2 \sum_{i \in N} \lambda_i \right] \\ \text{subject to constraints (6.12)–(6.15)} \end{cases}$$

where i_l and j_l are two terminal vertices of arc l. One should note here that for any given λ, problem $P_T(\lambda)$ is the same as problem P_T but with transformed costs $c_l = c_l + \lambda_{i_l} + \lambda_{j_l}$. The best values of the penalties, (λ^* say), are then those which maximize the expression:

$$V(P_T(\lambda^*)) = \max_\lambda [V(P_T(\lambda))] \quad (6.16)$$

where $V(\cdot)$ is used to denote the optimum value of problem $(\cdot)$.

$V(P_T(\lambda^*))$ is then a lower bound to $V(\text{TSP}_s)$. The above penalty procedure was first proposed by Held and Karp (1970) and Christofides (1970) for slightly different versions of the travelling salesman problem. In addition, Held and Karp gave a very effective iterative subgradient optimization procedure for solving equation (6.16). They also noted that if the λ_i are set initially to $-(u_i + v_i)/2$, where u_i and v_i are the optimal dual variables of the assignment problem P_A defined by equations (6.1)–(6.3), then $V(P_T(\lambda))$ is at least as large as $V(P_A)$. Improved methods for deriving λ^* were later introduced by Hansen and Krarup (1974).

A different scheme for solving equation (6.16), based on heuristically guided search techniques, and which proved to be computationally quite successful, was given by Camerini, Fratta, and Maffioli (1975).

It is simple to show that in general a positive gap g exists between the value of the bound $V(P_T(\lambda^*))$ and the optimum TSP solution $V(\text{TSP}_s)$. However, g is often less than 1% of the value of $V(\text{TSP}_s)$ and problems of up to 100 cities have been solved optimally by embedding bound $V(P(\lambda^*))$, described above, into branch and bound algorithms.

Asymmetric TSP

Consider the case when G is a directed graph and let $\hat{\Phi}$ be a Hamiltonian circuit which is the solution to the TSP. Again let v be a distinguished vertex on $\hat{\Phi}$. If an arc (i, v) incident at v is removed from $\hat{\Phi}$, the remaining graph is a directed path—with root at vertex v—passing through all the vertices of G. Clearly, if T_v is the directed shortest spanning tree (shortest spanning arborescence) with root at v, then the cost of T_v plus the cost of the shortest arc (i, v)—say (i_1, v)—is a lower bound on the cost of $\hat{\Phi}$.

Let the graph composed of the arcs in T_v and arc (i_1, v) be called G_v. The problem whose optimum solution is the graph G_v is:

$$P_{DT} \begin{cases} \min & \sum_{i=1}^{n} \sum_{j=1}^{n} c_{ij} x_{ij} \\ \text{s.t.} & \sum_{i=1}^{n} x_{ij} = 1 \qquad \forall j \in N \\ & \sum_{i \in S_t} \sum_{j \in \bar{S}_t} x_{ij} \geq 1 \qquad \forall S_t \subset N. \\ & x_{ij} \in \{0,1\} \; \forall i,j \in N \end{cases}$$

Let TSP$_a$ be the asymmetric TSP composed of (6.1)–(6.5a). It is then clear that problem P_{DT} above is a relaxation of TSP$_a$ where constraints (6.3) have been dropped.

Once more, a better bound than $V(P_{DT})$ on $V(TSP_a)$ can be derived by a lagrangian relaxation of constraints (6.3), thus producing the problem:

$$P_{DT}(\lambda) \begin{cases} \min_x & \sum_{i=1}^{n} \sum_{j=1}^{n} (c_{ij} + \lambda_i) x_{ij} - \sum_{i=1}^{n} \lambda_i \\ \text{s.t.} & \text{constraints (6.2), (6.4), and (6.5a).} \end{cases}$$

The value:

$$V(P_{DT}(\lambda^*)) = \max_{\lambda} [V(P_{DT}(\lambda))] \qquad (6.17)$$

is then a lower bound on $V(TSP_a)$.

For a given λ, $P_{DT}(\lambda)$ is essentially the problem of finding the minimum cost directed spanning tree rooted at v, of the graph G with modified costs c_{ij}. This problem can be solved by a simple polynomial bounded algorithm as shown by Edmonds (1967) and Fulkerson (1974).

The problem of solving equation (6.17) is similar to the problem of solving equation (6.16), and one of the previously mentioned techniques can be used to derive the penalties λ^* which are the solution to (6.17).

A few computational comments are in order here.

In the first instance, the problem of finding the directed SST of a (directed) graph is computationally more difficult than that of finding the SST of a nondirected graph, with the result that each solution of problem $P_{DT}(\lambda)$ requires 4–6 times longer than the solution of an equivalent size problem $P_T(\lambda)$ for a range of n up to about 100. Thus, the derivation of bound $V(P_{DT}(\lambda^*))$ from equation (6.17) is much more costly than the derivation of bound $V(P_T(\lambda^*))$ from equation (6.16).

Secondly, the quality of bound $V(P_{DT}(\lambda^*))$—although good (on average within 2% of $V(\text{TSP}_a)$—is appreciably inferior to the quality of bound $V(P_T(\lambda^*))$ for the corresponding symmetric problem.

As a result of the above two shortcomings asymmetric TSP's cannot be solved by the use of bound $V(P_{DT}(\lambda^*))$ for sizes of n much above 60, as opposed to symmetric TSP's which can be solved for sizes of n of up to about 100 by using bound $V(P_T(\lambda^*))$ (Smith and Thompson, 1975).

6.2.2 Bounds from the Assignment Problem (AP)

Consider the assignment problem P_A defined by expressions (6.1)–(6.4). This is a relaxation of problem TSP_a where constraints (6.5a), (6.5b), or (6.5c) have been dropped. Balas and Christofides (1976) considered the introduction in a lagrangian fashion of some of the violated constraints from (6.5a) and some others from (6.5c). Let λ_t be the multiplier associated with the tth constraint in (6.5a) which is not satisfied. The problem then becomes:

$$P_A(\lambda) \begin{cases} \min_x & \sum_{i=1}^n \sum_{j=1}^n c_{ij} x_{ij} - \sum_t \lambda_t \sum_{i \in S_t} \sum_{j \in \bar{S}_t} x_{ij} + \sum_t \lambda_t \\ \text{s.t.} & \text{constraints (6.2)–(6.4).} \end{cases}$$

In contrast to the previously introduced relaxations $P_T(\lambda)$ and $P_{DT}(\lambda)$, the number of multipliers λ_t in the objective function of $P_A(\lambda)$ can increase exponentially, rather than linearly, with n. Thus, if h is the number of circuits formed by the AP solution, then there are $2^{h-1} - 1$ possible sets S_t for which constraint (6.5a) could be violated; once for cut $(S_t, \bar{S}_t)$ and once for cut $(\bar{S}_t, S_t)$, thus producing a maximum total of $2^h - 2$ violated constraints of type (6.5a). Since h could be as high as $\lfloor n/2 \rfloor$, the number of violated constraints is exponential in n. Balas and Christofides (1976) considered a sequential procedure for choosing the constraints to enter into the objective function by means of a lagrange multiplier, and gave an approximate noniterative method for determining these multipliers. The procedure is as follows.

Let $[\bar{c}_{ij}]$ be the reduced cost matrix after the solution of the AP. Form the graph $G_0 = (N, A_0)$ where $A_0 = \{(i, j) \mid \bar{c}_{ij} = 0\}$. Choose any vertex $i \in N$ and form the set of vertices $R(i)$ which can be reached from i via arcs of G_0. If $R(i) = N$ choose another vertex from N and repeat. If $R(i) = N \; \forall i \in N$ stop: no more cuts will be generated. If $R(i) \neq N$ then a cut $K_t = (S_t, \bar{S}_t)$ with $S_t = R(i)$ is generated which violates (6.5a).

Once K_t is identified, an initial value of the corresponding multiplier λ_t is computed as:

$$\lambda_t^0 = \min_{(i,j) \in K_t} [\bar{c}_{ij}] \tag{6.18}$$

The Travelling Salesman Problem

and $[c_{ij}]$ is updated by:

$$\bar{c}_{ij} = \bar{c}_{ij} - \lambda_t^0 \quad \forall (i,j) \in K_t$$

$\bar{c}_{ij}$ unchanged otherwise.

Update G_0 to include the arcs for which $\bar{c}_{ij}$ has just become 0 and repeat until no more cuts can be generated. The number of cuts generated by this procedure is limited to be at the most $2(h-1)$.

At this stage it is no longer possible to increase $V(P_A(\lambda))$ by modifying the λ while keeping the modified costs $\bar{c}_{ij}$ nonnegative. However, the AP solution for problem P_A is still optimal for $P_A(\lambda)$ and it is possible to identify constraints of the form (6.5c) which are violated by this solution. Let the solution contain circuits Φ_p, where Φ_p is the set of arcs of forming that circuit. The lagrangian relaxation of the corresponding constraints (6.5c) then leads to the problem:

$$P_A(\lambda, \mu) \begin{cases} \min_x \sum_{i=1}^n \sum_{j=1}^n c_{ij} x_{ij} - \sum_t \lambda_t \sum_{i \in S_t} \sum_{j \in \bar{S}_t} x_{ij} \\ \quad + \sum_p \mu_p \sum_{(i,j) \in \Phi_p} x_{ij} - \sum_p \mu_p |\Phi_p| + \sum_t \lambda_t + \sum_p \mu_p \\ \text{s.t.} \quad \text{constraints (6.2)--(6.4).} \end{cases} \quad (6.19)$$

If the λ are assumed fixed at λ_t^0 as mentioned earlier, the problem of maximizing $V(P(\lambda, \mu))$ over all μ can be considered in a way similar to that used above to determine the λ, that is by setting μ_p to a value μ_p^0 which is the largest value of μ_p that will leave the AP solution to problem $P(\lambda, \mu)$ unchanged. This value μ_p^0 is simple to derive and can be computed in much the same way that the duals are computed for the AP.

With μ_p^0 determined, new dual variables u_i and v_j are computed for problem $P_A(\lambda^0, \mu_p^0)$ which can be used to update the costs $\bar{c}_{ij}$ corresponding to problem $P_A(\lambda^0)$ by $\bar{c}_{ij} = \bar{c}_{ij} - u_i - v_j$.

Another circuit Φ_p in the AP solution is now identified, and another value of μ_p^0 determined. New dual variables u_i and v_j are computed for problem $P_A(\lambda^0, [\mu_{p_1}^0, \mu_{p_2}^0])$, where p_1 and p_2 are the indices of the first and second circuits Φ_p considered, and $\bar{c}_{ij}$ again updated, etc., until all circuits Φ_p in the AP solution have been considered.

At the end of this procedure, it is clear that, since the initial AP solution is still optimal, the quantity:

$$V(P_A) + \sum_t \lambda_t^0 + \sum_p \mu_p^0$$

is a lower bound to the TSP and has been found to be a reasonable approximation to:

$$\max_{\lambda, \mu} [V(P_A(\lambda, \mu))].$$

Let B be an optimal solution to the assignment problem $P_A(\lambda^0, \mu^0)$. We know that the initial AP solution is optimal, but there may also be other alternative solutions. If B is a hamiltonian circuit, and if it satisfies the constraints for which $\lambda \neq 0$ and $\mu \neq 0$ as equalities, then B is an optimal solution to the TSP. If B is a hamiltonian circuit but some constraints with $\lambda, \mu \neq 0$ are not satisfied as equalities, then it may still be possible to modify the λ and μ and derive a better upper bound (and a better solution) than the value of the best known solution so far in the tree search. Some techniques for modifying λ and μ in such circumstances are given in Balas and Christofides (1976).

If B is not a hamiltonian circuit, let us again define the graph $G_0 = (X, A_0)$, $A_0 = \{(i, j) \mid \bar{c}_{ij} = 0\}$, where $\bar{c}_{ij}$ are the reduced costs of problem $P_A(\lambda^0, \mu^0)$. Figure 6.1 shows an example of a possible G_0.

Let $\hat{\Phi}$ be the optimum solution to the TSP. $\hat{\Phi}$ defines a graph $G(\hat{\Phi})$, consisting of a single hamiltonian circuit. If a vertex v (say) is removed from $G(\hat{\Phi})$, the resulting graph $G_v(\hat{\Phi})$ is a hamiltonian path through the remaining $(n-1)$ vertices and is, therefore, a unilaterally connected graph; that is for any two vertices i and j of $G_v(\hat{\Phi})$ there exists a path either from i to j or from j to i.

Let us now consider the graph G_0 with vertex v removed thus producing graph $G_0^v = (X^v, A_0^v)$. If G_0^v is not unilaterally connected, there must exist a pair of cuts K_v', K_v'' of G_0^v for which $A_0^v \cap K_v' = A_0^v \cap K_v'' = \emptyset$.

If we now define:

$$\pi_v = \min_{(i,j) \in K_v' \cup K_v''} [\bar{c}_{ij}]$$

and transform:

$$\bar{c}_{ij} = \bar{c}_{ij} - \pi_v \quad \forall (i, j) \in K_v' \cup K_v''$$

$\bar{c}_{ij}$ unchanged otherwise

a new graph G_0 can be defined which can again be tested for unilateral connectedness (after removing some other vertex v), and so on, until G_0^v remains unilaterally connected after the removal of any v.

Let $W = \{v_1, v_2, \ldots, v_k\}$ be the set of vertices whose removal has led to G_0^v not being unilaterally connected during this procedure. It is then clear that the quantity

$$B_{AP} = V(P_A) + \sum_t \lambda_t^0 + \sum_p \mu_p^0 + \sum_{v \in W} \pi_v$$

is a lower bound to the TSP.

Some computational comments are in order here. (The computational results mentioned below depend, to some extent, on the way that the problems are generated, but the relative magnitudes of the bounds and the conclusions drawn are, in general, correct.)

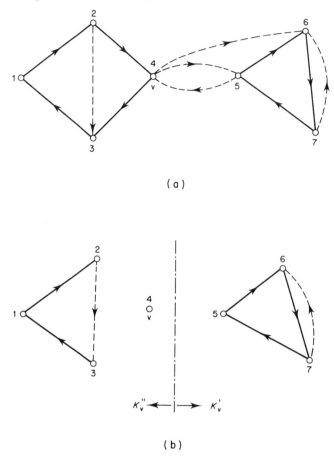

Figure 6.1 (a) The 0 graph G_0, ——— arcs in AP solution; - - - - other arcs.
(b) The graph G_0^v

The procedure can be applied to both symmetric and asymmetric problems but produces much better bounds for the latter type of problem.

For *symmetric* problems, the value of the assignment solution $V(P_A)$ is, on average, about 20% below $V(TSP_s)$, whereas B_{AP} is, on average, only 4% below $V(TSP_s)$. However, bound B_{AP} is very inferior to bound $V(P_T(\lambda^*))$—derived earlier from the SST—although on average it is 4 to 5 times faster to compute B_{AP} rather than $V(P_T(\lambda^*))$ for symmetric TSP's in the range up to $n = 100$.

For *asymmetric* problems, the value of the assignment solution $V(P_A)$ is, on average, about 3% below $V(TSP_a)$, whereas B_{AP} is on average only about

½% below $V(\text{TSP}_a)$. Thus, for asymmetric problems, bound B_{AP} is of approximately the same quality as bound $V(P_{\text{DT}}(\lambda^*))$—derived earlier from the directed SST—but requiring on average 10 to 20 times less computational effort.

Asymmetric TSP's with 250 or more cities can be solved by the use of AP-based bounds in tree search algorithms with 'subtour elimination' and other types of branching schemes. Balas and Christofides (1976)

6.2.3 Bounds from the Matching Problem (MP)

Although the AP bound described in the previous section can be computed for both symmetric and asymmetric TSP's, a much better bound can be derived for symmetric problems by the use of a matching problem (MP) relaxation which is very much in the same spirit as the AP relaxation described above.

Consider the following 2-matching problem.

$$P_M \begin{cases} \min & \sum_{l=1}^{m} c_l x_l & (6.20) \\ \text{s.t.} & \sum_{l \in A_i} x_l = 2 & (6.21) \\ & x_l \in \{0, 1\}. & (6.22) \end{cases}$$

This is the same as problem TSP_s—defined by (6.6), (6.10a), and (6.10b)—with constraint (6.10a) ignored. If (6.10a) is included into the objective function (6.20) via a lagrange multiplier, we get the problem:

$$P_M(\lambda) \begin{cases} \min_{x} & \sum_{l=1}^{m} c_l x_l - \sum_{t} \lambda_t \sum_{l \in K_t} x_l + \sum_{t} \lambda_t \\ \text{s.t.} & \text{constraints (6.21) and (6.22).} \end{cases}$$

This problem has essentially the same form as problem $P_A(\lambda)$, and the same techniques used in the last section to derive bound B_{AP} can be used here to derive an analogous bound B_M.

6.2.4 Bound from Shortest n-Paths (Houck et al., 1977)

Consider a path with n arcs starting from some given vertex v of the graph G and finishing at some given vertex w. Such a path is called an n-path from v to w. It is clear that the solution to the TSP is an n-path (circuit) from v back to v, where each vertex of G appears exactly once on this path.

The Travelling Salesman Problem

Clearly, the value of the shortest n-path from v back to v (where a vertex can appear an arbitrary number of times on the path) is a lower bound to the TSP.

The computation of the shortest n-path from v back to v is a simple path problem which can be solved by a dynamic programming recursion similar to that used by Bellman (1958) for finding shortest paths in graphs. Thus, let $f_k(i)$ denote the shortest k-path from vertex v to vertex i.

Starting from
$$f_1(i) = c_{1i} \quad i \neq v, i \in N$$
the recursion is
$$f_k(i) = \min_{j \neq i, vj \in N} [f_{k-1}(j) + c_{ji}] \quad i \neq v, i \in N, k = 2, \ldots, n-1 \quad (6.23)$$
and
$$f_n(v) = \min_{j \neq vj \in N} [f_{n-1}(j) + c_{jv}].$$

The n-path corresponding to $f_n(v)$ can be found by recording for each $f_k(i)$ the value of j which produced the minimum in (6.23). Obviously, if the n-path passes through each vertex exactly once it is a solution to the TSP. If not, then $f_n(v)$ is a lower bound on the value of the TSP.

It is easy to see that if a penalty λ_i is associated with each vertex i and the costs c_{ij} are transformed by $c'_{ij} = c_{ij} + \lambda_i + \lambda_j$, then the cost of any hamiltonian circuit in G is increased by the same constant amount $2\sum_i \lambda_i$, but n-paths that are not hamiltonian circuits are penalized differently. Let $f_n(v)$ be recomputed with the modified costs c'_{ij} and let this n-path pass through vertex i d_i times. Then
$$w(\lambda) = f_n(v) + 2\sum_i (d_i - 1)\lambda_i$$
is also a valid bound to the TSP. We therefore wish to choose that λ^* which corresponds to the maximum of the expression:
$$w(\lambda^*) = \max_\lambda [w(\lambda)] \quad (6.24)$$
and use $w(\lambda^*)$ as a bound for the TSP. Subgradient optimization is one possible procedure for solving equation 6.24.

In the path corresponding to $f_n(v)$, the same vertex (or arc) can appear more than once or not at all. It is not possible to constrain the path so that a vertex does not appear more than once on the path, but one type of vertex repetition, namely when the rth and $(r+2)$th vertices on the path correspond to the same vertex of G—for some value of r—can be excluded by a slight modification to recursion (23). If $f'_n(v)$ is the shortest n-path from v to v with the above vertex repetition excluded, a bound $w'(\lambda^*)$ can be computed in a way similar to that for $w(\lambda)$ but with $f_n(v)$ replaced with $f'_n(v)$.

Computational results reported in Houck et al. (1977) indicate that bound $w'(\lambda^*)$ is slightly worse than bound $V(P_T(\lambda^*))$ obtained from the SST for the symmetric case, and about the same as bound $V(P_{DT}(\lambda^*))$ obtained from the directed SST for the asymmetric case. A great disadvantage of bound $w'(\lambda)$ is that $O(n^3)$ operations are required to compute the shortest n-path of a graph (for a given root vertex v) as compared with $O(n^2)$ operations for the SST and $O(n^{2.5})$ for the AP. (Note, however, that in the last case only one solution of the AP is required.) On the other hand, an advantage of $w'(\lambda^*)$ is that it can be easily modified to include additional constraints which exist in problems closely related to the TSP, such as the vehicle routing problem and some machine scheduling problems.

6.3 Exact Solution Procedures Based on LP

Consider the TSP defined by expressions (6.1)–(6.4) and (6.5b). One possible procedure for solving the TSP is to solve the linear programming relaxation of TSP (say $\overline{\text{TSP}}$) by dropping constraints (6.4) and then imposing integrality either in a branch and bound algorithm or by a cutting plane procedure. An obvious problem that immediately arises is due to the very large number of constraints of type (6.5b) that exists, and which implies that such constraints must be introduced into the LP tableau if and when they are violated by an LP solution rather than in a single step.

The procedure was first suggested by Dantzig, Fulkerson, and Johnson (1954) and proceeds as follows. Relax the TSP as much as possible, for example by including only constraints (6.2), (6.3), and a very small subset of constraints (6.5b). Solve the LP to obtain a solution x. If x represents a hamiltonian circuit the problem is solved. If not, choose a set of inequalities from

(i) constraints of type 6.5b (or 6.5a) which are violated by x and/or

(ii) other constraints which are violated by x (for example constraints which are satisfied by any integer solution but which are violated if x turns out to be fractional).

Add the chosen inequalities to the LP tableau and reoptimize to obtain a new solution x, and repeat until an integer feasible solution is obtained.

The basic method described above has been adapted in a number of different ways.

Miliotis (1976) suggested and tested an algorithm in which integrality is first achieved—either by branching on fractional variables or by introducing Gomory cuts from the constraints in group (ii) above—and then adding constraints of type 6.5a from group (i).

Grötschel (1977) considered the simultaneous inclusion of some constraints from (i) and some constraints from (ii). For any (in general fractional) solution x to the LP, violated constraints of type 6.5b from group (i)

The Travelling Salesman Problem

were identified visually by plotting the solution on a map. (Obviously this is only possible when the TSP is a symmetric and—almost—Euclidean problem.) Constraints from group (ii) were chosen to be facets of the TSP polytope and, in particular, were chosen from a class of constraints known as the comb inequalities (Chvátal, 1973; Grötschel and Padberg, 1974). These inequalities (which are only defined on nondirected graphs) are similar to the inequalities introduced by Edmonds (1965) for the linear characterization of the matching polytope. They eliminate fractional vertices of the $\overline{\text{TSP}}$ polytope but may introduce other fractional vertices from their intersections with the constraints of type 6.5b. Once more these constraints from group (ii) were chosen visually.

Christofides and Whitlock (1978) considered the inclusion, first of constraints of type 6.5b from group (i), and only when all constraints of type 6.5b are satisfied then imposed constraints from group (ii) by branching. For a fractional solution x, a corresponding graph G^x was defined, and by using the Gomory–Hu algorithm (Christofides, 1975) for determining maximum flows between every pair of vertices of G^x, a number of violated constraints of type 6.5a were identified, or it was shown that all such constraints were satisfied. Thus, the generation of the constraints to be added is automatic.

Christofides and Whitlock also considered the *a priori* reduction in the size of the LP's to be solved, by calculating initially bound B_{AP} of Section 6.2.2 and making use of the resulting reduced costs $\bar{c}_{ij}$ at the end of the bound calculation procedure to eliminate variables x_{ij} from the LP formulation before the LP's are solved.

In general, LP-based methods are successful and at least competitive with pure branch and bound procedures for solving symmetric TSP's, although at least two of the procedures described above can also be described as LP's embedded in branch and bound schemes. Symmetric problems of up to at least 100 cities can be solved by LP-based algorithms without human intervention in subjective decision making. LP-based methods are not competitive with branch and bound methods for asymmetric TSP's.

6.4 Heuristic Procedures for the TSP

The TSP is an *NP*-complete problem (Garey, Graham, and Johnson, 1976) and all the methods previously described for its solution have a rate of growth of the computation time which is exponential in n (the number of cities in the TSP). There are several approximation algorithms whose rate of growth of the computation time is a low order polynomial in n and which have been experimentally observed to perform well. In this section we summarize some of these procedures, and restrict our attention to symmetric TSP's with cost matrices that satisfy the triangle inequality.

6.4.1 The Nearest Neighbour Rule (NNR)

With this procedure, one starts with an arbitrary vertex and proceeds to form a path by joining the vertex just added to its nearest neighbouring vertex which is not yet on the path, until all vertices are visited, in which case the two end vertices of the hamiltonian path are joined to form the TSP solution.

Rosenkrantz, Stearns, and Lewis (1974) have shown that:

$$\frac{V(\text{NNR})}{V(\text{TSP})} \leq \tfrac{1}{2}(\lceil \log n \rceil + 1) \tag{6.25}$$

where $V(\text{NNR})$ is the value obtained by NNR, $V(\text{TSP})$ is the value of the optimal solution to the TSP and $\lceil x \rceil$ is the smallest integer greater than or equal to x.

For $n \geq 15$, it is also shown in Rosencrantz, Stearns, and Lewis (1974) that:

$$\frac{V(\text{NNR})}{V(\text{TSP})} > \tfrac{1}{3}(\log(n+1) + 4/3) \tag{6.26}$$

and noted that the cause of the worst-case bound on $V(\text{NNR})$ (given by (6.26)) being logarithmically increasing with n is neither the fact that the last added arc in NNR is too long, nor that the starting vertex is chosen arbitrarily.

The NNR requies $O(n^2)$ operations to apply.

6.4.2 The Nearest Insertion Rule (NIR)

In this procedure are starts with a circuit Φ passing through a subset of the set of vertices and adds sequentially into Φ vertices not already in Φ until Φ becomes hamiltonian. The vertex (x say) to be added next at some stage can be chosen to be that vertex (not in Φ) nearest to any vertex already in Φ; having chosen x, it is inserted in that position of Φ which causes the least additional cost. The circuit Φ can be initialized to be a self-loop on an arbitrarily chosen vertex.

It is shown in Rosencrantz, Stearns, and Lewis (1974) that:

$$\frac{V(\text{NIR})}{V(\text{TSP})} < 2 \tag{6.27}$$

and that for every $n \geq 6$, there exists a graph for which

$$\frac{V(\text{NIR})}{V(\text{TSP})} = 2\left(1 - \frac{1}{n}\right) \tag{6.28}$$

that is, the worst-case bound on $V(\text{NIR})$—for arbitrary n—is $2V(\text{TSP})$.

The Travelling Salesman Problem

Two possible variants of NIR are:

(i) Given Φ, let us say that vertex x to be next inserted into Φ is chosen and that y is the vertex of Φ nearest to x. Instead of inserting x into the least-extra-cost position in Φ, we insert x either just before or just after y in Φ, whichever is the cheaper.

(ii) Given Φ, choose x to be that vertex which when inserted into its best position in Φ leads to the smallest extra cost, that is both x and the position into which it is to be inserted are chosen simultaneously.

Conditions (6.27) and (6.28) apply to both the above variants. NIR and variant (i) require $O(n^2)$ operations and variant (ii) requires $O(n^2 \log n)$ operations.

6.4.3 Lin's r-optimal Heuristic (Lin, 1965)

Starting from an arbitrary initial tour let r arcs be removed, thus producing r disconnected paths. These paths can be reconnected in one or more different ways to produce another tour, and this operation is called an r-change. A tour is r-optimal if no r-change produces a tour of lower cost.

It is shown in Rosencrantz, Stearns, and Lewis (1974) that for all $n \geq 8$ there exists a graph for which

$$\frac{V(r\text{-opt})}{V(\text{TSP})} = 2\left(1 - \frac{1}{n}\right) \qquad (6.29)$$

for all $r \leq n/4$, where $V(r\text{-opt})$ is the value of an r-optimal tour.

Although the problem of finding an r-optimal tour can be performed in a number of operations polynomial in n, this number is exponential in r and is bounded from below by n^r. Thus, only very small values of r can be used in any heuristic (Christofides and Eilon, 1972; Liu and Kernighan, 1973).

6.4.4 Christofides' Heuristic (CH) (Christofides, in press)

Let T^* be the solution to the SST of graph G. Relative to T^*, let X^0 be the set of vertices having odd degree. Solve the matching (1-matching) problem for graph $\langle X^0 \rangle$ and let M be the set of arcs in this matching. The graph $G' = (X, A')$, composed by the set X of vertices of G and having as arcs A' only those arcs in T^* and M, is eulerian, that is, has all vertices of even degree and can, therefore, be traversed by an eulerian circuit E so that every arc of G' is traversed once and once only by E. It is possible to construct a hamiltonian circuit Φ of G which serves as the heuristic solution to the TSP by making use of circuit E as follows.

Start constructing Φ by following arcs of E. If a vertex already visited by

Φ reappears in the vertex sequence of E, skip that vertex unless all vertices are in Φ in which case return to the starting vertex.

It is shown in Christofides (in press) that:

$$\frac{V(\text{CH})}{V(\text{TSP})} < \frac{3}{2} \tag{6.30}$$

and that CH requires $O(n^3)$ operations to be executed.

Recently, Cornuejols and Nemhauser (1978) have shown that for every $n \geq 3$, there exists a graph for which:

$$\frac{V(\text{CH})}{V(\text{TSP})} = \frac{3n-1}{2n}. \tag{6.31}$$

The results given for the heuristics in Sections (6.4.1)–(6.4.4) above are concerned with absolute performance guarantees. The average performance of these heuristics is much better than the worst case performance and is almost always within a few percent of the optimum solution. This is particularly true for the r-optimal heuristic of Lin and its variants (Liu, 1965; Liu and Kernighan, 1973; Christofides and Eilon, 1972) (for $r = 2$ or 3), which produces excellent results on average. The probabilistic analysis of any of the above heuristics (see Karp, 1976; Nemhauser et al., 1976; and Sahni and Gonzalez, 1976, for further results on the general analysis of heuristics) is too complex, but Karp (1977) has recently given a partitioning scheme for Euclidean TSP's which, for every $\varepsilon > 0$ produces a tour costing no more than $(1+\varepsilon)$ times the cost of the optimal tour (with probability that tends to 1 as $n \to \infty$) and which runs in time $k(\varepsilon)n + O(n \log n)$, where $k(\varepsilon)$ is a constant dependent on the choice of ε.

6.5 Some Solvable Cases of the TSP

There are a few particular cases of the TSP which can be solved exactly by algorithms in which the total number of operations involved is bounded from above by a polynomial function of n (polynomial algorithms).

The first example of a solvable TSP was found by Gilmore and Gomory (1964) in connection with the problem of sequencing items on a one-state-variable machine. Assume there are n items to be heated in a furnace and assume item i has to be taken from temperature A_i to B_i. Let us order the items so that $B_i \geq B_{i-1}$ for all i.

The cost of processing item j after item i is c_{ij} where:

$$c_{ij} = \int_{B_i}^{A_j} f(x)\,dx \quad \text{if} \quad A_j \geq B_i$$

$$c_{ij} = \int_{A_j}^{B_i} g(x)\,dx \quad \text{if} \quad A_j < B_i$$

The Travelling Salesman Problem

where $f(x)$ and $g(x)$ are any two functions of the temperature x so that $f(x) + g(x) \geq 0$.

The optimum solution to the TSP—of finding the minimum cost cyclic sequence of processing the items in the furnace—can be obtained by first solving the assignment problem (which in general will produce a number of subloops) and then finding best exchanges of two arcs in separate subloops by two other arcs which joint the subloops into a single loop, and repeating until a single hamiltonian circuit is produced. By a suitable choice of these arc exchanges it is shown in Gilmore and Gomory (1964) that this circuit is the solution to the TSP.

A second special case is due to Lawler (1971) and requires the cost matrix $[c_{ij}]$ to be upper triangular, that is:

$$c_{ij} = 0 \quad \text{for all} \quad i \geq j.$$

The assignment problem is now solved for a reduced matrix C' obtained from C by deleting column 1 and row n. The solution to this AP corresponds to a path from 1 to n and a number of circuits. Backward arcs (i, j) with $i \geq j$ (and which have zero cost) are now deleted from the circuits in the AP solution to produce a number of disconnected paths. Other backward arcs (also of zero cost) are now introduced to join together these paths into a single hamiltonian circuit. An example is shown in Figure 6.2.

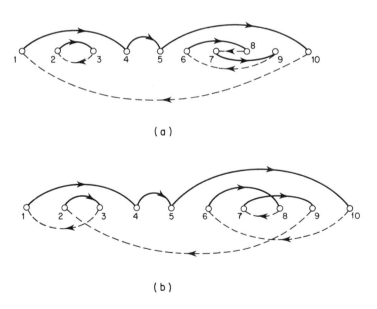

Figure 6.2 (a) AP solution. —— positive cost arcs; ---- zero cost arcs.
(b) TSP solution

A third special case was described by Syslo (1973), who noted that since the problem of finding the shortest Eulerian circuit in a graph can be solved easily, the special case of a TSP on a line graph can also be solved by the same algorithm, and in this case all hamiltonian circuits of the graph would have the same cost.

6.6 References

Balas, E. and N. Christofides (1976). A new penalty method for the travelling salesman problem. Presented at the *9th Math. Prog. Symposium, Budapest, 1976.*

Bellman, R. (1958). On a routing problem. *Quart. J. Appl. Math.*, 87.

Bellmore, M. and J. C. Malone (1971). Pathology of travelling salesman subtour elimination algorithms. *Ops. Res.*, **19**, 278.

Burkard, R. E. (1977). Travelling salesman and assignment problems—a survey. *Report 77–11, Mathematisches Institut, University of Cologne.*

Camerini, P., L. Fratta, and F. Maffioli (1975). Travelling salesman problem: Heuristically guided search and modified gradient techniques. *Report of Instituto di Elettronica, Politecnico di Milano.*

Christofides, N. (1970). The shortest hamiltonian chain of a graph. *J. SIAM (Applied Mathematics)*, **19**, 689.

Christofides, N. (1972). Bounds for the travelling salesman problem. *Ops. Res.*, **20**, 1044.

Christofides, N. (1975). *Graph Theory—An Algorithmic Approach*, Academic Press, London.

Christofides, N. (in press) Worst case analysis of a new heuristic for the travelling salesman problem. *Math. Prog.* (to be published).

Christofides, N. and S. Eilon (1972). Algorithms for large scale TSPs. *Opl. Res. Quart.*, **23**, 511.

Christofides, N. and C. Whitlock (1978). An LP-based TSP algorithm. *Imperial College Report* OR 78·14.

Chvátal, V. (1973). Edmonds polytopes and weakly hamiltonian graphs. *Math. Prog.*, **3**, 29.

Cornuejols, G. and G. L. Nemhauser (1978). Tight bounds for Christofides' TSP heuristic. *Math. Prog.*, **8**, 163.

D'Atri, G. (1977). Improved lower bounds to the travelling salesman problem. *IP77/3, Institut de Programmation.*

Dantzig, G., D. R. Fulkerson, and S. Johnson (1954). Solution of a large scale travelling salesman problem. *Ops. Res.*, **2**, 393.

Edmonds, J. (1965). Maximum matchings and a polyhedron with (0, 1) vertices. *J. Res., Nat. Bur. Stand.* **69B**, 125.

Edmonds, J. (1967). Optimum branchings. *J. Res., Nat. Bur. Stand.*, **71B**, 233.

Fulkerson, D. R. (1974). Packing routed directed cuts in a weighted directed graph. *Math. Prog.*, **4**, 1.

Gabovič, E. J. (1970). The travelling salesman problem. *Trudy Vyčisl. Centra Tartu. Gos. Univ.*, 52.

Garey, M. R., R. L. Graham, and P. S. Johnson (1976). Some NP-complete geometric problems. *Proc. 8th ACM Symp. on Theory of Computing, 1976.*

Gilmore, P. C. and R. E. Gomory (1964). A solvable case of the travelling salesman problem. *Proc. Nat. Acad. Sci.*, **12**, 178.

Grötschel, M. (1977). An optimal tour through 120 cities in Germany. *Report 7770, University of Bonn.*
Grötschel, M. and M. W. Padberg (1974). Linear characterisation of the symmetric travelling salesman polytope. *Report 7417, University of Bonn.*
Hansen, K. H. and J. Krarup (1974). Improvements of the Held–Karp algorithm for the symmetric travelling salesman problem. *Math. Prog.*, **4,** 87.
Held, M. and R. Karp (1970). The travelling salesman problem and minimum spanning trees. *Ops. Res.*, **18,** 1138.
Held, M. and R. Karp (1971). The travelling salesman problem and minimum spanning trees II. *Math. Prog.*, **1,** 6.
Houck, D., J.-C. Picard, M. Queyranne, and R. R. Vemuganti (1977). The travelling salesman problem and shortest n-paths. *University of Maryland.*
Karp, R. M. (1976). The probabilistic analysis of some combinatorial search algorithms. *Proc. Symposium on Algorithms and Complexity, Pittsburgh.*
Karp, R. M. (1977). Probabilistic analysis of partitioning algorithms for the TSP in the plane. *Math. Oper. Res.*, **2,** 209.
Lawler, E. L. (1971). A solvable case of the travelling salesman problem. *Math. Prog.*, **1,** 267.
Lin, S. (1965). Computer solution of the TSP. *Bell System Tech. J.*, **44,** 2245.
Lin, S. and B. W. Kernighan (1973). An effective heuristic algorithm for the travelling salesman problem. *Ops. Res.*, **21,** 498.
Miliotis, P. (1976). Integer programming approaches to the travelling salesman problem. *Math. Prog.*, **6,** 367.
Nemhauser, G. L., L. A. Wolsey, and M. L. Fischer (1976). An analysis of approximations for maximising submodular set functions. *CORE paper 7618, Louvain.*
Padberg, M. W. and S. Hong (1977). On the symmetric travelling salesman problem: A computational study. Presented at *Symp. on Discrete Opt., Banff.*
Rosenkrantz, D. J., R. E. Stearns, and P. M. Lewis (1974). Approximate algorithms for the TSP. *Proc. 15th IEEE Symp. on Switching and Automata Theory*, p. 33.
Rubinshtein, M. I. (1971). On the symmetric TSP. *Automatika i Telemekhanika*, **6,** 126.
Sahni, S. and T. Gonzalez (1976). P-complete approximation problems. *J. ACM*, **11,** 555.
Smith, T. H. C., and G. L. Thompson (1975). A comparison of two different lagrangean relaxations of the TSP. *MSR382, Carnegie-Mellon University.*
Syslo, M. M. (1973). A new solvable case of the travelling salesman problem. *Math. Prog.*, **3,** 347.

CHAPTER 7

Set Partitioning—A Survey†

EGON BALAS
Carnegie-Mellon University

MANFRED W. PADBERG
New York University

7.0 Introduction

This paper discusses the set partitioning or equality-constrained set covering problem. It is a survey of theoretical results and solution methods for this problem, and while we have tried not to omit anything important, we have no claim to completeness.

Section 7.1 gives some background material. It starts by discussing the uses of the set partitioning model; then it introduces the concepts to be used throughout the paper, and connects our problem to its close and distant relatives which play or may play a role in dealing with it: set packing and set covering, edge matching and edge covering, node packing and node covering, clique covering. The crucial equivalence between set packing/partitioning and node packing problems is introduced.

Section 7.2 deals with structural properties of the set packing and set partitioning polytopes. We discuss necessary and sufficient conditions for all vertices of the set packing polytope to be integer, and we describe the facial structure of this polytope to the extent that it is known. In this description, the one to one correspondence between graphs and set packing polytopes plays a central role: the facets of a set packing polytope are related to certain subgraphs of an associated graph. We review the various classes of facets and associated subgraphs that have been identified to date. Since facets of the set packing polytope can be arbitrarily complex and therefore computationally expensive to generate, we then discuss a class of inequalities derived from the disjunctive conditions of the set partitioning problem, which are facets of various relaxations of the set partitioning polytope, and can easily be computed from a fractional simplex tableau. These inequalities can be used in simplex-based fractional cutting plane algorithms. Another class of inequalities, also derived from the logical

† Work partially supported by the National Science Foundation and the U.S. Office of Naval Research. Reprinted with permission from the *SIAM Review*, **18**, 1976, p. 710–760.

implications of the set partitioning constraints, is unrelated to any particular simplex tableau. These inequalities have coefficients of 0, 1, or -1 and provide convenient all-integer cutting planes especially for a primal approach. Finally, we characterize adjacency relations between vertices of the set partitioning and set packing polytopes, on these polytopes as well as on their linear programming relaxations. The basic property, that every edge of the set partitioning (set packing) polytope is also an edge of the linear programming relaxation of the latter, is viewed in the context of the need for appropriate criteria to identify such edges that meet a given vertex. All this theory is relatively new: a product of the last five years. Proofs are in general omitted, but sources are referenced in each case.

Section 7.3 focuses on algorithms. We first discuss the two main types that are by now well established: implicit enumeration and (traditonal) cutting planes. While in the first category several specialized algorithms have been developed, of which we discuss the ones that to our knowledge have been tested and found successful, algorithms in the second category are basically nonspecialized. Nevertheless, since cutting planes are known to be relatively efficient on set partitioning problems, we discuss some features of the algorithms and codes in this class. For both of these approaches, we briefly review the published computational experience. Next we discuss some recently developed approaches which are either untested or tested to a very limited extent, but which are based on new ideas that seem to hold some promise. The first one is a column generating procedure, based on the adjacency properties discussed in Section 7.2. It uses a modified all-integer version of the primal simplex algorithm, and generates composite columns corresponding to edges of the feasible set which connect a given integer vertex to a better one. The second procedure is a hybrid algorithm which combines a primal cutting plane method based on all integer cuts with coefficients 0, 1, or -1, with implicit enumeration applied to subproblems so defined as to generate an improvement at each iteration. The third approach uses a new symmetric subgradient method to solve the set partitioning linear program, amended with cutting planes, in an attempt to eliminate the difficulties involved in solving these large, very constrained and very degenerate linear programs by the simplex method. Finally, the fourth one deals with the set partitioning problem via an equivalent weighted node covering problem, which it solves by a hybrid cutting-plane–branch-and-bound algorithm. The latter again avoids recourse to the simplex method, and uses a labelling technique instead.

We assume some familiarity on the part of the reader with the basic concepts of graph theory. For background material in this field, the reader is referred to Harary (1969), Berge (1970), Roy (1969, 1970), Christofides (1975). For background in the general areas of linear and integer programming, see Dantzig (1963) and Simmonard (1966) for the former, and Garfinkel and Nemhauser [1972] for the latter.

Set Partitioning—A Survey 153

7.1 Background

7.1.1 Set Partitioning and its Uses

Among all special structures in (pure) integer programming, there are three which have the most widespread applications: set paritioning, set covering, and the travelling salesman (or minimum length hamiltonian cycle) problem; if we were to rank the three, set partitioning would be a likely candidate for number one.

The (weighted) *set partitioning* (or equality-constrained set covering) problem is

SPP $\quad\quad\quad\quad\quad \min \{cx \mid Ax = e, x_j = 0 \text{ or } 1, \forall j \in N\}$

where A is an $m \times n$ matrix of zeros and ones, c is an arbitrary n-vector, $e = (1, \ldots, 1)$ is an m-vector, and $N = \{1, \ldots, n\}$. Its name comes from the following interpretation: if the rows of A are associated with the elements of the set $M = \{1, \ldots, m\}$ and each column a_j of A with the subset M_j of those $i \in M$ such that $a_{ij} = 1$, then SPP is the problem of finding a minimum-weight family of subsets M_j, $j \in N$, which is a partition of M, each subset M_j being weighted with c_j.

A partial list of applications described in the literature includes: railroad crew scheduling, truck deliveries, airline crew scheduling, tanker routing, information retrieval, switching circuit design, stock cutting, assembly line balancing, capital equipment decisions, location of offshore drilling platforms, some other facility location problems, and political districting. A special bibliography on applications is given as an Appendix to this chapter.

A great variety of scheduling problems can be formulated as follows. Given (i) a finite set M; (ii) a constraint set defining a family F of 'acceptable' subsets of M; and (iii) a cost (real number) associated with each member of F; find a minimum-cost collection of members of F which is a partition of M.

The usefulness and wide applicability of the set partitioning model follows from the simple observation that in most cases a problem of the above form can be solved to a satisfactory degree of approximation by the following two-stage procedure.

Stage 1 Using (ii), generate explicitly a subset $\bar{F} \subset F$, such that the probability of an optimal solution being contained in $\bar{F}$ is sufficiently high.

Stage 2 Replace the constraint set (ii) by a list of the members of $\bar{F}$ and solve the resulting SPP.

The most widely used application to date of the set partitioning model seems to be the airline crew scheduling problem, in which M corresponds to the set of flight legs (from city A to city B, at time t) to be covered during a planning period (usually a few days), while each subset M_i stands for a possible tour (sequence of flight legs with the same initial and terminal point) for a crew. In order to be acceptable, a tour must satisfy certain regulations. To set up the problem, one starts with a given set of (usually several hundred) flight legs, and one generates by computer a set of (usually several thousand) acceptable tours, with their respective costs. This produces A (of density usually ≤ 0.05) and $c > 0$, after which one attempts to solve the set partitioning problem. If the attempt is successful, the solution yields a minimum-cost collection of acceptable tours such that each flight leg is included in exactly on tour of the collection.

Airline crew scheduling problems with 300–500 constraints and 2500–4000 variables are sometimes solved (to optimality), but often much smaller problems (with several hundred variables) defy solution within reasonable time limits.

7.1.2 Set Packing and Set Covering

The set partitioning problem (SPP) has two seemingly close relatives; the *set packing* problem

SP $\qquad\qquad \max\{c'x \mid Ax \leq e, x_j = 0 \text{ or } 1, \forall j \in N\}$

and the *set covering* problem

SC $\qquad\qquad \min\{c''x \mid Ax \geq e, x_j = 0 \text{ or } 1, \forall j \in N\}$

where A, e, and N are defined as in SPP, while c' and c'' are arbitrary n-vectors.

At a closer look, however, it turns out that SC is a much more distant relative than SP. Intuitively, one can guess this from the fact that SP, like SPP, is a 'tightly constrained' problem (each constraint requires at most one, or exactly one, of many variables to be 1), whereas SC is a 'loosely constrained' problem (at least one of many variables is required to be 1). More precisely, the relationship is as follows.

SPP can be brought to the form SC by writing

$$\min\{cx + \theta ey \mid Ax - y = e, y \geq 0, x_j = 0 \text{ or } 1, j \in N\}$$

and then, using $y = Ax - e$,

$$\min\{-\theta m + c'x \mid Ax \geq e, x_j = 0 \text{ or } 1, j \in N\}$$

with $c' = \theta eA + c$. For sufficiently large θ (e.g. $\theta > \sum_{j \in N} c_j$), this problem has the same set of optimal solutions as SPP whenever the latter is feasible (see Lemke, Salkin, and Spielberg, 1971). The converse, however, is not true, i.e. SC cannot be brought to the form SPP.

Set Partitioning—A Survey

On the other hand, SP is a special case of SPP; conversely, SPP can be restated as

$$\max\{\theta m + c''x \mid Ax \leq e, x_j = 0 \text{ or } 1, j \in N\}$$

with $c'' = \theta eA - c$; and again, for θ sufficiently large, this problem has the same set of optimal solutions as SPP, whenever the latter is feasible.

The equivalence of SP and SPP is crucial to some of the results to be discussed.

Next we turn to a family of four interrelated problems defined on an undirected graph, two of which are special cases of SP.

7.1.3 Edge Matching and Covering, Node Packing and Covering

Let $G = (N, E)$ be a finite undirected graph with $n = |N|$ nodes and $q = |E|$ edges. Let A_G be the $n \times q$ node–edge incidence matrix of G, e_n, and e_q the n-vector and q-vector respectively, whose components are all 1.

An *edge matching* in G is a subset E' of edges such that every node of G is incident with at most one edge in E'. If every node of G is incident with exactly one edge in E', then E' is a *perfect* matching. An *edge cover* (a covering of nodes by edges) in G is a subset E'' of edges such that every node of G is incident with at least one edge in E''. The edge matching problem, or the problem of finding a maximum-cardinality edge matching in G, is then

EM $\qquad \max\{e_q y \mid A_G y \leq e_n, y_i = 0 \text{ or } 1, i = 1, \ldots, q\}$

while the edge covering problem, or the problem of finding a minimum-cardinality edge cover in G, is

EC $\qquad \min\{e_q y \mid A_G y \geq e_n, y_i = 0 \text{ or } 1, i, \ldots, q\}.$

A *node packing* (vertex packing) in G is a subset N' of nodes such that every edge of G is incident with at most one node in N'''. The node packing problem, or the problem of finding a maximum-cardinality node packing (internally stable node set, independent node set) in G, is then

NP $\qquad \max\{e_n x \mid A_G^T x \leq e_q, x_j = 0 \text{ or } 1, j = 1, \ldots, n\}$

while the node covering problem, or the problem of finding a minimum cardinality node cover in G, is

NC $\qquad \min\{e_n x \mid A_G^T x \geq e_q, x_j = 0 \text{ or } 1, j = 1, \ldots, n\}.$

If $\alpha_0, \beta_0, \alpha_1$, and β_1 are the cardinality of a maximum node packing, minimum node cover, maximum edge matching, and minimum edge cover in G, respectively, then these four numbers are connected by the following simple formula.

Theorem 7.1 (*Gallai* 1958) *For any nontrivial connected graph G with n nodes,*

$$\alpha_0 + \beta_0 = n = \alpha_1 + \beta_1.$$

A maximum edge matching and a minimum edge cover are easily obtained from each other; the same is true of a maximum node packing and a minimum node cover.

Clearly, EM and NP are special cases of SP. The problem of finding a perfect edge matching, obtained from EM by replacing the inequality with equality, is on the other hand a special case of SPP.

When G is bipartite, A_G is totally unimodular and the four integer programs listed above can be replaced by the associated linear programs. In the general case, however, the optimal solutions to these linear programs need not be integer. On the other hand, a basic result on convex polyhedra, due to Weyl (1935), implies that there always exists a finite system of linear inequalities whose solution set is the convex hull of feasible integer points. Identifying such a defining linear system is not an easy task in general. For the edge matching problem, this task was solved by Edmonds (1965a) (see also Edmonds, 1965b), who has also given an algorithm of complexity $O(n^3)$ for solving EM or its weighted version (in which e_q is replaced by an arbitrary integer vector), based on Berge's (1957) theorem of alternating chains and using the above mentioned defining linear system to prove optimality (for related work see also Balinski, 1969, 1972).

More recently, Pulleyblank and Edmonds (1973) have sharpened the characterization of the edge mathing polytope by identifying its (unique) minimal defining linear system.

An optimal solution to EM also (trivially) yields an optimal solution to EC.

The pair of problems NP, NC, is considerably more difficult. Balinski (1969) has characterized maximum node packings in terms of alternating subgraphs (see also Edmonds, 1962), but no polynomially bounded algorithm is known for the solution of this problem. More recently, several classes of facets of the node packing polytope have been identified (see Padberg, 1971, 1973a, 1975b; Chvátal, 1972; Nemhauser and Trotter, 1974; Trotter, 1974; Balas and Zemal, 1976a,b). Some algorithms have also been proposed (see Trotter, 1973; Balas and Samuelsson, 1973, 1974a), none of which is polynomially bounded.

7.1.4 Node Packing, Set Packing, Clique Covering

Denote by a_j the jth column of the matrix A of SP. The *intersection graph* $G_A = (N, E)$ of A has one node for every column of A, and one edge for every pair of nonorthogonal columns of A [i.e. $(i, j) \in E$ if and only if $a_i a_j \geq 1$]. Let A_G be the node–edge incidence matrix of G_A, and denote by

Set Partitioning—A Survey

NP the weighted node packing problem whose weights c_j are the same for each node as those of SP, i.e.

NP $\qquad \max\{cx \mid A_G^T x \leq e_q, x_j = 0 \text{ or } 1, j = 1, \ldots, n\}.$

Remark 7.1 x is a feasible (optimal) solution to SP if and only if it is a feasible (optimal solution to NP.

Thus, one way of solving set packing (and set partitioning) problems is to solve the associated node packing problem. Obviously, A_G^T has a more special structure than A (exactly two ones per row). Note, however, that while the two integer programs are equivalent, the two associated linear programs are not.

Remark 7.2

$$\max\{cx \mid Ax \leq e_m, x \geq 0\} \leq \max\{cx \mid A_G^T x \leq e_q, x \geq 0\}$$

It can easily be seen that the linear program associated with SP is more tightly constrained and that, apart from very special situations (e.g. when G is bipartite), the above relation holds with strict inequality.

A *clique* in G is a maximal complete subgraph. A *clique covering of the edges* of G is a set K of cliques such that each edge of G belongs to the edge set of some clique in K.

Remark 7.3 (Padberg, 1973a) Let K be a clique covering of the edges of G, and let A_K be the incidence matrix of the cliques in K (rows of A_K) versus the nodes of G (columns of A_K). Then SP is equivalent to (has the same set of feasible and optimal solutions, respectively, as) the set packing problem obtained from SP by replacing A with A_K.

Thus, many (seemingly different) set packing problems are equivalent to the same (unique) node packing problem NP.

A *clique covering of the nodes* of G is a set K' of cliques such that each node of G belongs to the node set of some clique in K'.

The clique matrix A_C of G is the incidence matrix of the set C of *all* cliques in G (rows of A_C) versus the nodes of G (columns of A_c). In view of the last remark, the unweighted node packing problem is equivalent to

SP$_C$ $\qquad \max\{e_n x \mid A_C x \leq e_C, x_j = 0 \text{ or } 1, j = 1, \ldots, n\},$

where $e_C = (1, \ldots, 1)$ is dimensioned compatibly with A_C.

Consider now the problem of finding a minimum-cardinality clique covering of the nodes of G:

KC $\qquad \min\{e_C y \mid A_C^T y \geq e_n, y_i = 0 \text{ or } 1, i = 1, \ldots, |C|\}.$

If $\alpha_0(G)$ is the value of an optimal solution to (SP$_C$), i.e., the cardinality of a maximum independent node set in G, and $\omega_0(G)$ the value of an optimal solution to KC, i.e. the cardinality of a minimum clique covering of the nodes of G, then $\alpha_0(G) \leq \omega_0(G)$, since the linear programs associated with the two problems are dual to each other. Further, $\alpha_0(G) = \omega_0(G)$ if and only if both linear programs have integral optimal solutions.

The subgraph of $G = (N, E)$ *induced* by a subset N' of the nodes of G, is $G' = (N', E')$, where $(i, j) \in E' \Leftrightarrow i \in N', j \in N', (i, j) \in E$. The *complement* $\bar{G}$ of a graph $G = (N, E)$ is the graph $\bar{G} = (N, \bar{E})$, where $(i, j) \in \bar{E} \Leftrightarrow (i, j) \notin E$. A *chordless cycle* C in G is a cycle each of whose nodes is adjacent to exactly two other nodes of C. A cycle is called odd or even according to whether it is of odd or even length. A cycle of length 3 is obviously chordless, and is a clique. A chordless cycle of length greater than 3 is called a *hole*, its complement an *anti-hole*.

A graph $G = (N, E)$ such that $\alpha_0(G') = \omega_0(G')$ for all induced subgraphs G' of G is called *perfect*. The perfect graph theorem, conjectured by Berge (1961) and proved by Lovász (1972) (see also Fulerson, 1971, 1973) asserts that G is perfect if and only if its complement $\bar{G}$ is perfect. Further, it is known that G is perfect if and only if all vertices of the polytope

$$P = \{x \in R^n \mid A_C x \leq e_C, x \geq 0\}$$

are integral (Chvátal, 1972). Finally, the strong perfect graph conjecture (Berge, 1970) asserts that G is perfect if and only if it contains on odd holes or anti-holes.

An alternative way of looking at arbitrary zero–one matrices is via the theory of *hypergraphs* (see Berge, 1970). A hypergraph is a pair $H = (N, \mathscr{E})$, where N is a set of elements called nodes, and $\mathscr{E}$ is a family of nonempty subsets of N, called edges. The incidence matrix $A = (a_{ij})$ of the hypergraph H has a column for every node and a row for every edge of H, with $a_{ij} = 1$ if node j is contained in edge i, $a_{ij} = 0$ otherwise. Every 0–1 matrix which has no zero rows or zero columns is the incidence matrix of some hypergraph. Thus SP and SPP can be formulated in terms of hypergraphs (see Berge, 1970; Lovász, 1972).

7.2 Theory

7.2.1 Facets of the Set Packing Polytope and Associated Graphs

Throughout this chapter, we assume that A has no zero rows or columns. Then the constraint sets of the linear programs associated with SP and let SPP are bounded.

Let

$$P = \{x \in R^n \mid Ax \leq e, x \geq 0\}$$

where A is the coefficient matrix of SP, and let P_I be the *set packing polytope*, i.e. the convex hull of points satisfying the constraints of SP:

$$P_I = \text{conv}\{x \in P \mid x \text{ integer}\}.$$

Set Partitioning—A Survey

We first note that $\dim P = \dim P_I = n$. A *facet*, or $(n-1)$-dimensional face, of P_I is a set $P_I \cap \{x \in R^n \mid \pi x = \pi_0\}$ such that $\pi x \leq \pi_0$, $\forall x \in P_I$, and $\pi x = \pi_0$ for exactly n affinely independent points $x \in P_I$. As is customary in the literature, an inequality $\pi x \leq \pi_0$ defining a facet will itself be called a facet. The facets of P_I are the inequalities of the (unique) minimal defining linear system of P_I. It is easily seen that each inequality $x_j \geq 0$, $j \in N$, is a (trivial) facet of P_I. From the nonnegativity of A it follows easily that every facet $\pi x \leq \pi_0$ of P_I that is different from the above trivial facets satisfies $\pi_j \geq 0$ for $j \in N$ and $\pi_0 > 0$. We first examine the conditions under which some of the constraints $Ax \leq e$ themselves are facets of P_I. As before, G_A denotes the intersection graph of A. The next theorem is from Padberg (1974a), though it also follows from Theorem 8 of Fulkerson (1971).

Theorem 7.2 *The inequality*

$$\sum_{j \in K} x_j \leq 1$$

where $K \subseteq N$, is a facet of P_I if and only if K is the node set of a clique in G_A.

As a direct consequence, the polytope

$$P_C = \{x \in R^n \mid A_C x \leq e_C, x \geq 0\}$$

where A_C is, as before, the clique matrix of G_A, satisfies $P_I \subseteq P_C \subseteq P$. In general, P_C is different from P_I as well as from P. There is, however, a large class of matrices A for which the three polytopes P, P_C, and P_I coincide, i.e. $P = P_C = P_I$. Zero–one matrices with this property are called *perfect* and can be characterized in terms of certain 'forbidden' submatrices.

Let A' be a $m \times k$ zero–one matrix, with $m \geq k$. A' is said to have the property $\pi_{\beta,k}$ if the following conditions are met:

(i) A' contains a $k \times k$ nonsingular submatrix A'_1 whose row and column sums are all equal to β.

(ii) Each row of A' which is not a row of A'_1, either is componentwise equal to some row of A'_1, or has row sum strictly less than β.

Theorem 7.3 (Padberg, 1974a). *The following two conditions are equivalent for an arbitrary $m \times n$ zero–one matrix A:*

(i) *A is perfect, i.e. $P = P_C = P_I$.*

(ii) *For $\beta \geq 2$ and $3 \leq k \leq n$, A does not contain any $m \times k$ submatrix A' having the property $\pi_{\beta,k}$.*

Perfect matrices are closely related to perfect graphs (Berge, 1970), and normal hypergraphs (Lovász, 1972). They properly subsume nonnegative totally unimodular matrices, as well as balanced matrices (Berge, 1972). (For a discussion of their interrelationships see Padberg 1974b, c.) As mentioned in Section 7.1.4, the clique-matrices of perfect graphs are all perfect. A unified treatment of perfect graphs and perfect matrices, from an algebraic point of view, is to be found in Padberg (1975a).

Whenever $P_C \neq P$, a facet of P_I of the type characterized by Theorem 7.2 is absent from the constraint set defining P. Facets of P_I of that type are easy to detect. In general, however, more complicated types of inequalities are needed in order to characterize P_I. Since these inequalities are associated with certain subgraphs of G_A, the intersection graph of A, we will from now on use a notation which relates G_A directly to the (unique) node packing/set packing polytope P_I defined by G_A. Thus we will denote G_A and P_I by G and $P(G)$ respectively.

The next few theorems are concerned with classes of graphs whose associated set packing polytopes have a facet with all (left-hand side) coefficients equal to 1. From Theorem 7.2, a first such class is that of complete graphs. The next theorem (whose part (i) is due to Padberg 1971, 1973a; part (ii) to Nemhauser and Trotter, 1974) specifies two further classes.

Theorem 7.4 *If G is either (i) an odd hole, or (ii) an odd anti-hole, then*

$$\sum_{j \in N} x_j \leq k$$

is a facet of $P(G)$, with $k = (n-1)/2$ in case (i), and $k = 2$ in case (ii).

Two additional classes of graphs defining facets with coefficients equal to 1 are webs and their complements (Trotter, 1974). A *web* $W_{(n,k)} = (N, E)$ is defined by $|N| = n \geq 3$ and

$$(i, j) \in E \Leftrightarrow j = i+k, i+k+1, \ldots, i+n-k$$

(where sums are taken modulo 1), with $1 \leq k \leq [n/2]$. The web $W_{(n,k)}$ is regular of degree $n - 2k + 1$, and has exactly n maximum node packings of size k. The anti-web $\bar{W}_{(n,k)}$, or complement of the web $W_{(n,k)}$, is regular of degree $2(k-1)m$ and has exactly n maximum cliques of size k.

Theorem 7.5 (Trotter 1974)
(i) *If $G = W_{(n,k)}$ is a web, the inequality*

$$\sum_{j \in N} x_j \leq k$$

is a facet of $P(G)$ if and only if n and k are relatively prime.

(ii) *If $G = \bar{W}_{(n,k)}$ is an anti-web with n and k relatively prime, $k \geq 2$, the inequality*

$$\sum_{j \in N} x_j \leq [n/k]$$

is a facet of $P(G)$.

Note that for all the facets with left-hand side coefficients equal to 1 considered so far, the right-hand side coefficient was equal to the maximum cardinality of an independent node set (independence number) of the graph G associated with $P(G)$, namely: 1 in the case of the complete graph or clique K_n; $(n-1)/2$ and 2 in the case of the odd hole H_n and odd anti-hole $\bar{H}_n$, respectively; k and $[n/k]$ in the case of the web $W_{(n,k)}$ and anti-web $\bar{W}_{(n,k)}$, respectively. This raises the more general question as to when an inequality with left-hand side coefficients equal to 1, and right-hand side equal to $\alpha(G)$, the independence number of some graph G, is a facet of $P(G)$.

We first state a sufficient condition due to Chvátal (1972). An edge $u \in E$ of $G = (N, E)$ is called α-*critical* if its removal produces a graph with an independence number greater than $\alpha(G)$.

Theorem 7.6 (Chvátal, 1972) *Let E^* be the set of α-critical edges of G. If $G^* = (N, E^*)$ is connected, then*

(i) $$\sum_{j \in N} x_j \leq \alpha(G)$$

is a facet of $P(G)$.

It is easy to check that all the graphs examined so far which give rise to facets of the above form (i.e. cliques, odd holes and odd anti-holes, webs and anti-webs with n and k relatively prime), satisfy the condition of Theorem 7.6. Nevertheless, Chvátal's condition is not necessary, as the following example shows.

Example 7.1 Let $G = H_5^1 + H_5^2$ be the join of two (disjoint) odd holes, H_5^1 and H_5^2 respectively; i.e. G consists of $H_5^1 \cup H_5^2$ and edges joining each node of H_5^1 to each node of H_5^2 (see Figure 7.1). Then the inequality

$$\sum_{j=1}^{10} x_j \leq 2$$

is a facet of $P(G)$, in spite of the fact that none of the edges joing H_5^1 to H_5^2 is critical, and hence G^* is disconnected.

Next we give a necessary condition for (i) to be a facet of $P(G)$, based on the concept of an α-*critical cutset* (Balas and Zemel, 1976b).

A *cutset* $C \subset E$ in a graph $G = (N, E)$ is defined relative to a partition of N. Given any proper subset N_1 of N, the cutset $C = (N_1, N_2)$ is the (possibly empty) set of edges joining the nodes in N_1 to those in $N_2 = N - N_1$. A cutset $C = (N_1, N_2)$ is called α-*critical* if $\alpha(G_1) + \alpha(G_2) \geq \alpha(G) + 1$, where G_1 and G_2 are the subgraphs of G induced by N_1 and N_2 respectively.

Theorem 7.7 (*Balas and Zemel, 1976b*) *If the inequality*

(i) $$\sum_{j \in N} x_j \leq \alpha(G)$$

is a facet of $P(G)$, then every cutset of G is α-critical.

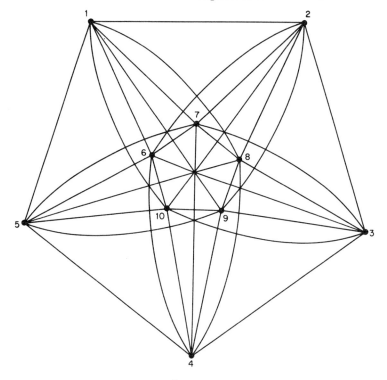

Figure 7.1

The following are a few immediate consequences of Theorem 7.7.

Remark 7.4 If (i) is a facet of $P(G)$, then
  (α) G is connected;
  (β) every vertex of G has degree at least equal to 2;
  (γ) for every clique K of G there exists a maximum-cardinality independent node set of G, not containing any node of K.

The necessary condition of Theorem 7.7 is not sufficient for (i) to be a facet of $P(G)$, as the following example shows.

Example 7.2 Consider the graph G' obtained from $G = H_5^1 + H_5^2$ (of Example 7.1) by inserting a chord (i.e. an edge joining two nodes of H_5^1 or two nodes of H_5^2). It is easy to check that every cutset of G' is α-critical, yet (i) is not a facet of G' since G' has only 9 (i.e. $n-1$) independent node sets of size 2.

Next we give a sufficient condition for (i) to be a facet of $P(G)$, weaker than the condition of Theorem 7.6.

Set Partitioning—A Survey

Theorem 7.8 (Balas and Zemel, 1976b) *Let $G = (N, E)$ be a graph such that either $G^* = (N, E^*)$ is connected, or G has an α-critical cutset $C = (N_1, N_2)$ satisfying the following conditions for $i = 1, 2$:*

(α) the inequality

$$\sum_{j \in N_i} x_j \leq \alpha(G_i)$$

is a facet of $P(G_i)$;

(β) every maximum-cardinality independent node set of G_i is contained in some maximum-cardinality independent node set of G.

Then the inequality (i) is a facet of $P(G)$.

Example 7.3 The condition of Theorem 7.8 is satisfied by the graph G of Example 7.1, since (α), (β) hold for $G_1 = H_5^1$, $G_2 = H_5^2$, and, indeed, (i) is a facet of $P(G)$. On the other hand the graph G' of Example 7.2, where (i) is not a facet of $P(G')$, violates the condition of Theorem 7.8, since there is no partition of N for which (α), (β) hold.

The theorem to follow states a necessary and sufficient condition for an inequality with 0–1 coefficients to be a facet of $P(G)$.

Theorem 7.9 (Balas and Zemel, 1976b) *Let $G_1 = (N_1, E_1)$ be the subgraph of $G = (N, E)$ induced by some proper subset N_1 of N, and let*

(i) $$\sum_{j \in N_1} x_j \leq \alpha(G_1)$$

be a facet of $P(G_1)$. For every $k \in N_2 = N - N_1$, let G^k be the subgraph of G induced by $N_1 \cup \{k\}$.

Then (i) is a facet of $P(G)$ if and only if, for every $k \in N_2$, the cutset (k, N_1) of G^k is not α-critical.

We now discuss a more general result which relates the facets of $P(G)$ to those of a lower dimensional polytope $P(G')$ defined by some induced subgraph G' of G.

Theorem 7.10 *Let G' be the subgraph of $G = (N, E)$ induced by $S \subset N$. If*

(i) $$\sum_{j \in S} \alpha_j x_j \leq s$$

is a facet of $P(G')$, then there exist integers β_j, $0 \leq \beta_j \leq s$, such that

(ii) $$\sum_{j \in S} \alpha_j x_j + \sum_{j \in N-S} \beta_j x_j \leq s$$

is a facet of $P(G)$.

The existence of facets of the form (ii) was first stated by Pollatschek (1970). (We are indebted to Uri N. Peled for providing us with a translation of the relevant portion of Pollatschek's thesis.) Independently, Padberg (1971, 1973a) proved the theorem for the case when G' is an odd hole, and gave a (sequential) procedure for calculating the coefficients β_j. Nemhauser and Trotter (1974) showed that the result (and the procedure) extends to arbitrary G'.

Thus, given any induced subgraph G' of G, the facets of $P(G')$ can be 'lifted' into the space of $P(G)$, to yield facets of the latter. The lifting procedure, i.e. the procedure for calculating the coefficients β_j of (ii), consists of solving a sequence of set packing problems, one for each $j \in N - S$, in the variables $j \in S \cup S'$, where $S' \subseteq N - S$ is the index set for the coefficients already computed. The sequence in which the coefficients β_j are computed does matter, and different sequences may give rise to different facets. We call this procedure *sequential lifting*.

Generalizations of the sequential lifting procedure to other than set packing polytopes are discussed in Padberg (1973b), Nemhauser and Trotter (1974), Balas (1975a), Hammer, Johnson, and Peled (1974, 1975), Wolsey (1974, 1975), Johnson (1974), Zemel (1974), and Balas and Zemel (1974, 1975).

The class of facets defined in Theorem 7.4 does not exhaust the family of facets of $P(G)$ obtainable by lifting the facet (i) of $P(G')$. In particular (see Zemel, 1974), there exists facets of $P(G)$ of the form (ii) whose coefficients β_j are not necessarily integer, and which cannot be obtained by sequential lifting, but can be obtained by a generalization of the latter, called *simultaneous lifting*. Furthermore, it was shown (see Balas and Zemel, 1975) that every facet of $P(G)$ can be obtained from a facet of $P(G')$, where G' is some induced subgraph of G, by a procedure which combines simultaneous lifting with projection on a certain subspace.

7.2.2 Facet-Producing Graphs

It should be clear by now that the task of characterizing all facets of the set packing polytope $P(G)$ associated with a graph G is closely related to that of characterizing all 'facet-producing' subgraphs of G. We call a graph $G = (N, E)$ *facet-producing* if there exists a facet $\pi x \leq \pi_0$ of $P(G)$ which is not a facet of $P(G')$ for any subgraph G' of G induced by a node set $N' \subset N$ of cardinality $n - 1$; furthermore, if in this definition N' is allowed to be of *any* cardinality less than n, G will be called *strongly* facet-producing. (This definition slightly differs from that of Trotter, 1974, who introduced the concept.) In other words, G is called facet-producing (strongly facet-producing) if $P(G)$ has a facet which cannot be obtained by *sequentially* (*simultaneously*) lifting a facet of a lower dimensional polytope.

Set Partitioning—A Survey

One possible attack on the problem of finding a linear characerization of $P(G)$ is to enumerate the facet-producing subgraphs of G. This, however, turns out to be a very difficult task, since there are procedures, to be discussed below, for constructing arbitrarily complex graphs with the facet-producing property.

Of the graphs discussed in Section 7.2.1, odd holes and odd anti-holes are strongly facet-producing, while cliques are not facet-producing. The web $W_{(n,k)}$ is facet-producing (strongly and otherwise) if and only if $k \geq 2$ and n and k are relatively prime (Trotter, 1974), while the anti-web $\bar{W}_{(n,k)}$ is strongly facet-producing (for n and k relatively prime) if and only if $k[n/k] = n - 1$ (Padberg, 1975b).

One of the first questions one may want to ask in this context is whether the complement of a facet-producing graph is also facet-producing. Contrary to what intuition might suggest, the answer to this question, contained in the next theorem, is in the negative (see also Figure 7.2, which displays a facet-producing graph G whose complement is not facet-producing, as the clique matrix of G is of rank 9).

A square matrix M is called irreducible if there exists no permutation matrices P and R such that

$$PMR = \begin{pmatrix} N & 0 \\ L & S \end{pmatrix}$$

where N and S are square matrices and 0 is a zero matrix.

Theorem 7.11 (Padberg, 1975b) *Let $G = (N, E)$ be any graph, $P(G)$ the associated set packing polytope, and $P_C = \{x \in R^n \mid A_C x \leq e_C, x \geq 0\}$, where A_C is the clique matrix of G.*

If $P(G)$ has a facet $\pi x \leq \pi_0$, such that $\pi_j > 0$, $j \in N$, and πx assumes its maximum over P_C at a vertex $\bar{x}$ of P_C satisfying $0 < \bar{x}_j < 1$, $\forall j \in N$, then $P(\bar{G})$ has a facet $\pi' y \leq \pi'_0$ such that $\pi'_j > 0$, $j \in N$. Furthermore, if the submatrix A_1 of A_C for which the constraints of $A_C \bar{x} \leq e$ are tight is square and irreducible, then $\bar{G}$ is strongly facet-producing.

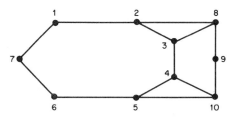

Figure 7.2

We now discuss several procedures for constructing facet-producing graphs. From the nature of these procedures it will be seen that such graphs can be of arbitrary size and of an enormous diversity.

We start with a construction due to Chvátal (1972), called graph substitution. Let $G = (N, E)$ and $H = (Q, F)$ be graphs with node sets $N = \{v_1, \ldots, v_n\}$, $Q = \{v'_1, \ldots, v'_q\}$ ($N \cap Q = \varnothing$), and edge sets E and F respectively. The graph G_k^H obtained from G by *substituting* H *for* v_k is defined as the (disjoint) union of $G - \{v_k\}$ and H, together with edges joining each node of H to those nodes of G adjacent to v_k. Figure 7.3 illustrates the concept.

Theorem 7.12 (*Chvátal, 1972*) *Let*

$$\sum_{j \in N} \alpha_{ij} x_j \leq 1 \quad i \in I$$
$$x_j \geq 0 \quad j \in N$$

be a defining linear system for $P(G)$, *and let*

$$\sum_{j \in Q} \beta_{kj} y_j \leq 1 \quad k \in K$$
$$y_j \geq 0 \quad j \in Q$$

be a defining linear system for $P(H)$. *Then*

$$\sum_{j \in N-\{n\}} \alpha_{ij} x_j + \alpha_{in}^+ \sum_{j \in Q} \beta_{kj} y_j \leq 1 \quad i \in I, k \in K$$
$$x_j \geq 0 \quad j \in N-\{n\}, \quad y_j \geq 0 \quad j \in Q$$

is a defining linear system for $P(G_n^H)$, *where* $\alpha_{in}^+ = \max\{0, \alpha_{in}\}$, $i \in I$.

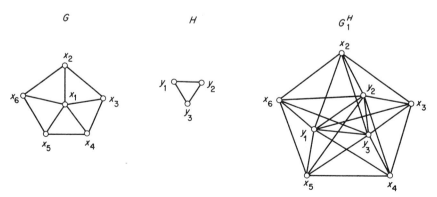

Figure 7.3

Set Partitioning—A Survey

This theorem characterizes the set packing polytope $P(G_n^H)$ in terms of the lower dimensional polytopes $P(G)$ and $P(H)$. It leaves open, however, the problem of when the defining systems in question are minimal. This problem is settled in the next theorem.

Theorem 7.13 (Balas and Zemel, 1976a) *If the inequalities*

(i) $$\sum_{j \in N} \alpha_j x_j \leq 1$$

and

(ii) $$\sum_{j \in Q} \beta_j y_j \leq 1$$

are facets of $P(G)$ and $P(H)$ respectively, then the inequality

(iii) $$\sum_{j \in N-\{n\}} \alpha_j x_j + \alpha_n \sum_{j \in Q} \beta_j y_j \leq 1$$

is a facet of $P(G_n^H)$.

Conversely, if the inequality

(iv) $$\sum_{j \in N-\{n\}} \alpha_j x_j + \sum_{j \in Q} \gamma_j y_j \leq 1$$

is a facet of $P(G_n^H)$, then (i) is a facet of $P(G)$, with

$$\alpha_n = \max \left\{ \sum_{j \in Q} \gamma_j y_j \,\middle|\, y \in P(H) \right\};$$

and if $\gamma_k > 0$ for at least one $k \in Q$, then (ii) is a facet of $P(H)$, with $\beta_j = \gamma_j \alpha_n^{-1}$, $j \in Q$.

Theorem 7.13 implies that if the defining systems of $P(G)$ and $P(H)$ introduced in Theorem 7.12 are minimal, then so is the defining system derived for $P(G_n^H)$. Conversely, if the minimal defining linear system of $P(G_n^H)$ is of the form stated in Theorem 7.12, with

$$\max \left\{ \sum_{j \in Q} \beta_{kj} y_j \,\middle|\, y \in P(H) \right\} = 1 \quad k \in K$$

then the corresponding defining systems for $P(G)$ and $P(H)$ are also minimal, with $\alpha_{in} = \alpha_{in}^+$, $i \in I$.

Example 7.4 Consider the graphs G, H, and G_1^H illustrated in Figure 7.3. The inequalities

$$2x_1 + x_2 + x_3 + x_4 + x_5 \leq 2$$

and

$$y_1 + y_2 + y_3 \leq 1$$

are facets of $P(G)$ and $P(H)$ respectively. Hence the inequality

$$2y_1 + 2y_2 + 2y_3 + x_2 + x_3 + x_4 + x_5 \leq 2$$

is a facet of $P(G_1^H)$.

Next we give necessary and sufficient conditions for a graph obtained by substitution to have the facet-producing property. G is said to be k-*almost facet producing* if $P(G)$ has a facet $\pi x \leq \pi_0$ which is not a facet of $P(G \setminus \{v_i\})$ for any $i \in N \setminus \{k\}$.

Theorem 7.14 (Balas and Zemel, 1976a) G_n^H *is a facet-producing graph if and only if H is facet-producing and G is n-almost facet-producing.*

Example 7.5 In Example 7.4 (see Figure 7.3), the wheel $G = W_6$ is not facet-producing, but the odd hole $G - \{v_1\}$ is. However, since the triangle H is not facet-producing, neither is G_1^H. On the other hand, the graph $G = H_5^1 + H_5^2$ of Example 7.1 can be obtained by substituting the odd hole H_5^2 for node v_1 of the wheel $W_6 = H_5^1 + \{v_1\}$ (here W_6 has 6 nodes, v_1 being the node adjacent to all the others). W_6 is not facet-producing, but the odd hole $H_5^1 = W_6 - \{v_1\}$ is, and so is H_5^2. Therefore the inequality

$$\sum_{j=1}^{10} x_j \leq 2$$

of Example 7.1 is a facet produced by G.

We now consider another procedure for constructing facet-producing graphs. The star $K_{1,n}$ with $n+1$ nodes is the complete bipartite graph whose n edges are $(i, n+1)$, $i = 1, \ldots, n$. For a graph $G = (N, E)$, let $G^\#$ be the graph consisting of $G \cup K_{1,n}$ together with edges joining node i of G to node i of $K_{1,n}$, $i = 1, \ldots, n$ (see Figure 7.4).

Theorem 7.15 (Padberg, 1975b) *If $P(G)$ has a facet $\pi x \leq \pi_0$ such that $\pi_j > 0$, $j \in N$, then the inequality*

$$\pi x + \pi y + (\pi e - \pi_0) y_{n+1} \leq \pi e$$

where $e \in R^n$, $e = (1, \ldots, 1)$, $y \in R^n$, and y_i is the variable associated with the

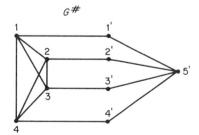

Figure 7.4

Set Partitioning—A Survey

i-th node of $K_{1,n}$, $i = 1, \ldots, n+1$, is a facet of $P(G^{\#})$, and $G^{\#}$ is strongly facet-producing.

Furthermore, if $P_C = \{x \in R^n \mid A_C x \le e_C, x \ge 0\}$, where A_C is the clique matrix of G, and if πx assumes its maximum over P_C at a vertex $\bar{x}$ of P_C such that $0 < \bar{x}_j < 1$, $\forall j \in N$, then

$$\bar{x}x + (e - \bar{x})y + \bar{x}_{j*}y_{n+1} \le 1$$

is a facet of $P(\bar{G}^{\#})$, where $\bar{x}_{j*} = \min_{j \in N} \bar{x}_j$, and $\bar{G}^{\#}$ is the complement of $G^{\#}$.

Note that the facet $\pi x \le \pi_0$ of Theorem 7.15 need not be *produced* by G. In this context, the following procedure for constructing graphs which give rise to facets of a certain form is of interest. By the insertion of two nodes v_{n+1}, v_{n+2} into an edge (v_i, v_j) of a graph $G = (N, E)$, we mean the substitution of the path $(v_i, v_{n+1}, v_{n+2}, v_j)$ for the edge (v_i, v_j). If $\pi x \le \pi_0$ is a facet of $P(G)$, and edge $u = (v_i, v_j)$ is called π-*critical* if there exists an independent node set S in $G - \{u\}$, such that

$$\sum_{j \in S} \pi_j > \pi_0 \quad \text{but} \quad \sum_{j \in S - \{v_h\}} \pi_j = \pi_0 \quad \text{for} \quad h = i \text{ or } j.$$

Theorem 7.16 (Padberg, 1975b) *Let $C = (N, E)$ be a graph with $|N| = n \ge 3$, and let $\pi x \le \pi_0$ be a facet of $P(G)$ such that $\pi_j > 0$, $j \in N$. If $G' = (N', E')$ is the graph obtained from G by inserting two nodes v_{n+1}, v_{n+2} into a π-critical edge (v_i, v_k) of G, then*

$$\pi x + \pi_{j*}(x_{n+1} + x_{n+2}) \le \pi_0 + \pi_{j*}$$

is a facet of $P(G')$, with $\pi_{j*} = \min\{\pi_i, \pi_k\}$.

Example 7.6 Consider the graph $G^{\#}$ of Figure 7.4 with the vertices numbered consecutively from 1 to 9 (with vertex 5' numbered 9). Then the inequality

$$\sum_{j=1}^{8} x_j + 3x_9 \le 4$$

is a facet of $P(G^{\#})$, with $\pi_j > 0$, $\forall j$. It is easy to check that every edge of $G^{\#}$ is π-critical. Inserting two nodes, 10 and 11, into the edge $(8, 9)$, gives rise to a graph G' such that

$$\sum_{j=1}^{8} x_j + 3x_9 + x_{10} + x_{11} \le 5$$

is a facet of $P(G')$.

The results discussed above show that, unlike the facets of the edge matching polytope, which are all of the same type and easy to describe, the facets of the node packing/set packing polytope are of an enormous variety.

This raises the question whether knowing certain classes of facets is of any practical use, when a *complete* characerization of the facial structure of set packing polytopes seems so hopelessly difficult. The answer to this question is in the positive, since knowledge of the more frequently occurring classes of facets often enables one either to solve or come close to solving the set packing problem. We illustrate the issue in the following example.

Example 7.7 Consider the maximum-cardinality node packing problem defined on the odd anti-hole $\bar{H}_n$ of size n, and its linear programming relaxation. The constraints of the latter are $x_i + x_j \leq 1$ for each edge (i, j), and $x_j \geq 0$, $j \in N$. Setting $x_j = \frac{1}{2}$, $j \in N$, is obviously a feasible solution, and easily seen to be optimal with a value of $n/2$. Thus, the linear programming optimum increases with n. The integer optimum, however, is 2 for any n, no matter how large.

Suppose now that the simplest class of facets, namely the one associated with the maximum-cardinality cliques of $\bar{H}_n$, is used as a set of additonal inequalities in solving the linear program (the addition of these inequalities in fact makes the original ones redundant). Since each such clique of $\bar{H}_n$ is of size $(n-1)/2$ and $\bar{H}_n$ is symmetric with each node belonging to $(n-1)/2$ cliques, the clique matrix of $\bar{H}_n$ (the coefficient matrix of the newly added inequalities) is the circulant matrix of order n, with $(n-1)/2$ ones in each row and each column. Setting $x_j = 2/(n-1)$, $j \in N$, is clearly a feasible solution to the linear program with the added constraints, and it is not hard to see that this solution is optimal, with a value of $2n/(n-1) = 2 + 2/(n-1)$. As n increases, this value approaches 2, the value of the integer optimum; actually for any $n \geq 4$ it can be rounded down to yield the integer 2.

This example illustrates how dramatically the addition of some facets can sometimes influence the strength of the linear programming relaxation of a set packing polytope. In the case of an odd anti-hole $\bar{H}_n$, the straightforward linear programming formulation using the edge–node incidence matrix produces a bound which is not only bad, but gets worse as n increases, whereas the addition of the n facets associated with the maximum-cardinality cliques of $\bar{H}_n$ produces, also via linear programming, a bound equal to the value of the integer optimum. Note that the n maximum-cardinality cliques of $\bar{H}_n$ needed to obtain such a strong bound constitute only a very small fraction of all cliques of $\bar{H}_n$ (for a sufficiently large n).

In this and the preceding section we have reviewed the main classes of facets of the set packing polytope known to date. The set packing polytope itself is a relaxation of the set partitioning polytope $\bar{P}_I$. In the next section we discuss a class of inequalities which are facets of some other relaxations of $\bar{P}_I$. These inequalities are usually weaker than the facets of the set packing polytope, but they can readily be derived from a fractional simplex tableau and used to cut off the vertex of the linear programming feasible set

Set Partitioning—A Survey

associated with the tableau. Compared with other cutting planes derived from a fractional simplex tableau, they have the advantage of using the struture of the set partitioning constraints.

7.2.3 Facets of Relaxed Polytopes: Cuts from Disjunctions

Let $\bar{P}$ be the feasible set of the linear program associated with SPP, that is

$$\bar{P} = \{x \in R^n \mid Ax = e, x \geq 0\}$$

and let

$$\bar{P}_I = \text{conv}\{x \in \bar{P} \mid x \text{ integer}\}$$

be the set paritioning polytope.

In this section we use the disjunctive programming approach (see Balas, 1975b) to generate valid inequalities (cutting planes) which are facets of some relaxation $\bar{P}'_1$ of $\bar{P}_I$. By disjunctive programming we mean linear programming with disjunctive constraints. Integer programs (pure or mixed) and a host of other nonconvex programming problems can be stated as linear programs with logical conditions, and the latter can always be expressed as a disjunction between sets of linear inequalities. Special cases of this class of problem have been examined by several authors in the past (see Glover and Klingman, 1973a,b; Owen, 1973; Zwart, 1972). However, a theoretical study of the general disjunctive programming problem has only recently been undertaken (see Balas, 1974a,b, 1975b; Jeroslow, 1974; for a somewhat different but closely related approach see Glover, 1975).

Not only can integer programs be formulated as disjunctive programs, but for special structures, like SPP and SP, as well as other combinatorial problems, such a formulation offers definite advantages. Cutting planes can be obtained which are computationally cheap, and have the important feature of displaying coefficients of different signs, unlike the traditional cuts; a property which tends to mitigate the tendency towards dual degeneracy, common to cutting plane algorithms (for background on the latter, see Garfinkel and Nemhauser, 1972).

The *disjunctive normal form* of a disjunctive program is

DP $\qquad\qquad\qquad\qquad \min\{cx \mid x \in F\}$

where

$$F = \left\{x \in R^n \;\middle|\; \bigvee_{h \in Q} (A^h x \geq b^h, x \geq 0)\right\}$$

and where each A^h is an $m_h \times n$ matrix, each b^h is an m_h-vector, while $\vee$ stands for 'or' i.e. at least one of the $|Q|$ systems $A^h x \geq b^h$, $x \geq 0$, must hold. It is usually convenient to express the problem in the nonbasic variables associated with an optimal solution to the linear program, for in that case one can view the inequalities $\alpha x \geq \alpha_0$ implied by the constraints of F as potential cutting planes, which cut off the current solution $x = 0$ if and only if $\alpha_0 > 0$.

Let Q^* be the set of those $h \in Q$ such that

$$\{x \in R^n \mid A^h x \geq b^h, x \geq 0\} \neq \emptyset.$$

In the next theorem, the forward direction of the main statement is from Balas (1974a), while the reverse direction is from Jeroslow (1974). Statements (i) and (ii) are from Balas (1974b).

Theorem 7.17 *The inequality $\alpha x \geq \alpha_0$ is satisfied by all $x \in F$ if and only if there exist vectors $\theta^h \in R^{m_h}$, $h \in Q^*$, such that*

$$\alpha \geq \theta^h A^h \qquad \alpha_0 \leq \theta^h b^h \qquad \theta^h \geq 0 \qquad h \in Q^*.$$

Further, if conv F is closed and full dimensional, then (i) and (ii) hold.

(i) If $\alpha_0 \neq 0$, $\alpha x \geq \alpha_0$ is a facet of conv F if and only if $\alpha \neq 0$ is a vertex of $F^{\#} = \{y \in R^n \mid y \geq \theta^h A^h; \theta^h b^h \geq \alpha_0; \theta^h \geq 0, \forall h \in Q^*\}$.

(ii) If $\alpha x \geq 0$ is a facet of conv F, then $\alpha \neq 0$ is an extreme direction vector of $F^{\#}$.

Remark 7.5 If some of the inequalities of F are replaced by equalities, Theorem 7.17 holds without the nonnegativity constraint on the corresponding components of the vector θ.

The vertices of the set $F^{\#}$, which define the facets of conv F, can be obtained by solving a linear program. When Q is large, this may be hopeless, though the linear program is strongly structured. However, by relaxing part of the constraints of SPP and retaining only a convenient subset, one can take advantage of the structure imposed by the latter so as to solve the resulting linear program trivially.

For instance, let the linear program associated with SPP have an optimal solution of the form

$$x_i = \bar{a}_{i0} + \sum_{j \in J} \bar{a}_{ij}(-t_j) \qquad i \in I \cup J$$

where I and J are the basic and nonbasic index sets respectively (i.e. for $i \in J$, $\bar{a}_{i0} = 0$ and $\bar{a}_{ij} = 0$ for $i \neq j$, $\bar{a}_{ij} = -1$ for $i = j$), and let

$$\sum_{j \in Q} x_j = 1$$

be one of the constraints of SPP, such that $Q' \neq \emptyset$, where $Q' = \{i \in Q \mid 0 < \bar{a}_{i0} < 1\}$. Note that this equality is satisfied by the current solution, so it cannot be used as a cut. However, the logical condition that it expresses, namely that exactly one of the variables x_j, $j \in Q$, must be one, and all the others zero, i.e.

$$\bigvee_{i \in Q} \left(x_i = 1, \sum_{j \in Q - \{i\}} x_h = 0 \right)$$

is violated by the current solution and can be used to generate a cut.

Let $\sigma = (\sigma_i)$ be a q-vector, where $q = |Q|$, such that $0 \leq \sigma_i \leq 1$, $\forall i \in Q$. If the numbers σ_i, $1 - \sigma_i$ are used to take convex combinations of the two equations within each member of the above disjunction, the latter becomes

$$\bigvee_{i \in Q} \left[(1 - \sigma_i) x_i + \sigma_i \left(\sum_{h \in Q - \{i\}} x_j \right) = 1 - \sigma_i \right].$$

Consider now the family of relaxations of $\bar{P}_I$ whose members are of the form conv $F(\sigma)$, for some $\sigma \in R^q$ such that $0 \leq \sigma_i \leq 1$, $i \in Q$, where

$$F(\sigma) = \left\{ t \in R^n \;\middle|\; \begin{array}{l} x_i = \bar{a}_{i0} + \sum_{j \in J} \bar{a}_{ij}(-t_j), \; i \in I \cap Q \\ t_j \geq 0, \quad j \in J, \\ \bigvee_{i \in Q} \left[(1 - \sigma_i) x_i + \sigma_i \left(\sum_{h \in Q - \{i\}} x_h \right) = 1 - \sigma_i \right] \end{array} \right\}.$$

The family of cuts defined in the next theorem is from Balas (1974a). The fact that it represents the family of those facets of conv $F(\sigma)$ which cut off the current solution follows from Theorem 4.5 of Balas (1974b).

Theorem 7.18 *For every $\sigma \in R^q$ such that $0 \leq \sigma_i \leq 1$, $\forall i \in Q$, the unique facet of conv $F(\sigma)$ which cuts off the solution $t_j = 0$, $j \in J$, is*

$$\sum_{j \in J} \alpha_j(\sigma) t_j \geq 1$$

where

$$\alpha_j(\sigma) = \max_{i \in Q} \left(\sigma_i \sum_{h \in Q} \bar{a}_{hj} - \bar{a}_{ij} \right)(1 - \bar{a}_{i0})^{-1} \quad j \in J.$$

These cuts are computationally not expensive, and one has some freedom in choosing the parameters σ_i so as to make the cut stronger in one direction or another. A very convenient choice of the parameters, which makes the cut particularly easy to compute, is $\sigma_i = (1 - \bar{a}_{i0})$, $\forall i \in Q$, which yields

$$\alpha_j = \sum_{h \in Q} \bar{a}_{hj} - \min_{i \in Q} \bar{a}_{ij} (1 - \bar{a}_{i0})^{-1}.$$

Note that this cut is likely to have some negative entries, since the coefficients $\bar{a}_{hj}$ of the current simplex tableau are of arbitrary signs.

Another disjunctive cut, computationally even cheaper, yet of considerable strength, is the following.

Theorem 7.19 (Balas, 1976) *Let Q and Q' be as in Theorem 7.17, but $i_1, i_2 \in Q'$, and*

$$S_1 = \left\{ j \in J \;\Big|\; \frac{\bar{a}_{i_1 j}}{\bar{a}_{i_1 0}} \geq \frac{\bar{a}_{i_2 j}}{\bar{a}_{i_2 0}} \right\} \qquad S_2 = J - S_1.$$

Then the inequality

$$\sum_{j \in J} \alpha_j t_j \geq 1$$

is a valid cut, with

$$\alpha_j = \begin{cases} \max\left\{ \dfrac{\bar{a}_{i_1 j} - 1}{\bar{a}_{i_1 0}}, \dfrac{\bar{a}_{i_2 j}}{\bar{a}_{i_2 0}} \right\} & j \in Q \cap S_1 \\[1ex] \max\left\{ \dfrac{\bar{a}_{i_1 j}}{\bar{a}_{i_1 0}}, \dfrac{\bar{a}_{i_2 j} - 1}{\bar{a}_{i_2 0}} \right\} & j \in Q \cap S_2 \\[1ex] \min\left\{ \dfrac{\bar{a}_{i_1 j}}{\bar{a}_{i_1 0}}, \max\left\{ \dfrac{\bar{a}_{i_1 j} - 1}{\bar{a}_{i_1 0}}, \dfrac{\bar{a}_{i_2 j} + 1}{\bar{a}_{i_2 0}} \right\} \right\} & j \in (J \backslash Q) \cap S_1 \\[1ex] \min\left\{ \dfrac{\bar{a}_{i_2 j}}{\bar{a}_{i_2 0}}, \max\left\{ \dfrac{\bar{a}_{i_1 j} + 1}{\bar{a}_{i_1 0}}, \dfrac{\bar{a}_{i_2 j} - 1}{\bar{a}_{i_2 0}} \right\} \right\} & j \in (J \backslash Q) \cap S_2. \end{cases}$$

This cut can be shown to be a strengthening of the (unique) facet of conv F' which cuts off the current solution, where

$$F' = \left\{ t \in R^n \;\Big|\; \begin{array}{l} x_i = \bar{a}_{i0} + \sum_{j \in J} \bar{a}_{ij}(-t_j) \quad i \in I \cap Q \\ t_j \geq 0 \quad j \in J \\ \left(\sum_{i \in Q_1} x_i \leq 0 \right) \vee \left(\sum_{i \in Q_2} x_i \leq 0 \right) \end{array} \right\}$$

and where $Q_k = \{i_k\} \cup (Q \cap S_k)$, $k = 1, 2$.

In choosing i_1 and i_2, it is reasonable to give preference to the pair with the largest number of negative coefficients in both rows; this yields a cut with at least that number of negative coefficients.

Each of the above two cutting planes can further be strengthened at a reasonable computational cost by using the integrality constraints on the nonbasic variables, via the procedure of Balas and Jeroslow (1975). Whether strengthened or not, these cutting planes can be used in the context of a simplex-based dual cutting plane algorithm of the type to be discussed in Section 7.3.2, in the same way as the Gomory cuts are used. Since the above cuts use the structure of the set partitioning polytope, their use can be expected to enhance the convergence rate of the algorithm.

Set Partitioning—A Survey

In the next section we discuss a different class of valid inequalities, which are all-integer and unrelated to any solution to the linear programming relaxation of SPP.

7.2.4 Other Valid Inequalities

In this section we discuss the class of valid inequalities introduced by Balas (1975c). These inequalities are derived from the logical implications of the set partitioning constraints, and they often dominate some facet of the set packing polytope, and thus cut off one or more vertices of the latter. The results of this section are from the above-mentioned paper.

As before, let $\bar{P}_I$ be the set partitioning polytope, P_I the set packing polytope, and let M and N be the row and column index sets of $A = (a_{ij})$. Denote

$$M(k) = \{i \in M \mid a_{ik} = 1\} \quad \bar{M}(k) = M - M(k) \quad k \in N$$
$$N(i) = \{j \in N \mid a_{ij} = 1\} \quad i \in M$$

and for $k \in N$, $i \in \bar{M}(k)$, let

$$N(k, i) = \{h \in N(i) \mid a_h a_k = 0\}.$$

Theorem 7.20 *For every $k \in N$ and $i \in \bar{M}(k)$, the inequality*

$$(\alpha) \qquad x_k - \sum_{j \in N(k,i)} x_j \leq 0$$

is satisfied by all $x \in \bar{P}_I$.

Further, every nonzero vertex of P_I, not contained in $\bar{P}_I$, is cut off by at least one inequality (α); each such inequality cuts off some $x \in P_I - \bar{P}_I$.

Remark 7.6 The number of inequalities of the above form is

$$\sum_{k \in N} |\bar{M}(k)|.$$

Theorem 7.21 *For every $k \in N$, $i \in \bar{M}(k)$, and $Q(k, i) \subseteq N(k, i)$, the inequality*

$$(\beta) \qquad x_k - \sum_{j \in Q(k,i)} x_j \leq 0$$

is valid if and only if $x \in \text{vert } \bar{P}_I$, $x_k = 1$ implies $x_j = 0$, $\forall j \in N(k\ i) - Q(k, i)$.

A valid inequality of the form $\pi x \leq \pi_0$ is called *maximal* if there exists no valid inequality $\pi' x \leq \pi_0$ such that $\pi'_j = \pi_j$, $j \in N - \{k\}$, $\pi'_k > \pi_k$ for some $k \in N$.

Theorem 7.22 Assume that for every $j \in N$ there exists $x \in \bar{P}_I$ such that $x_j = 1$, and that (β) is a valid inequality. Then (β) is maximal if and only if
  (i) for every $j \in Q(k, i)$ there exists $x \in \text{vert } \bar{P}_I$ such that $x_j = x_k = 1$;
  (ii) for every $j \in \bar{N}(i) - \{k\}$ there exists $x \in \text{vert } \bar{P}_I$ such that $x_j = 1$ and $x_k \geq x_h$, $\forall h \in Q(k, i)$.

Furthermore, if the assumption holds and (β) is valid but not maximal, and if S_1 and S_2 are the index sets for which conditions (i) and (ii) respectively are violated, then the inequality

$$x_k - \sum_{j \in Q(k,i) - S_1} x_j \leq 0$$

is valid, and so are the inequalities

$$x_k + \sum_{j \in S_2 \cap T} x_j - \sum_{j \in Q(k,i)} x_j \leq 0$$

for every $T \subseteq \bar{N}(i) - \{k\}$ *such that* $a_h a_j \neq 0$, $\forall h, j \in T$.

Based on these results, one can start with any of the inequalities of Theorem 7.20, which are always valid, and use certain strengthening procedures (two of which are described in Balas, 1975c) to obtain new valid inequalities.

Since the dimension of $\bar{P}_I$, unlike that of P_I, is always less than n (in fact, at most $n - m$), an inequality $\pi x \leq \pi_0$ valid for $\bar{P}_I$ may define an *improper face* (this is the case when $\pi x = \pi_0$ for all $x \in \bar{P}_I$). Sufficient conditions for a maximal valid inequality (β) to define a facet or an improper face of $\bar{P}_I$ can be found in Balas (1975c).

Homogeneous inequalities of the above type have their nonhomogeneous equivalents. Namely, for $i \in \bar{M}(k)$, $k \in N$, let $Q(k, i) \subseteq N(k, i)$. Further, let

$$V(k, i) = \{k\} \cup N(i) \backslash Q(k, i) \qquad i \in \bar{M}(k), k \in N$$
$$W(k, i, h) = Q(k, i) \cup N(h) \backslash \{k\} \qquad i \in \bar{M}(k), h \in M(k), k \in N.$$

Theorem 7.23 A point $x \in \bar{P}_I$ satisfies any one of the following inequalities (for some $k \in N$ and $i \in \bar{M}(k)$) if and only if it satisfies each of them:

$$x_k - \sum_{j \in Q(k,i)} x_j \leq 0$$

$$\sum_{j \in V(k,i)} x_j \leq 1$$

$$\sum_{j \in W(k,i,h)} x_j \geq 1 \qquad h \in M(k).$$

Though the second inequality of Theorem 7.23 is of the set packing type, it is not in general valid for the set packing polytope P_I associated with $\bar{P}_I$.

Set Partitioning—A Survey 177

Theorem 7.24 *The inequality*

$$\sum_{j \in V(k,i)} x_j \leq 1$$

cuts off some $x \in P_I - \bar{P}_I$ *if and only if* $Q(k, i) \neq N(k, i)$.

The intersection graph G_A of a 0–1 matrix A, defined in Section 7.1.4, has a node for each column of A, and an edge for each pair of columns a_i, a_j such that the set packing polytope P_I has no point x for which $x_i = x_j = 1$.

By analogy, one can define (see Balas, 1975c) the *strong intersection graph* $G(A)$ of A as having a node for each column of A, and an edge for each pair of columns a_i, a_j such that the set partitioning polytope $\bar{P}_I$ has no point x for which $x_i = x_j = 1$. G_A is always a proper subgraph of $G(A)$. The inequalities discussed above can then be interpreted on the strong intersection graph $G(A)$ as follows.

Theorem 7.25 *The inequality*

$$\sum_{j \in V} x_j \leq 1$$

is satisfied by all $x \in \bar{P}_I$ *if and only if* V *is the node set of a complete subgraph* G' *of* $G(A)$; *and, provided that for each* $j \in N$ *there exists* $x \in \bar{P}_I$ *such that* $x_j = 1$, *it is maximal if and only if* G' *is a clique.*

Example 7.8 Consider the matrix

$$A = \begin{pmatrix} 1 & 0 & 0 & 0 & 0 & 0 & 1 & 0 & 0 & 0 \\ 1 & 1 & 0 & 0 & 0 & 0 & 0 & 0 & 0 & 0 \\ 0 & 1 & 1 & 0 & 0 & 0 & 0 & 1 & 0 & 0 \\ 0 & 0 & 1 & 1 & 0 & 0 & 0 & 0 & 0 & 0 \\ 0 & 0 & 0 & 1 & 1 & 0 & 0 & 0 & 0 & 1 \\ 0 & 0 & 0 & 0 & 1 & 1 & 0 & 0 & 0 & 0 \\ 0 & 0 & 0 & 0 & 0 & 1 & 1 & 0 & 0 & 0 \\ 0 & 0 & 0 & 0 & 0 & 1 & 1 & 0 & 0 & 0 \\ 0 & 0 & 0 & 0 & 0 & 0 & 0 & 1 & 1 & 0 \\ 0 & 0 & 0 & 0 & 0 & 0 & 0 & 0 & 1 & 1 \end{pmatrix}$$

Figure 7.5 shows G_A and $G(A)$. The thin lines are the edges of both G_A and $G(A)$, while the heavy lines are those edges of $G(A)$ not in G_A: (1, 6), (1, 8), (1, 9), (2, 4), (2, 5), (2, 9), (3, 7), (4, 9), (5, 8).

Clearly, the facet-producing subgraphs of $G(A)$ give rise to valid inequalities for $\bar{P}_I$, stronger than those obtained from the facet-producing subgraphs of G_A.

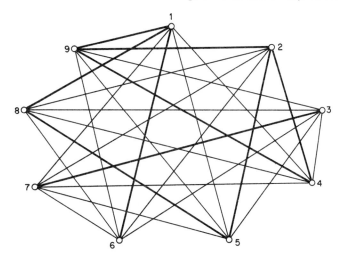

Figure 7.5

The homogeneous inequalities discussed above, called *elementary*, can be combined according to certain composition rules (see Balas, 1975c, for details) into homogeneous inequalities of a more general type, stronger than the sum of the elementary inequalities used to obtain them (but still of a canonical form, i.e. having coefficients equal to 1, −1, or 0). Alternatively, some of these *composite* inequalities can be derived directly. One such instance is shown in the next theorem.

Theorem 7.26 Let $K \subset N$ be such that

$$\sum_{j \in K} x_j \leq 1, \forall x \in \text{vert } \bar{P}_I$$

and let

$$L(K) = \{j \in N - K \mid a_j a_k = 0 \text{ for some } k \in K\}.$$

Then the inequality

$$\sum_{j \in K} x_j - \sum_{j \in S} x_j \leq 0$$

where $S \subseteq L(K)$, is satisfied by all $x \in \bar{P}_I$ if and only if

$$\sum_{j \in Q} a_j \neq e - a_k \quad \forall k \in K, \quad \forall Q \subseteq L(K) - S.$$

The inequalities of Theorem 7.26 are particularly well suited for use as primal all-integer cutting planes. This is discussed in Section 7.3.4.

Set Partitioning—A Survey 179

7.2.5 Adjacent Vertices of the Set Partitioning and Set Packing Polytopes

Integer programmers have often been puzzled by the frequency of the occurrence of integer solutions among the basic solutions to the linear program associated with SPP, which we will call LSPP. This is one of the features which makes SPP relatively easily accessible to cutting plane methods: a few cuts applied to a fractional tableau, even very large, often (though not always) yield an integer solution. The results to be discussed below throw some light on this phenomenon.

On the other hand, the linear program associated with SPP is often difficult to handle, because of the large size, massive degeneracy, and the lack of any specialized technique to take advantage of the structure at hand. Thus, solving the linear program usually becomes the bottleneck of cutting plane algorithms.

Therefore, it is potentially of great practical interest to know that, due to certain adjacency properties of the vertices of $\bar{P}$ and $\bar{P}_I$, the set partitioning problem can in principle be solved by a modified version of the simplex method, generating only integer solutions and not using any cutting planes. More precisely, the following is true, where two bases of LSPP are called *adjacent* if they differ by only one column, i.e. can be obtained from each other by one pivot.

Theorem 7.27 (Balas and Padberg, 1972a) Let x^1 be a feasible integer (but not optimal) solution to LSPP associated with the basis B_1. If x^2 is an optimal solution to SPP, then there exists a sequence of adjacent bases $B_{10}, B_{11}, B_{12}, \ldots, B_{1p}$, such that $B_{10} = B_1$, $B_{1p} = B_2$ is a basis associated with x^2, and:

(i) the basic solutions $x^{1i} = B_{1i}^{-1} e$, $i = 0, 1, \ldots, p$, are all feasible and integer;

(ii) $cx^{10} \geq cx^{11} \geq \cdots \geq cx^{1p}$; and

(iii) $p = |J_1 \cap Q_2|$, where J_1 is the index set of nonbasic variables associated with B_1, while $Q_2 = \{j \in N \mid x_j^2 = 1\}$.

Statements (i) and (ii) of the above theorem can also be deduced from a result of Trubin (1969). On the other hand, statement (iii) proves the famous Hirsch conjecture (see Dantzig, 1963, p. 168) for the special class of linear programs discussed here if it is restricted to integer solutions only, since clearly $p \leq m$.

Two vertices of a polytope are *adjacent* if they lie on an edge, i.e. are contained in a one-dimensional face, of the polytope. Theorem 7.27 implies that, given a basic feasible integer solution x^1 to LSPP, there is a better integer solution if and only if there is one which is a vertex of $\bar{P}$, the feasible polytope of LSPP. adjacent to x^1. The difficulty lies in identifying such adjacent vertices. Since set partitioning problems tend to be highly degenerate, there are many bases associated with the same solution, and there is a very large number of vertices of LSPP adjacent to a given vertex. Furthermore, lexicographic or similar techniques are of no avail in coping with degeneracy, since the sequence of pivots required to reach an adjacent vertex may include pivots on a negative entry in a degenerate row (i.e. a row with $a_{i0} = 0$). What is needed, therefore, is a detailed knowledge of adjacency relations among integer vertices of $\bar{P}$.

Next we give a constructive characterization of such adjacency relations, based on Balas and Padberg (1973b), which makes it possible to generate all edges of $\bar{P}$ connecting a given integer vertex to integer adjacent vertices.

We start out with a general characterization of an integer vertex of $\bar{P}$ in terms of any other integer vertex, and an associated basis.

Given a basic feasible integer solution x^1 to LSPP, with associated basis B_1 and (basic and nonbasic) index sets $I_1 = \{1, \ldots, m\}$, J_1 ($I_1 \cup J_1 = N$), we will denote

$$\bar{a}_j = B_1^{-1} a_j, \quad \bar{a}^j = \begin{pmatrix} \bar{a}_j \\ -e_j \end{pmatrix}$$

where e_j is the $(n-m)$-dimensional unit vector with 1 in position j, and

$$Q_1 = \{j \in N \mid x_j^1 = 1\} \quad \bar{Q}_1 = N - Q_1.$$

Theorem 7.28 (Balas and Padberg, 1973b) *Let x^1 be a basic feasible integer solution to LSPP. Then x^2 is a basic feasible integer solution to LSPP if and only if there exists $Q \subseteq J_1$ such that*

$$\sum_{j \in Q} \bar{a}_{kj} = \begin{cases} 0 \text{ or } 1 & k \in Q_1 \\ 0 \text{ or } -1 & k \in I_1 \cap \bar{Q}_1 \end{cases}$$

and

$$x_j^2 = \begin{cases} 1 & j \in Q_2 = Q \cup S \\ 0 & \text{otherwise} \end{cases}$$

where

$$S = \left\{ k \in Q_1 \,\middle|\, \sum_{j \in Q} \bar{a}_{kj} = 0 \right\} \cup \left\{ k \in I_1 \cap \bar{Q}_1 \,\middle|\, \sum_{j \in Q} \bar{a}_{kj} = -1 \right\}.$$

When this condition holds, then

$$x^2 = x^1 - \sum_{j \in Q} \bar{a}^j.$$

The next theorem characterizes adjacent integer vertices of $\bar{P}$. From the last part of Theorem 7.28, it follows that x^2 is an integer vertex adjacent to x^1 on $\bar{P}$ if and only if

$$\Lambda = \left\{ x \in R^n \,\middle|\, x = x^1 - \left(\sum_{j \in Q} \bar{a}^j \right) \lambda, \, 0 \leq \lambda \leq 1 \right\}$$

Set Partitioning—A Survey 181

is an edge (1-dimensional face) of $\bar{P}$ which is also an edge of $\bar{P}_I$. The necessary and sufficient condition for this to be true is given in terms of a certain property of the set of columns indexed by Q.

Given a basic feasible integer solution x^1, with J_1 defined as above, a set $Q \subset J_1$ which satisfied the condition of Theorem 7.28 is called *decomposable* if it can be partitioned into two subsets Q^* and Q^{**} such that the condition remains true when Q is replaced by Q^* and Q^{**}, respectively. This concept of decomposability is used in the next theorem to characterize pairs of adjacent integer vertices of $\bar{P}$ in terms of a basis associated with one member of the pair. While this characterization can be used directly to generate the integer vertices adjacent to a given vertex from any simplex tableau associated with the latter, it is desirable also to have an equivalent characterization in terms of the matrix A, without reference to a specific basis. This requires a concept analogous to decomposability, defined in terms of the columns of A.

For any $S \subset N$, $T \subset N$, such that

$$\sum_{j \in S} a_j = \sum_{j \in T} a_j \leq e$$

we say that (S, T) is *pairwise decomposable* if there exists a pair (S', T') of proper nonempty subsets $S' \subset S$, $T' \subset T$ such that the above condition remains true when S and T are replaced by S' and T' respectively.

The central result of the next theorem is the equivalence of statements (i) and (iii) (see Balas and Padberg, 1973b). This constitutes the key to a procedure for generating integer vertices adjacent to a given integer vertex, and also implies the equivalence between (i) and (ii), obtained earlier by Trubin (1969). The result on the equivalence of (iii) and (iv) is due to Padberg and Rao (1973).

Theorem 7.28 *Let x and y be any two vertices of $\bar{P}_I$, and for $z = x, y$ let*

$$Q(z) = \{j \in N \mid z_j = 1\} \quad \bar{Q}(z) = \{j \in N \mid z_j = 0\}.$$

Further, let $J(x)$ be the nonbasic index set for some (arbitrarily chosen) basis associated with x. Then the following four statements are equivalent:

(i) *x and y are adjacent on $\bar{P}$*
(ii) *x and y are adjacent on $\bar{P}_I$*
(iii) *$Q(y) \cap J(x)$ is not decomposable*
(iv) *$[Q(x) \cap \bar{Q}(y), \bar{Q}(x) \cap Q(y)]$ is not pairwise decomposable.*

It is known that all vertices of the feasible set (convex hull of feasible 0–1 points) of a 0–1 program are vertices of the associated linear programming polytope. The equivalence of (i) and (ii) above puts SPP into the class of 0–1 programs having the much stronger geometric property, that all edges of the feasible set (convex hull of feasible 0–1 points) are edges of the associated linear programming polytope.

For the set packing polytope P_I, which is a special case of $\bar{P}_I$, a necessary and sufficient condition for the adjacency of two vertices (on P_I only) was given independently, in graph-theoretical terms, by Chvátal (1972).

Theorem 7.30 *Let x and y be two vertices of P_I and let G' be the subgraph of (the intersection graph) G_A induced by the node set*

$$[Q(x) \cap \overline{Q(y)}] \cup [\overline{Q(x)} \cap Q(y)].$$

Then x and y are adjacent on P_I if and only if G' is connected.

This theorem is closely related to an earlier result on weighted node packings, due to Balinski. Given any node packing S, define an *alternating subgraph* $H(S) = (N', E')$ of G_A relative to S to be a bipartite graph whose edges only connect nodes in S to nodes in $N - S$, and such that if $i \in N'$, $j \in N - N'$, and $(i, j) \in E$, then $j \notin S$. Define an *augmenting subgraph* to be an alternating subgraph such that the weight-sum of the nodes in $N' - S$ exceeds that of the nodes in S. Let the weight of a node packing be the weight-sum of its nodes. Then Balinski's result can be stated in terms of the node packing polytope as follows (see also Edmonds, 1962).

Theorem 7.31 (Balinski, 1970) *A vertex x of P_I is of maximum weight if and only if G_A admits no connected augmenting subgraph relative to $Q(x)$.*

In other words, the graphs G' of Theorem 7.30 are precisely the connected alternating subgraphs defined above. Note however, that Balinski's and Chvátal's results only concern the adjacency of vertices of P_I; they are not concerned with adjacency on the associated linear programming polytope P.

Returning now to the more general case of the set partitioning polytope $\bar{P}_I$, Theorem 7.29 above implies a necessary and sufficient condition for two vertices of the set partitioning polytope $\bar{P}_I$ to be *nonadjacent*. This condition leads to an interesting geometric characterization of $\bar{P}_I$. For an arbitrary polyhedron P, a *path* between two vertices x and y of P is a sequence of vertices $x^1, x^2, \ldots, x^k$ with $x^1 = x$, $x^k = y$, such that every pair of vertices x^i, x^{i+1}, $i = 1, \ldots, k-1$, is connected by an edge of P, the length of the path being $k - 1$. The *edge-distance* $d(x, y)$ between x and y is then defined as the length of a shortest path on P between x and y. The diameter $\delta(P)$ of P is the longest edge-distance between any pair of vertices of P, i.e.

$$\delta(P) = \max_{x, y \in \text{vert} P} d(x, y).$$

If we require the matrix A in the definition of $\bar{P}_I$ not to have identical columns, we then have Theorem 7.32.

Set Partitioning—A Survey

Theorem 7.32 (Balas and Padberg, 1973) *For any 0–1 matrix A, the diameter of the associated set partitioning polytope $\bar{P}_I$ satisfies*

$$\delta(\bar{P}_I) \leq \left[\frac{z^*}{2}\right] \leq \left[\frac{m}{2q}\right]$$

where

$$q = \min_{j \in N} \sum_{i=1}^{m} a_{ij}, \quad \text{and} \quad z^* = \max\left\{\sum_{j=1}^{n} x_j \,\Big|\, x \in \bar{P}_I\right\}.$$

In fact, the upper bound on the diameter of $\bar{P}_I$ provided by the above theorem is a best possible one, and is actually attained if $A = (A_G, I)$ in the definition of $\bar{P}_I$, where A_G is the $m \times \binom{m}{2}$ node–edge incidence matrix of a complete graph with m nodes, and I is the identity of order m.

A further geometric property is contained in the following theorem, which has some interesting algorithmic implications.

Theorem 7.33 (Balas and Padberg, 1973b) *Let x^1 be a nonoptimal vertex of $\bar{P}_I$, let x^{1i}, $i = 1, \ldots, k$, be those vertices of $\bar{P}_I$ adjacent to x^1, and such that $cx^{1i} < cx^1$, $i = 1, \ldots, k$. Then the convex polyhedral cone*

$$C = \left\{x \in R^n \,\Big|\, x = x^1 + \sum_{i=1}^{k} \lambda_i (x^{1i} - x^1), \lambda_i \geq 0, i = 1, \ldots, k\right\}$$

contains an optimal vertex of $\bar{P}_I$.

The property stated in Theorem 7.33 is not true for arbitrary integer programs, as shown by the trivial counter-example of Figure 7.6. In this example, $cx^1 > cx^2 > cx^3$, and the cone C (here just a halfline) clearly does not contain the unique optimal point x^3.

Since the set packing polytope P_I is a special case of the set partitioning polytope $\bar{P}_I$, all the results of this section stated for SPP are also valid for SP. Furthermore, the main results, including Theorems 7.28, 7.29 without (i),

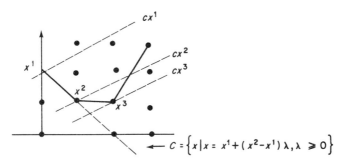

Figure 7.6

and 7.33, can be generalized to arbitrary 0–1 programs (see Balas and Padberg, 1973).

7.3 Algorithms

Before discussing any particular algorithm it should be mentioned that problem size can often be substantially reduced by using one or more of the rules listed below, prior to applying any solution method.

Let M and N be the row and column index sets of A; let a^i be the ith row, and a_j the jth column, of A.

The following three 'reduction rules' are trivial but useful and easy to implement.

1. If for some $i \in M$ and $k \in N$, $a_{ik} = 1$, $a_{ij} = 0$, $\forall j \in N - \{k\}$, then
   (a) set set $x_k = 1$ and remove column k
   (b) remove all columns $j \in N - \{k\}$ such that $a_k a_j \geq 1$
   (c) remove all rows $h \in M$ such that $a_{hk} = 1$.
2. If for some $i, k \in M$, $a^i \leq a^k$, then one can
   (a) remove row k
   (b) remove all columns $j \in N$ such that $a_{ij} = 0$, $a_{kj} = 1$.
3. If for some $k \in N$ and some subset $N' \subset N$, $\sum_{j \in N'} a_j = a_k$ and $\sum_{j \in N'} c_j \leq c_k$, one can remove column k.

The next 'reduction rule' is somewhat more expensive to implement.

4. For each $i \in M$, let $N_i = \{j \in N \mid a_{ij} = 1\}$. Then any column a_k such that

$$a_k a_j \geq 1 \quad \forall j \in N_i \text{ for some } i \in M$$

can be removed.

7.3.1 Implicit Enumeration

Several specialized versions of the implicit enumeration approach of Balas (1964, 1965) (see also Glover, 1965; Geoffrion, 1967) have been proposed for the set partitioning problem and implemented. The ones that seem to have been successful include Pierce (1968), Garfinkel and Nemhauser (1969), Pierce and Lasky (1973), Marsten (1974).

The first two of these algorithms are very similar, and we will discuss the Garfinkel–Nemhauser version. The solution space is systematically searched by generating partial solutions (assigning 0–1 values to variables taken one at a time) and exploring the logical implications of these value assignments. The special structure simplifies the logical tests and lends them great power. Like the general implicit enumeration procedures, these algorithms are additive, i.e. require no divisions and therefore pose no numerical stability problems. Furthermore, since in this case all constraint coefficients are

Set Partitioning—A Survey

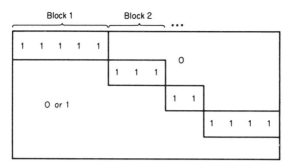

Figure 7.7

binary, they can be stored and manipulated as bits. Also, the use of the logical 'and' and logical 'or' statements permits the efficient execution of many of the required operations.

To start with, the matrix A is brought by row and column permutations to the staircase form shown in Figure 7.7.

In other words, the columns of A *are partitioned into* t non-empty subsets ('blocks') B_j, $j = 1, \ldots, t$, such that block B_j satisfies $a_{ik} = 1$ for all $k \in B_j$ and $a_{ik} = 0$ for

$$k \in \bigcup_{l=j+1}^{t} B_l$$

for some row i of the matrix A. The rows of A are then ordered so that the row defining block B_j becomes the jth row for $j = 1, \ldots, t$. If $t < m$, empty blocks B_j are defined for $j = t+1, \ldots, m$. Within each block the columns are ordered by some heuristic criterion: according to increasing costs (Garfinkel–Nemhauser), or increasing costs per number of rows covered (i.e. increasing $c_j/\sum_{i=1}^{m} a_{ij}$) (Pierce), or increasing reduced costs, obtained by solving LSPP (Pierce and Lasky; Heurgon, 1972).

Denote by S (the index set of) a partial solution, by S^+ its subset of variables fixed at 1, by z its value, and by R the set of rows (constraints) satisfied by S. Further, let $\bar{z}$ be the value of the best solution found so far (if any).

Then the algorithm proceeds as follows.

Step 1 (Initialization) Set up the initial tableau, let $S = \emptyset$, $R = \emptyset$ and $\bar{z} = \infty$. Go to Step 2.

Step 2 (Choose next block) Let $r = \min\{i \mid i \notin R\}$. Set a marker at the top (lowest cost element) of block r. Go to Step 3.

Step 3 (Test for an augmenting variable) Beginning at the marked position in block r, examine all columns of A in block r in order. If a column j is found such that $a_{ij} = 0$, $\forall i \in R$, and $z + c_j < \bar{z}$, go to Step 4. If a column j is reached such that $z + c_j \geq \bar{z}$, or if block r is exhausted, go to Step 5.

Step 4 (*Test for a new solution*) Redefine S^+ to be $S^+ \cup \{j\}$, R to be $R \cup \{i \mid a_{ij} = 1\}$ and z to be $z + c_j$. If $R = \{1, \ldots, m\}$, a better solution has been found, which is recorded, and $\bar{z}$ is updated. Go to Step 5. Otherwise, go to Step 2.

Step 5 (Backtrack) If $S^+ = \emptyset$ (i.e. block 1 has been exhausted, terminate with the best solution found (if any). Otherwise, let k be the last index included in S^+. Redefine S^+ to be $S^+ - \{k\}$. Let B_r be the (unique) block to which column a_k belongs. Put a marker at the next column in block r, remove the previous marker in block r, and go to Step 3.

As one would expect, this algorithm performs better on high density than on low density problems (where density means the number of nonzero entries of A divided by mn). Computational results reported by Garfinkel and Nemhauser include the solution of 4 problems with 26–100 constraints and 385–1790 variables, and a density of 0.16–0.25, in 25–900 seconds of IBM 7094 time; however, a fifth problem (37×1790, 0.25) could not be solved in 15 minutes. Comparable results are reported by Pierce.

One of the means of enhancing the efficiency of implicit enumeration is to calculate for each partial solution a lower bound on the cost of its completion. This can be done, for instance, by solving the linear program in the free variables, as done by Geoffrion (1967) for general 0–1 programs, by Lemke, Salkin, and Spielberg (1971) in the context of an algorithm primarily designed for set covering problems, or by Michaud (1972) in his implicit enumeration algorithm for the SPP. Pierce, and Pierce and Lasky, avoid solving a linear program for each partial solution by solving instead, or finding a lower bound on the value of, a knapsack problem obtained by adding up all the constraints. Ming-Te Lu (1970) weights the constraints before adding them up. A stronger bound, which brought a considerable improvement in computing time, was obtained by Christofides and Korman (1973) by solving an auxiliary problem via dynamic programming techniques. While this was achieved in the context of a set covering algorithm the bound applies to set partitioning problems as well.

Another important improvement was obtained, first by Pierce and Lasky, then also by others (see Gondran and Laurière, 1974; Heurgon, 1972), by replacing the cost vector c with the reduced cost vector $c - c^B B^{-1} A$ of an optimal simplex tableau for LSPP. This is permissible, since $(c - c^B B^{-1} A)x = cx - c^B B^{-1} e$ for any feasible x, i.e. the value of the reduced cost function differs from that of the original cost function by a constant, for any feasible solution. This requires of course the solution of LSPP prior to implicit enumeration, which for large problems is quite a high price to pay for hopefully speeding up the subsequent enumeration. However, the gain seems to be worth the price, for Pierce and Lasky have solved implicit enumeration, without making any other use of the linear program except for solving it once to obtain an optimal reduced cost vector, 7 out of 10 problems with 15–60 constraints and 1200–3500 variables in 15–300 seconds of CPU time on an IBM 360/67. Their code could not, however, solve the remaining 3 problems within 9, 23, and 23 minutes respectively. A cheaper way of improving the cost structure (see Gondran and Laurière, 1974) is the following. Choose any row a^i of A, and replace c by $c - c_{j_i} a^i$, where $c_{j_i} = \min\{c_j \mid a_{ij} = 1\}$. Then choose another row and repeat the procedure. Since no row can yield an improvement twice, it is sufficient to consider each row once. The legitimacy of the procedure follows from the same argument as that of replacing the initial costs by the reduced costs of the linear program.

An implicit enumeration algorithm based on a different enumerative scheme, which can be viewed as a specialization of the one proposed for general 0–1 programming by Graves and Whinston (1968), was developed by Marsten (1974). The columns of A are partitioned into blocks, just as in the above procedure, but the enumeration is based on assigning rows to blocks. If $M_i = \{k \in N \mid a_{ik} = 1\}$ and $M_{ij} = M_i \cap B_j$, where i is the index of any row and j that of any block, then assigning row i to block j amounts to assigning the value 1 to the *sum* of variables x_k, $k \in M_{ij}$, rather than to a single variable. The logical implications of such value assignments are then explored in the usual manner. This enumerative scheme gives rise to a different search tree, whose number of nodes usually has a smaller bound than the number of nodes in a usual search tree. This does not necessarily have anything to do with the number of nodes actually generated in the two cases, and at the moment no convincing evidence is available in favour of one search tree or another. Marsten also solves LSPP first and uses the reduced costs instead of the original ones. He reports the solution on an IBM 360/91 of four large problems, of size 90×303, 63×1641, 111×4826, and 200×2362, in 40.64, 168.76, 479.77, and 418.86 seconds, respectively (excluding the time required to bring A to the proper format). Marsten's code seems to perform better than the Garfinkel–Nemhauser code on low density problems, while the latter has the lead for higher densities. Unfortunately, however, it is not clear to what extent these differences can be ascribed to the different search trees used, since the two codes have a number of other differences. A limited comparative study of the two search strategies, undertaken by Nemhauser, Trotter, and Nauss (1972), indicates better performance for the usual type of search tree.

One important fact that has emerged from Marsten's experience, and confirms earlier computational experience by Martin and others, is the difficulty of solving the linear programs associated with large set covering problems. The reasons for this were already mentioned and will be taken up again later.

7.3.2 Simplex-Based Cutting Plane Methods

As mentioned earlier, the properties of the set partitioning polytope discussed in Section 7.2.3 produce a high frequency of integer solutions among the basic solutions of LSPP. This makes SPP relatively easily accessible via cutting plane methods. Martin (1969) reports in his, unfortunately unpublished, talk, the solution of several large set partitioning (airline crew scheduling) problems with 300–500 constraints and 1500–400 variables by the use of Gomory's (1963a) cutting planes, generated in a special way described in Martin (1963). In spite of the large size of the problems involved, the number of cuts used was reportedly less than 10 in all but one of the cases (when it was 16). However, other problems, in some cases with less than 300 variables, could not be solved within reasonable time limits.

In a recent study on solving SPP by cutting plane methods, Délorme (1974) describes a computer code implementing a version of Gomory's (1963a) algorithm, and reports on his computational experience with several fairly large set partitioning problems, used in the scheduling of crews for the Paris bus lines (see Heurgon, 1972, on the treatment of some of the same problems by implicit enumeration). The problems are of a special type, namely SPP with two additional constraints, one of which is an upper bound on a positively weighted sum of all the variables, while the other one is $ex = k$, $e = (1, \ldots, 1)$ i.e. a constraint that fixes the cardinality of a feasible partition. These two constraints drastically reduce the feasible set, and thus make the problem easier than a general SPP.

The cuts are generated from the row with largest fractional constant term. After each pivot, a new cut is generated (unlike in Gomory's algorithm) until the solution (though not necessarily the whole tableau) becomes integer, which rarely seems to take more than 3–4 cuts. Once integer, however, the solution is not necessarily primal feasible, and in the process of reoptimization, integrality may be lost again. The main data of the study are summarized in Table 7.1.

Densities are not shown, but judging from the above data they must have been very low. The time and number of pivots used to solve the LP are given only for the first two problems, and they are 8'42" and 577 pivots for problem 1, 2'51" and 343 pivots for problem 2.

The procedure of generating sequences of cuts (one cut after each pivot) is compared with the usual one of re-establishing primal feasibility after the addition of each cut. It is found that the number of cuts required is about the same in the two cases, whereas the number of pivots required is considerably higher in the second case. Also, the first procedure is compared

Table 7.1 1, problem number; 2, constraints multiplied by variables; 3, maximum number of cuts in one sequence; 4, total number of cuts; 5, time spent on post-optimization; 6, number of pivots in post-optimization; 7, basis determinant at LP optimum; 8, optimal LP value; 9, optimal integer value

1	2	3	4	5	6	7	8	9
No.	$m \times n$	Cuts in sequence	Cuts total	Time in post-opt	Pivots in post opt	Basis det	Z_{LP}	Z_{INT}
1	111×2186	4	6	2'32"	59	4	3320	3327
2	77× 969	4	7	46"	27	122	3240	3271
3	78× 693	4	5	35"	66	12	2942	2945
4	91×1019	3	3	28"	42	4	3161	3162
5	51× 913	3	3	15"	49	3	450	479
6	88× 953	1	1	2"	1	68	3305	2216
7	58× 405	4	5	23"	35	24	2870	2930
8	185×1043	1	1	12"	5	3	3061	3080

with the one which 'consolidates' the sequence of cuts into a single cut as recommended by Martin (1963) (see also Gondran, 1973), and found superior to the latter. A number of questions of general interest for solving large linear programs are discussed. The program has reportedly been compared with the version of the Garfinkel–Nemhauser algorithm described by Heurgon (1972) and found to be able to solve a larger number of problems, in shorter average times. An evaluation of the effect of the two extra constraints was not attempted, and no information is given on the performance of the code on 'pure' set partitioning problems (i.e. without the two extra constraints). The study contains much other useful information. For related computational experience see also Délorme and Heurgon (1975).

Another simplex-based cutting plane method that was tested on SPP is Gomory's (1963b) all-integer algorithm. In a computational study undertaken by Salkin and Koncal (1973), this algorithm was applied, basically unchanged, to the set partitioning and set covering problems. A set of 13 problems with 30 constraints and 30–90 variables, having unit costs and density of 0.07, were solved on a Univac 1108 in less then 3 seconds each. The average number of cuts generated per problem was 22, with a maximum of 45, and the average number of pivots was 131; these performances are quite good. The code failed to solve any of a set of 8 larger problems with about 400–900 variables within the time limit of 2 minutes, but apparently the time limit was set too low since the maximum number of pivots performed on any of the problems was $3m$ (where m is the number of constraints), which is about the number of pivots it usually takes to solve a linear program, and we cannot reasonably expect to be able to solve integer programs by the same number of pivots as one needs to solve the corresponding linear program.

The cutting plane methods discussed above have three characteristics in common: they are all based on the simplex method (primal and dual, or only dual), they all use the traditional cutting planes introduced by Gomory, and they are all general-purpose algorithms which take little if any advantage of problem structure. The fact that, in spite of this last feature, they perform reasonably well on SPP shows that the cutting plane approach holds great promise for this class of problems. One possible direction of improvement seems to be at hand in the form of cutting planes derived from the special structure of SPP. These cuts, discussed in Section 7.2.3, have not yet been extensively tested. What they are likely to accomplish is to reduce substantially the number of cuts needed. While this could render solvable those problems which were not solved by the above procedures because they required too many cuts, it cannot of course affect the effort required to solve the linear program itself, and the latter is currently the main bottleneck in solving large SPP's. In this respect, two directions of improvement seem to be open; specializations of the simplex method which use the structure of LSPP, and non-simplex-type methods for solving LSPP. One development in the first of these directions is discussed in the next section (though not in a cutting plane context), and another one, in the second direction, is discussed in Section 7.3.5.

While the methods discussed in this and the previous section have achieved a certain maturity in the sense of having been implemented and tested by different persons in different versions, the approaches discussed in the next three sections are new and have only been tested to a very limited extent.

7.3.3 A Column Generating Algorithm

Next we discuss a column generating primal simplex algorithm for SPP, which produces a sequence of integer solutions that converges on an optimal one. The algorithm is based on the characterization of adjacency relations among vertices of $\bar{P}_I$ described in Section 7.2.3. This is a modified version of the algorithm by Balas and Padberg (1973b), discussed in Balas, Gerritsen, and Padberg (1974). The algorithm performs nondegenerate primal simplex pivots on $+1$ entries as long as this is possible. When this cannot be continued, degenerate pivots are performed on positive or negative entries, as long as they decrease total dual infeasibility. (In the present context, a pivot in row i is called degenerate if $a_{i0}=0$, nondegenerate if $a_{i0}>0$.) When neither type of pivoting can be continued, a column generating procedure is used to produce a composite column defining an edge of $\bar{P}_I$ which connects the current vertex of $\bar{P}_I$ to a better one, or to establish the absence of any better vertex. This procedure differs from the one described in Balas and Padberg (1973b), mainly in that it is performed entirely on the original A matrix instead of the current simplex tableau, and thus it is all-binary (see Balas, Gerritsen, and Padberg, 1974). The idea of allowing for degenerate pivots on -1 entries is due to Andrew, Hoffman, and Krabek (1968), who implemented it and obtained remarkably good computational results in the sense of being able to get to, or close to, the integer optimum fairly often. Experience has shown, however, that it pays to go further and allow for degenerate pivots on any nonzero entry (i.e. to give up the integrality of the tableau, though not that of the solution) as long as improvements can be obtained according to some reasonable criterion.

We start by describing the binary column generating procedure, then imbed it into the rest of the algorithm.

Let $\bar{x}$ denote a feasible integer solution to LSPP with basis B, and reduced cost vector $\bar{c} = c - c^B B^{-1} A$, and let $Q = \{j \in N \mid \bar{x}_j = 1\}$. Let T designate initially the binary tableau consisting of those columns a_h of A such that $h \in S = N - Q$. Each composite column is of the form $a_j = \sum_{h \in Q_j} a_h$, where the index j represents the set Q_j. For the original columns, $Q_j = \{j\}$. T and S always refer to the current tableau, T' and S' to the next one.

BCGP (*Binary Column Generating Procedure*)

Step 0 Initialization: $S = N - Q$, $T = \{a_j\}_{j \in S}$.

Step 1 If $\bar{c}_j \geq 0$, $\forall j \in S$, stop: the current tableau is optimal. Otherwise go to Step 2.

Step 2 Choose $t \in S$ such that $\bar{c}_t = \min_{j \in S} \bar{c}_j$. Define $R(t) = \{j \in Q \mid a_j a_t \geq 1\}$ and let $a(t) = \sum_{j \in R(t)} a_j$. If $a_t = a(t)$, stop: an improving adjacent vertex x^* is given by $x_j^* = 1$ for all $j \in (Q - R(t)) \cup \{t\}$, $x_j^* = 0$ otherwise. If $a_t \neq a(t)$, let $C(t) = \{j \in S \mid a_j a(t) \geq 1, a_j a_t = 0\}$ and go to Step 3.

Step 3 Find all $N' \subseteq C(t)$ such that
(i) $a_t + \sum_{j \in N'} a_j \geq a(t)$
(ii) $a_j a_k = 0$, $\forall j, k \in N'$
(iii) $\bar{c}_t + \sum_{j \in N'} \bar{c}_j + z(t, N') < 0$

where $z(t, N')$ is a lower bound on the value of an optimal solution to the problem obtained by setting $x_k = 1$, $\forall k.\{t\} \cup N'$.

(iv) $Q_t \cup (\bigcup_{j \in N'} Q_j)$ cannot be paritioned into Q_h, $h = j_1, \ldots, j_p$, with $j_i \in S$, $j_i \neq t$, for $i = 1, \ldots, p$.

Step 3a If there exists no $N' \subseteq C(t)$ satisfying (i)–(iv), remove a_t permanently from the tableau; set $S' = S - \{t\}$, $T' = T - \{a_t\}$, and go to Step 4.

Step 3b If there exists $N' \subseteq C(t)$ satisfying (i)–(iv), such that (i) holds with equality and $z(t, N') = 0$, stop: pivoting into the basis the composite column

$$a_{j*} = \sum_{j \in Q_{i*}} \bar{a}_j$$

where $Q_{j*} = \{t\} \cup N'$, or each of the columns $\bar{a}_j$, $j \in Q_{j*}$, yields an integer solution x^*, with $x_j^* = 1$, $\forall j \in [Q - R(t)] \cup Q_{j*}$, $x_j^* = 0$ otherwise, which is adjacent to and better than $\bar{x}$.

Step 3c Otherwise, adjoin a new column to T and a composite index to S for every (t, N') satisfying (i)–(iv) but not satisfying (i) with equality; discard column a_t and index t; call the resulting tableau T' and index set S', and go to Step 4.

Step 4 Define T' and S' to be the current tableau T and index set S, respectively, and go to Step 1.

We now turn to the execution of Step 3 of BCGP. First, we notice that a crude lower bound $z(t, N')$ for test (iii) can be obtained as follows. Let

$$Z = \left\{ j \in S \mid j \notin \{t\} \cup N', a_j \left(a_t + \sum_{h \in N'} a_h \right) = 0, \bar{c}_j < 0 \right\}.$$

Then $z(t, N') = \sum_{j \in Z} \bar{c}_j$ is a usable bound.

Set Partitioning—A Survey

Further, finding the sets $N' \subseteq C(t)$ which satisfy (i)–(iv) involves solving an auxiliary set partitioning problem defined on a submatrix of A. To be more specific, let A_R^C be the submatrix of A with column set $C = C(t)$ and row set $R = \{i \in \{1, \ldots, m\} \mid a_i(t) = 1, a_{it} = 0\}$. Then identifying the sets $N' \subseteq C(t)$ which satisfy (i)–(iv) amounts to finding those solutions to

APP $\qquad A_R^C \eta = e_r, \qquad \eta_j = 0 \text{ or } 1, j = 1, \ldots, |C|$

that satisfy the additional tests to qualify for a feasible improving solution to SPP, where $e_r = (1, \ldots, 1)$ has $r = |R|$ components. This is best done by a modified implicit enumeration procedure which includes in its test those conditions (ii)–(iv) not present in APP.

Having discussed the column generating procedure, we can now state the algorithm in its entirety.

Step 1 NPIP (Nondegenerate primal integer pivot) Apply the primal simplex method to LSPP as long as you can pivot on +1 in a nondegenerate row. Whenever this becomes impossible, let $\bar{x}$ be the current (integer) solution, B the associated basis, and $\bar{c} = c - c^B B^{-1} A$. If $\bar{c} \geq 0$, stop: $\bar{x}$ is optimal. Otherwise go to Step 2.

Step 2 DP (Degenerate pivot) Let $N_0 = \{j \in N \mid \bar{c}_j < 0\}$, and $c_0 = \sum_{j \in N_0} \bar{c}_j$. If there exists $\bar{a}_{ij} \neq 0$ in a degenerate row (i.e. $\bar{x}_i = 0$), such that
 (i) pivoting on $\bar{a}_{ij}$ replaces $\bar{c}, N_0$, and c_0 by $\bar{c}', N_0'$, and c_0' respectively
 (ii) $c_0' < c_0$

then pivot on $\bar{a}_{ij}$. If $\bar{c}' \geq 0$, stop: the solution $\bar{x}'$ obtained from the pivot is optimal. Otherwise go to Step 1. If no such $\bar{a}_{ij}$ exists, go to Step 3.

Step 3 BCGP (Binary column generating procedure) This was described above. It either shows the current solution to be optimal, or yields an index set $Q_{j_*} = \{t\} \cup N'$. In the latter case, go to Step 4.

Step 4 BP (Block pivot) Pivot into the basis each $a_j, j \in Q_{j_*}$, and go to Step 1.

The version of the algorithm described here is not yet tested. An earlier version, tested on a set of problems with unit costs, 30 rows, and 40–90 columns, density 0.07, showed behaviour somewhat similar to implicit enumeration, which is not surprising. Thus, finding a good (often optimal) solution usually takes a small fraction of the time required to prove optimality. The importance of Step 2 emerged clearly, in that its inclusion or exclusion has affected computing times by a factor of about 4.

7.3.4 A Hybrid Primal Cutting Plane/Implicit Enumeration Algorithm

In this section we discuss a hybrid algorithm (see Balas, 1975c), which combines a primal cutting plane approach based on cuts of the type discussed in Section 7.2.4, with implicit enumeration applied to subproblems so defined as to generate an improvement at each iteration.

We first describe the particular cutting plane to be used with the algorithm. Let $\bar{x}$ be a basic feasible *integer* solution to the linear programming relaxation of the set partitioning problem, possibly amended with some cuts of the type to be described below, and let

$$x_i = \bar{a}_{i0} + \sum_{j \in J} \bar{a}_{ij}(-t_j) \qquad i \in I \cup \{0\}.$$

Denote

$$J^- = \{j \in J \mid \bar{a}_{0j} < 0\}$$

and assume that $\emptyset \neq J^- \subseteq N$ and $I \subseteq N$, i.e. no slack variable (possibly introduced with the cuts) is basic or has a negative reduced cost. These conditions can easily be met and maintained throughout the procedure, as explained below. Assume also that

$$\min_{i \in I} \left\{ \frac{\bar{a}_{i0}}{\bar{a}_{ij}} \,\middle|\, \bar{a}_{ij} > 0 \right\} < 1 \qquad \forall j \in J^-$$

i.e. pivoting into the basis any single nonbasic variable with a negative reduced cost would make the solution fractional.

Let $j_* \in J^-$ and let $i_* \in I$ be such that $J^- \cap N_{(i_*)} \neq \emptyset$. Define

$$f_j = \min\{0, \bar{a}_{0j} - \bar{a}_{0j_*}\} \qquad j \in J^- \cap N(i_*)$$
$$g_j = \min\{0, \bar{a}_{0j}\} \qquad j \in J \cap N \setminus N(i_*)$$
$$h_j = \min\{0, \bar{a}_{0j} + \bar{a}_{0j_*}\}, \qquad j \in J \cap N \setminus N(i_*)$$

and let $J \cap N \setminus N(i_*) = \{j(1), \ldots, j(r)\}$ be ordered so that $h_{j(k)} \leq h_{j(k+1)}$ $k = 1, \ldots, r-1$. Finally, define $g_{j(0)} = 0$, and let $p \in \{0\} \cup \{1, \ldots, r\}$ be any integer such that

(i) $$\sum_{j \in J^- \cap N(i_*)} f_j + \sum_{j=j(0)}^{j(p)} g_j + \sum_{j=j(p+1)}^{j(r)} h_j > \sum_{j \in J^-} \bar{a}_{0j}.$$

Such p always exists, since (i) holds for $p = r$. The left-hand side of (i) is the sum of negative reduced costs after a pivot in column j_* and the row provided by the cut (iv) below.

If there exists $k \in J^- \cap N(i_*)$ and $y \in \{0, 1\}^q$, where $q = |Q(p)|$, $Q(p) = I \cup \{j(1), \ldots, j(p)\}$, satisfying

(ii) $$\sum_{j \in Q(p)} a_j y_j = e - a_k$$

and

(iii) $$\sum_{j \in Q(p)} c_j y_j < c\bar{x} - c_k,$$

Set Partitioning—A Survey

then $\hat{x} \in R^n$ such that $\hat{x}_j = y_j$, $j \in Q(p)$, $\hat{x}_j = 0$, $j \in N \setminus Q(p)$, is obviously a feasible integer solution better than $\bar{x}$.

On the other hand, if this is not the case, then we have the following.

Theorem 7.34 (Balas, 1975c) *If there exists no $k \in J^- \cap N(i_*)$ and $y \in \{0, 1\}^q$ satisfying (ii) and (iii), then the inequality*

(iv) $$\sum_{j \in J^- \cap N(i_*)} t_j - \sum_{j=j(p+1)}^{j(r)} t_j \leq 0$$

is satisfied by all $x \in P$ such that $cx < c\bar{x}$; and pivoting in the row provided by (iv) and in column j_* produces a simplex tableau with nonbasic index set $\hat{J}$ and reduced costs $\hat{a}_{0j}$ such that

$$\sum_{j \in J^-} |\hat{a}_{0j}| < \sum_{j \in J^-} |\bar{a}_{0j}|$$

and $\hat{a}_{0j} \geq 0$, $\forall j \in J \setminus N$, while the solution $\hat{x}$ remains unchanged.

The inequality (iv) is a valid cut for any integer p for which (i) holds. The strength of the cut increases with the size of p, but so does the computational effort involved in the search for a pair (k, y) satisfying (ii), (iii) (which can be done by implicit enumeration). Let $p_{\min}$ be the smallest value of p for which condition (i) holds. Note that when $p_{\min} = 0$, (ii) cannot be satisfied for any k. Therefore in this case the inequality (iv) (with $p = 0$) is always a valid cut, and there is no need for implicit enumeration to establish this fact.

Since implicit enumeration is highly efficient on small sets, but its efficiency tends to decline rapidly with the increase of the set size, a reasonable choice for p is

(v) $$p = \max\{p_0, p_{\min}\}$$

where p_0 is the largest integer sufficiently small to keep the cost of the implicit enumeration acceptably low.

An algorithm based on Theorem 7.34 can be described as follows.

Step 0 Choose a value of p_0. Start with the linear programming relaxation of the set partitioning problem and go to Step 1.

Step 1 Perform simplex pivots which leave the solution primal feasible and integer, and either reduce the objective function value or leave the latter unchanged and reduce the absolute value of the sum of negative reduced costs. (Note that this does not exclude pivots on negative entries, or pivots which make the tableau fractional, provided they occur in degenerate rows. The algorithm remains valid, however, if such pivots are excluded.) When this cannot be continued, if $\bar{a}_{0j} \geq 0$, $\forall j \in J$, stop: the current solution is optimal. Otherwise go to Step 2.

Step 2 Define i_* and j_* by

$$|J^- \cap N(i_*)| = \max_{i \in I} |J^- \cap N(i)|$$

and

$$|\bar{a}_{0j_*}| = \min_{j \in J^-} |\bar{a}_{0j}|$$

respectively, order the set $J \cap N \setminus N(i_*)$, choose p according to (v), and define $Q(p)$. Then use implicit enumeration (if necessary) to find $k \in J^- \cap J(i_*)$ and $y \in \{0, 1\}^q$ satisfying (ii), (iii) (case α), or to establish that no such pair (k, y) exists (case β); and go to Step 3 (case α) or Step 4 (case β).

Step 3 Pivot into the basis with a value equal to 0 each nonbasic slack variable, and remove from the simplex tableau the correspond row. Then pivot into the basis each nonbasic variable t_j such that $y_j = 1$, and go to Step 1.

Step 4 Generate the cutting plane (iv), add it to the simplex tableau, pivot in the new row and column j_*, and go to Step 1.

Every iteration of this algorithm either decreases the objective function value z, or leaves z unchanged and decreases the absolute value σ of the sum of negative reduced costs. Since both z and σ are bounded from below, the procedure is finite. Its efficiency depends to a considerable extent on a judicial choice of the parameter p_0 in (v), which governs the ratio of pivoting versus implicit enumeration.

7.3.5 A Symmetric Subgradient Cutting Plane Method

In Section 7.3.2 we mentioned that in solving large set partitioning problems by cutting plane techniques the main bottleneck is the handling of the linear program LSPP. This, and the special structure of LSPP, suggests the idea of trying to solve the latter by other methods than the (primal or dual) simplex algorithm. One such attempt (see Balas and Samuelsson, 1974b) uses some of the cutting planes discussed in Section 7.2, while solving the linear program by a subgradient method of the general type used by Held and Karp (1971) for the travelling salesman problem and discussed by Held, Wolfe, and Crowder (1974). The subgradient approach used here (see Samuelsson, 1974b differs from that of the above mentioned authors in that it works simultaneously with the primal and the dual, which makes it possible to achieve faster convergence. Among the positive features of this approach, we mention a high degree of numerical stability, low memory requirements, and an ability to use data in compactly stored form. Next we briefly outline the procedure.

Set Partitioning—A Survey

First SPP is restated as SP, a set packing problem, via the cost transformation mentioned in Section 7.1.2. Since it is essential to keep the costs as low as possible, a tight bound is derived for the transformation parameter θ. Let

(P) $$\max\{cx \mid Ax \leq b, x \geq 0\}$$

be the linear program associated with SP, possibly amended with some cutting planes; and

(D) $$\min\{yb \mid yA \geq c, y \geq 0\}$$

its dual. Let A be $q \times n$. The constraints of SP imply $x \leq e$, where $e = (1, \ldots, 1)$ is an n-vector, but the procedure requires upper bounds on the dual variables too. Whenever c and A are nonnegative,

$$\bar{y}_i = \max_{j \in N \mid a_{ij} > 0} (c_j / a_{ij})$$

is easily seen to be a valid upper bound on y_i, $\forall i$. The rows of A corresponding to the coefficient matrix of SP are by definition nonnegative, and the cutting planes to be generated can be shown to have only nonnegative coefficients when expressed in the structural variables x_j, $j \in N$. Thus the above bounds on the dual variables are valid.

The pair of dual linear programs can now be restated as

(P') $$\max_{0 \leq x \leq e} z(x)$$

and

(D') $$\min_{0 \leq y \leq \bar{y}} w(y)$$

where $z(x)$ and $w(y)$ are defined by

$L_P(x)$ $$z(x) = \min_{0 \leq \eta \leq \bar{y}} \{\eta(b - Ax) + cx\}$$

and

$L_D(y)$ $$w(y) = \max_{0 \leq \xi \leq e} \{(c - yA)\xi + yb\}$$

respectively.

For any given $x \in R^n$ and $y \in R^q$, all optimal solutions $\eta^*(x)$, $\xi^*(y)$ to the pair of linear programs $L_P(x)$, $L_D(y)$ satisfy

$$\eta_i^*(x) = \begin{cases} 0 & \text{if } a^i x < b_i \\ 0 \text{ or } \bar{y}_i & \text{if } a^i x = b_i \\ \bar{y}_i & \text{if } a^i x > b_i \end{cases}$$

and

$$\xi_i^*(y) = \begin{cases} 0 & \text{if } ya_j > c_j \\ 0 \text{ or } 1 & \text{if } ya_j = c_j \\ 1 & \text{if } ya_j < c_j \end{cases}$$

where a^i and a_j are the ith row and jth column of A, respectively. Clearly, $\eta^*(x)$ and $\xi^*(y)$ are trivially easy to compute for any given x and y.

Consider now the problem

PD
$$\min_{\substack{0 \leq x \leq e \\ 0 \leq y \leq \bar{y}}} v(x, y) = w(y) - z(x).$$

This problem is amenable to subgradient optimization; unlike other formulations, however, it has the advantage of a known optimal value, 0, which helps choosing an appropriate step length.

At any point $(\hat{x}, \hat{y})$, each pair of optimal solutions $\eta^* = \eta^*(\hat{x})$, $\xi^* = \xi^*(\hat{y})$, to $L_P(\hat{x})$ and $L_D(\hat{y})$ respectively, defines a subgradient $(s_x, s_y) \in R^n \times R^q$ of $v(x, y)$, given by

$$(s_x, s_y) = (\eta^* A - c, b - A\xi^*)$$

which is again trivially easy to compute; the convex hull of these subgradients for all optimal η^*, ξ^*, is

$$\partial v(x, y) = \text{conv} \bigcup_{\eta^*, \xi^*} (\eta^* A - c, b - A\xi^*)$$

the subdifferential of $v(x, y)$ at the point $(\hat{x}, \hat{y})$.

For any point $(x, y) \in R^n \times R^q$, let $P_F(x, y)$ denote the projection of (x, y) on the set

$$F = \{(x, y) \in R^n \times R^q \mid 0 \leq x \leq e, 0 \leq y \leq \bar{y}\}.$$

Projecting a point $(x, y) \notin F$ on F consists of simply replacing those components of (x, y) which exceed one of their bounds with the respective bound, while leaving the other components unchanged.

The procedure can now be summarized as follows.

Step 0 (Initialize Let A be the coefficient matrix of SPP, and $b = e$. Define F by computing the upper bounds $\bar{y}_i$ for each component of y. Start with any $(x^0, y^0) \in F$, set $i = 0$, and go to Step 1.

Step 1 Compute $v(x^i, y^i)$. If $v(x^i, y^i) < \varepsilon$, where $\varepsilon > 0$ is a given parameter, define $N_0 = \{j \in N \mid y^i a_j = c_j\}$ and go to Step 3. Otherwise go to Step 2.

Step 2 Compute a subgradient $(s_x^i, s_y^i) \in \partial v(x^i, y^i)$ and set

$$(x^{i+1}, y^{i+1}) = P_F[(x^i, y^i) - t^i(s_x^i, s_y^i)]$$

where the step length t^i is given by

$$t^i = \lambda \frac{v(x^i, y^i)}{\|s_x^i\|^2 + \|s_y^i\|^2}$$

for some λ, $0 < \lambda < 2$, and where $\|s\|$ is the Euclidean norm of s. Set $i \leftarrow i+1$ and go to Step 1.

Step 3 Search N_0 for a subset N_0' such that

$$\sum_{j \in N_0'} a_j \leq b$$

and equality holds for those rows h such that $y_h^i > 0$. If such N_0' exists, stop: x^* given by $x_j^* = 1$, $j \in N_0'$, $x_j^* = 0$ otherwise, is an optimal solution to SPP. Otherwise go to Step 4.

Step 4 Generate a set of cuts, expressed in the structural variables as

$$Dx \leq d$$

where $D = (d_{ij}) \geq 0$, $d > 0$. Redefine PD and F by replacing A and b with

$$\binom{A}{D} \quad \text{and} \quad \binom{b}{d}$$

respectively, and setting

$$\bar{y}_h = \max_{j \in N | d_{hj} > 0} (c_j / d_{hj})$$

for each new component of $\bar{y}$. Define the new components of y^i to be $y_h^i = \bar{y}_h$ for each new constraint h, and go to Step 1.

If all the cuts were present in the tableau at the start, the rate of convergence of the above procedure would be geometrical, as it is known to be in the case of the general class of algorithms to which it belongs. Since they are not, the actual convergence rate will, of course, depend on the number of cuts that have to be introduced. Convergence seems to depend crucially on the step length t^i. The value $\lambda = 1$ appears to give most of the time the best performance, though sometimes switching from $\lambda = 1$ to $\lambda = \frac{1}{2}$ at some stage in the process improves the convergence.

The main advantage of the symmetric procedure over the asymmetric subgradient method discussed in Held, Wolfe, and Crowder (1974) lies in the fact that the latter requires an estimate of the optimal objective function value for a proper choice of the step length. Experience shows that when the estimate is replaced by the actual value of the optimum, convergence is very fast. In the symmetric procedure such an estimate is not needed, since the function that one minimizes is known to have 0 as its optimal value. To compare the two procedures, an adaptation of the asymmetric method described by Held, Wolfe, and Crowder (1974) to LSPP was implemented, with three different estimates of the optimal objective function value, namely the true optimum, an overestimation of 20%, and one of 50%. Both the symmetric and the asymmetric procedure were run on 6 set partitioning linear programs, with 30 constraints, 70–90 variables, and density 0.07 (problems 9–14 of Salkin and Koncal, 1973). When the exact optimum was assumed known in the one-sided procedure, the two methods performed about equally. When the estimate was 20% off, for 4 of the 6 problems the one-sided procedure required between 1.3 and 2.1 times the number of iterations needed for the symmetric procedure, while for the remaining 2 problems it converged to the wrong value. When the estimate was 50% off, the asymmetric procedure converged to the right value for only one of the 6 problems. The average number of iterations required with the symmetric procedure to reach a value of $v(x, y)$ less than 1.0 was between 133 and 759, with an average of 299. The computational effort involved in an iteration of this procedure can be estimated (very roughly) to be about m times less than for a usual revised simplex iteration, where m is the number of constraints. These initial tests are of course insufficient to draw any conclusion, but the method certainly looks very promising.

7.3.6 Set Partitioning Via Node Covering

As discussed in Section 7.1.4, a set packing problem SP with coefficient matrix A is equivalent to the weighted node packing problem defined on the intersection graph G_A of A, where the weights are the same as in SP. On the other hand, the weighted node packing problem is equivalent to the weighted node covering problem, since the complement of any node packing is a node covering; if the node packing is of maximum weight, its complement must certainly be of minimum weight. Thus, one way of solving set packing and set partitioning problems is via solving the corresponding (weighted) node packing or node covering problems.

In this section we discuss one such method (see Balas and Samuelsson, 1973, 1974a). We first discuss the procedure for the unweighted node covering problem, then its extension to the weighted case.

The unweighted node covering problem can be stated as

NC $\qquad \min \{e_n x \mid A^T x \geq e_q, x_j = 0 \text{ or } 1, j \in N\}$

where $N = \{1, \ldots, n\}$, A is the $n \times q$ node–edge incidence matrix of a graph $G = (N, E)$, T means transpose, and e_n, e_q are vectors of ones of appropriate dimensions.

A *minimum partial cover* $\tilde{N}$ is a subset of nodes which is a minimum cover in the subgraph obtained from G by removing all edges not incident with $\tilde{N}$. For the purposes of this section, a *clique* will be defined as a set of pair wise adjacent nodes (i.e. a set of nodes inducing a complete subgraph; however, the set does not have to be maximal, and the term refers to the node set instead of the subgraph itself).

The algorithm starts with a minimum partial cover and uses a labelling procedure to increase the number of edges that are covered. This procedure ends after at most $n(q+1)$ steps. If all edges are covered, the solution is optimal. Otherwise the problem is partitioned, i.e. G is replaced by two proper subgraphs, and the procedure is applied to the latter.

The key concept behind the labelling procedure is that of a *dual node–clique set*. If K is the set of all cliques (in the above sense) in G, then $(\tilde{N}, \tilde{K})$, where $\tilde{N} \subset N$, $\tilde{K} \subset K$, is defined to be a dual node–clique set if

(i) the cliques of $\tilde{K}$ are pairwise node-disjoint;
(ii) each node in $\tilde{N}$ belongs to some clique in $\tilde{K}$;
(iii) each clique in $\tilde{K}$ contains exactly one node not in $\tilde{N}$.

It can be shown that if $(\tilde{N}, \tilde{K})$ is a dual node–clique set, then $\tilde{N}$ is a minimum partial cover, which defines a dual-feasible solution to the linear program obtained from NC by removing the integrality requirement and adding a constraint of the form

$$\sum_{j \in Q} x_j \geq |Q| - 1$$

for each clique in G. (When the cliques are maximal, this constraint is a facet—the complement of the corresponding clique-facet for the node packing or set packing problem, discussed in Section 7.2.1.)

These properties are used in the algorithm to perform implicitly, via a clique-labelling procedure, what in fact amounts to sequences of pivots in all-integer dual cutting plane method, where the cuts are the above inequalities. Dual feasibility is preserved by the fact that the partial covers generated by the labelling procedure are all associated with dual node–clique sets, i.e. are all minimal. A statement of the algorithm follows.

Step 0 Finding an initial solution Any edge matching (set of disjoint 2-cliques) can be used to generate an associated minimal partial node cover (and dual node–clique set). The larger the size of the edge-matching, the better the starting solution. An heuristic is used to find a good matching rapidly.

Let $(\tilde{N}, \tilde{K})$ be the current dual node–clique set ($\tilde{K}$ is the set of *labelled* cliques).

Step 1 Improving the solution (Labelling procedure) Scan E for edges not covered by $\tilde{N}$, and for each such edge (i, j) attempt to perform one of the following steps in order:

(a) (First labelling step) If neither i nor j belongs to a labelled clique (i.e. a clique in $\tilde{K}$), label the 2-clique $\{i, j\}$, and put into $\tilde{N}$ either i or j, whichever covers more new edges.

(b) (Reassignment step) If either i or j can replace in $\tilde{N}$ one of the members of the labelled clique to which it belongs without uncovering an edge, make the switch to cover (i, j).

(c) (Second labelling step) Find a largest unlabelled clique Q_* containing i and j, if it exists, such that $|Q_*| \geq 3$ and
 (i) if any $h \in Q_*$ belongs to some $Q \in \tilde{K}$, then $Q \subset Q_*$ or Q is a 2-clique;
 (ii) Q_* contains a labelled clique or a node $j \in \tilde{N}$.

If found, then

(α) label (put into $\tilde{K}$) Q_*, and 'unlabel' (remove from $\tilde{K}$) all labelled cliques contained in Q_* and labelled 2-cliques incident with Q_*;

(β) put into $\tilde{N}$ all but one of the nodes in Q_*, and remove from $\tilde{N}$ all nodes not in Q_*, belonging to labelled 2-cliques incident with Q_*. If, when no more applications of (a), (b), (c) are possible, the current $\tilde{N}$ is a cover, then it is optimal (for its subproblem). Otherwise go to Step 2.

Step 2 Partitioning (Branching and bounding) Choose $i_* \in \tilde{N}$ having a maximum number of adjacent nodes not in $\tilde{N}$, and partition the feasible set by

$$(x_{i_*} = 1) \bigvee [x_{i_*} = 0 \text{ and } x_j = 1, \forall j : (i_*, j) \in E].$$

This gives rise to two subproblems, whose associated graphs are the subgraphs of G induced by the node sets $N - \{i_*\}$ and $N - [\{i_*\} \cup \{j \in N \mid (i_*, j) \in E\}]$, respectively. For each of the two subgraphs, a dual node–clique set can be obtained from the dual node–clique set of the parent graph, by local modifications only.

Calculate a lower bound on the value of each subproblem (the cardinality of the minimal partial cover $\tilde{N}$ defined by the new dual node–clique set is one such bound; solving a structured linear program yields another, often stronger bound).

Select a subproblem and go to Step 1.

A first version of a computer code implementing the above algorithm was tested on 9 problems with 50 nodes and 64–329 edges that were randomly generated by Trotter (1973) for testing his implicit enumeration algorithm for the same problem. For 6 of the 9 problems (the sparser ones) the number of partitions required was between 1 and 7, while for the remaining 3 problems (the denser ones) it was between 81 and 102. The results for the sparser problems is needed before any conclusions can be drawn.

Set Partitioning—A Survey

The algorithm for the weighted node covering problem (Balas and Samuelsson, 1974a) is very similar in spirit to the one just described above. It is based on the equivalence of the weighted problem to an unweighted problem on an associated graph. If the weighted problem, with positive integral weight-vector c, is defined on a graph $G = (N, E)$, the node set of the associated graph $G(c) = [N(c), E(c)]$ contains c_i copies $i_1, \ldots, i_{c_i}$ of each node $i \in N$, i.e.

$$N(c) = \bigcup_{i \in N} N_i$$

where $N_i = \{i_1, \ldots, i_{c_i}\}$, while the edge set of $G(c)$ contains an edge for each pair of nodes i_k, j_l in $N(c)$ that are copies of nodes $i, j \in N$ connected by an edge of E; i.e.

$$E(c) = \{(i_k, j_l) \mid (i, j) \in E\}.$$

It can then be shown that the weighted node covering problem on G is equivalent to the unweighted node covering problem on $G(c)$, in the sense that for each optimal solution to one, there is an optimal solution with the same value to the other. This would make it possible to apply the algorithm discussed above to the unweighted problem defined on $G(c)$, but the latter problem is unwieldy, and one can do better. Indeed, the steps of the above algorithm on $G(c)$ can be carried out implicitly by a labelling procedure that operates only on the original graph G. This is accomplished via the concept of a *weighted dual node–clique set* in G, defined to be a pair $(\hat{N}, \hat{K}) \subseteq (N, K)$ (where K, as before, is the set of all cliques in G, while $\hat{K}$ is a subset of labelled cliques), with an associated positive weight-vector w, such that

(i) $\hat{N}$ contains exactly one node i_Q of each labelled clique $Q \in \hat{K}$ ($i_{Q_1} = i_{Q_2}$ not excluded);
(ii) w has exactly one component for each $Q \in \hat{K}$ and

$$\sum_{Q \in K \mid i \in Q} w(Q) \leq c_i \quad \forall i \in N.$$

These weighted dual node–clique sets can be shown to have properties similar to their unweighted counterparts. In particular, one can associate with each such set a soluton x to the linear programming relaxation of the weighted NC, for which there exists an easily computable function $f(x)$ such that if $f(x) = 0$ then x is integer and optimal. This makes it possible to transform a given weighted dual node–clique set into a 'better' one with the same properties (i) and (ii), by a labelling procedure on G. As in the unweighted case, this procedure terminates in a polynomially bounded number of steps, after which, if the terminating solution is not optimal, the problem is partitioned. Strong bounds on the resulting subproblems can be obtained via a procedure proposed by Samuelsson (1974a) which uses Lagrangian duality together with a relaxation of NS to a network flow problem. Computational experience with the weighted version of the node covering algorithm is not yet available.

References

Andrew, G., T. Hoffman, and C. Krabek (1968). On the Generalized Set Covering Problem. *CDC, Data Centers Division, Minneapolis.*

Balas, E. (1964). Un algorithme additif pur la résolution des programmes linéaires a variables bivalentes. *Comptes Rendus de l'Acadmie des Sciences, Paris,* **258,** 3817–3820.

Balas, E. (1965). An Additive Algorithm for Solving Linear Programs With Zero-One variables. *Operations Research,* **13,** 517–546.

Balas, E. (1974a). Intersection Cuts from Disjunctive Constraints. *MSRR No.* 330, *Carnegie-Mellon University, February* 1974.

Balas, E. (1974b). Disjunctive Programming: Properties of the Convex Hull of Feasible Points. *MSRR No.* 348, *Carnegie-Mellon University, July* 1974.

Balas, E. (1975a). Facets of the Knapsack Polytope. *Mathematical Programming,* **8,** 146–164.

Balas, E. (1975b). Disjunctive Programming: Cutting Planes from Logical Conditions. In O. L. Mangasarian, R. R. Meyer, and S. M. Robinson (Eds), *Nonlinear Programming,* Vol. 2, Academic Press, New York. pp. 279–312.

Balas, E. (1975c). Some Valid Inequalities for the Set Partitioning Problem. *MSRR No.* 368, *Carnegie-Mellon University, July* 1975. (*Revised January* 1976.)

Balas, E. (1976). A Disjunctive Cut for Set Partitioning. *W.P.* 57–75–76, *Carnegie-Mellon University, January* 1976.

Balas, E., R. Gerritsen, and M. W. Padberg (1974). An All-Binary Column Generating Algorithm for Set Partitioning. Paper presented at *ORSA-TIMS, Boston, April* 1974.

Balas, E. and R. G. Jeroslow (1975). Strengthening Cuts for Mixed Integer Programs. *MSRR No.* 359, *Carnegie-Mellon University, February* 1975.

Balas, E. and M. W. Padberg (1972a). On the Set Covering Problem. *Operations Research,* **20,** 1152–1161.

Balas, E. and M. W. Padberg (1973). Adjacent Vertices of the Convex Hull of Feasible 0–1 Points. *MSRR No.* 298, *Carnegie-Mellon University, November–April* 1973.

Balas, E. and M. W. Padberg (1975). On the Set Covering Problem, II. An Algorithm for Set Partitioning. *Operations Research,* **23,** 74–90.

Balas, E. and H. Samuelsson (1973). Finding a Minimum Node Cover in an Arbitrary Graph. *MSRR No.* 325, *Carnegie-Mellon University, November* 1973.

Balas, E. and H. Samuelsson (1974a). Finding a Minimum Node Cover in an Arbitrary Graph. II: The Weighted Case. *MSRR No.* 336, *Carnegie-Mellon University, April* 1974.

Balas, E. and H. Samuelsson (1974b). A Symmetric Subgradient Cutting Plane Method for Set Partitioning. *W.P.* 5–74–75, *Carnegie-Mellon University August.* 1974.

Balas, E. and E. Zemel (1974). Facets of the Knapsack Polytope from Minimal Covers. *MSRR No.* 352, *Carnegie-Mellon University, December* 1974.

Balas, E. and E. Zemel (1975). All the Facets of 0–1 Programming Polytopes with Positive Coefficients. *MSRR No.* 374, *Carnegie-Mellon University, October* 1975.

Balas, E. and E. Zemel (1976a). Graph Substitution and Set Packing Polytopes. *MSRR No.* 384, *Carnegie-Mellon University, January* 1976.

Balas, E. and E. Zemel (1976b). Critical Cutsets of Graphs and Canonical Facets of Set Packing Polytopes. *MSRR No.* 385, *Carnegie-Mellon University, February* 1976.

Balinski, M. L. (1969). Labelling to Obtain a Maximum Matching. In R. C. Bose and T. A. Dowling (Eds), *Combinatorial Mathematics and Its Applications*, University of North Carolina Press. pp. 585–601.

Balinski, M. L. (1970). On Maximum Matching, Minimum Covering and Their Connections, In H. W. Kuhn (Ed.), *Proceedings of the Princeton Symposium on Mathematical Programming*, Princeton University Press, Princeton, N.J.

Balinski, M. L. (1972). Establishing the matching Polytope. *Journal of Combinatorial Theory*, **13**, 1–13.

Berge, C. (1970). *Graphes et Hypergraphes*, Dunod, Paris. (English translation North Holland, Amsterdam, 1973.)

Berge, C. (1961). Färbung von Graphen, deran sämtliche, beur. deren ungerade Kreise starr sind. Zusammenfassung Wiss. Z. Martin-Luther-Univ., Halle-Wittenberg. *Math.–Natur. R*; 119.

Berge, C. (1970). *Graphes et Hypergraphes*, Dunod, Paris. (English translation North Holland, Amsterdam, 1973.)

Berge, C. (1972). Balanced Matrices. *Mathematical Programming*, **2**, 19–31.

Christofides, N. and S. Korman (1973). A Computational Survey of Methods for the Set Covering Problem. *Report No.* 73/2, *Imperial College of Science and Technology, April* 1973.

Christofides, N. (1975). *Graph Theory; An Algorithmic Approach*, Academic Press, New York.

Chvátal, V. (1972). On Certain Polytopes Associated with Graphs. *CRM*-238, *Universite de Montreal, October* 1972. Forthcoming in *Journal of Combinatorial Theory*.

Dantzig, G. B. (1963). *Linear Programming and Extensions*, Princeton University Press, Princeton, N.J.

Délorme, J. (1974). Contribution à la résolution du problème de recouvrement: méthodes de troncatures. *Thèse de Docteur Ingénieur, Université Paris VI.*

Délorme, J. and E. Heurgon (1975). Problèmes de partionnement: exploration arborescente ou méthodes de troncatures?, *Revue Francaise d'Automatique, Informatique et Recherche Opérationnelle*, **9**, V-2, 53–65.

Edmonds, J. (1962). Covers and Packings in a Family of Sets. *Bulletin of the American Mathematical Society*, **68**, 494–499.

Edmonds, J. (1965a). Maximum Matching and a Polyhedron with 0, 1 Vertices. *Journal of Research of the National Bureau of Standards*, **69B**, 125–130.

Edmonds, J. (1965b). Paths, Trees and Flowers. *Canadian Journal of Mathematics*, **17**, 449–467.

Fulkerson, D. R. (1971). Blocking and Anti-Blocking Pairs of Polyhedra. *Mathematical Programming*, **1**, 168–194.

Fulkerson, D. R. (1973). On the Perfect Graph Theorem. In T. C. Hu and S. M. Robinson (Eds), *Mathematical Programming*, Academic Press, New York.

Gallai, T. (1958). Über Extreme Punkt-und Kantenmengen. *Ann. Univ. Sci. Budapest, Eötvös, Sect. Math.*, **2**, 133–138.

Garfinkel, R. S. and G. L. Nemhauser (1969). The Set Partitioning Problem: Set Covering with Equality Constraints. *Operations Research*, **17,** 848–856.

Garfinkel, R. S. and G. L. Nemhauser (1972). *Integer Programming*, Wiley, New York.

Geoffrion, A. M. (1967). Integer Programming by Implicit Enumeration and Balas' Method. *SIAM Review*, **7,** 178–190.

Glover, F. (1965). A Multiphase Dual Algorithm for the Zero–One Integer Programming Problem. *Operations Research*, **13,** 94–120.

Glover, F. (1975). Polyhedral Annexation in Mixed Integer Programming. *Mathematical Programming*, **9.**

Glover, F. and D. Klingman, (1973a). The Generalized Lattice Point Problem. *Operations Research*, **21,** 141–156.

Glover, F. and D. Klingman (1973b). Improved Convexity Cuts for Lattice Point Problems. *CS133, University of Texas, April 1973*.

Gomory, R. E. (1963a). An Algorithm for Integer Solutions to Linear Programs. In R. L. Graves and P. Wolfe (Eds), *Recent Advances in Mathematical Programming*, McGraw-Hill, New York.

Gomory, R. E. (1963b). All-Integer Integer Programming Algorithm. In J. F. Muth and G. L. Thompson (Eds), *Industrial Scheduling*, Addison-Wesley, New York.

Gondran, M. (1973). Un outil pour la programmation en nombres entiers: la méthode des congruences décroissantes. *RAIRO*, **3,** 35–54.

Gondran, M. and J. L. Laurière (1974). Un algorithme pour le problème de partitionnement. *RAIRO*, **8,** 27–70.

Graves, G. W. and A. B. Whinston (1968). A new Approach to Discrete Mathematical Programming. *Management Science*, **15,** 177–190.

Hammer, P. L., E. L. Johnson, and U. N. Peled (1975). Facets of Regular 0–1 Polytopes. *Mathematical Programming*, **8,** 179–206.

Hammer, P. L., E. L. Johnson and U. N. Peled (1974). The Role of Master Polytopes in the Unit Cube. *CORR 74-25, University of Waterloo, October 1974*.

Harary, F. (1969). *Graph Theory*, Addison-Wesley, New York.

Held, M. and R. M. Karp (1971). The Traveling-Salesman Problem and Minimum Spanning Trees: Part II. *Mathematical Programming*, **1,** 6–25.

Held, M., P. Wolfe, and H. D. Crowder (1974). Validation of Subgradient Optimization. *Mathematical Programming*, **6,** 62–88.

Heurgon, E. (1972). Un problème de recouvrement exact: l'habillage des horaires d'une ligne d'autobus. *RAIRO*, **6,** p. 177.

Jeroslow, R. G. (1974). Principles of Cutting Plane Theory: Part I. *Carnegie-Mellon University, February 1974*.

Johnson, E. L. (1974). A Class of Facets of the Master 0–1 Knapsack Polytope. *IBM, Watson Research Center, November 1974*.

Lemke, C. E. H. M. Salkin, and K. Spielberg (1971). Set Covering by Single Branch Enumeration with Linear Programming Subproblems. *Operations Research*, **19,** 998–1022.

Lovász L. (1972). Normal Hypergraphs and the Perfect Graph Conjecture. *Discrete Mathematics*, **2,** 253–267.

Marsten, R. E. (1974). An Algorithm for Large Set Partitioning Problems. *Management Science*, **20,** 779–787.

Martin, G. T. (1963). An Accelerated Euclidean Algorithm for Integer Programming. In R. L. Graves and P. Wolfe (Eds), *Recent Advances in Mathematical Programming*, Wiley, New York.

Martin, G. T. (1969). Gomory Plus Ten. Paper presented at the ORSA Meeting in Miami, November 1969 (Unpublished).

Michaud, P. (1972). Exact Implicit Enumeration Method for Solving the Set Partitioning Problem. *IBM Journal of Research and Development*, **16**, 573–578.
Ming-Te Lu (1970). A Computerized Airline Crew Scheduling System. *Ph.D. Thesis, University of Minnesota.*
Nemhauser, G. L. and L. E. Trotter (1974). Properties of Vertex Packing and Independence System Polyhedra. *Mathematical Programming*, **6**, 48–61.
Nemhauser, G. L. L. E. Trotter, and R. M. Nauss (1972). Set Partitioning and Chain Decomposition. *Technical Report No. 161, Cornell University.*
Owen, G. (1973). Cutting Planes for Programs with Disjunctive Constraints. *Journal of Optimization Theory and Applications*, **11**, 49–55.
Padberg, M. W. (1972). Essays in Integer Programming. *Ph.D. Thesis, Carnegie-Mellon University, May 1971.*
Padberg, M. W. (1973a). On the Facial Structure of Set Packing Polyhedra. *Mathematical Programming*, **5**, 199–215.
Padberg, M. W. (1973b). A Note on Zero–One Programming. *Operations Research*, **23**, 833–837.
Padberg, M. W. (1974a). Perfect Zero–One Matrices. *Mathematical Programming*, **6**, 180–196.
Padberg, M. W. (1974b). Perfect Zero–One Matrices—II. *Proceedings in Operations Research*, **3**, Physica-Verlag, Wurzburg-Wien, 75–83.
Padberg, M. W. (1974c). Characterizations of Totally Unimodular, Balanced and Perfect Matrices. In B. Roy (Ed.), *Combinatorial Programming: Methods and Applications*, D. Reidel, Dordrecht/Boston. pp. 275–284.
Padberg, M. W. (1975a). Almost Integral Polyhedra Related to Certain Combinatorial Optimization Problems. *Working paper No. 75-25, GBA, New York University.* Forthcoming *Linear Algebra and its Applications.*
Padberg, M. W. (1975b). On the Complexity of Set Packing Polyhedra. *Working paper No. 75-105, GBA, New York University, New York.*
Padberg, M. W. and M. R. Rao (1973). The Travelling Salesman Problem and a Class of Polyhedra of Diameter Two. *IIM Preprint NO. I/73-5, International Institute of Management, Berlin.* Forthcoming in *Mathematical Programming.*
Pierce, J. F. (1968). Application of Combinatorial Programming to a Class of All Zero–One Integer Programming Problems. *Management Science*, **15**, 191–209.
Pierce, J. F. and J. S. Lasky (1973). Improved Combinational Programming Algorithms for a Class of All Zero–One Integer Programming Problems. *Management Science*, **19**, 528–543.
Pollatschek, M. A. (1970). Algorithms on Finite Weighted Graphs. *Ph.D. thesis, Technion, Haifa (in Hebrew, with an English summary).*
Pulleyblank, W. and J. Edmonds (1973). Facets of 1-Matching Polyhedra. *CORR 73-3, University of Waterloo, March 1973.*
Roy, B. (1969). *Algèbre moderne et Théorie des Graphes, I*, Dunod, Paris.
Roy, B. (1970). *Algèbre moderne et Théorie des Graphes, II*, Dunod, Paris.
Salkin, H. M. and R. D. Koncal (1973). Set Covering by an All Integer Algorithm: Computational Experience. *ACM Journal*, **20**, 189–193.
Samuelsson, H. (1974a). Integer Programming Duality for Set Packing–Partitioning Problems. *MSRR No. 337, Carnegie-Mellon University, March 1974.*
Samuelsson, H. (1974b). A Symmetric Ascent Method for Finitely Regularizable Linear Programs. *W.P. 6–74/75, Carnegie-Mellon University, August 1974.*
Simmonard, M. (1966). *Linear Programming*, Prentice-Hall, Englewood Cliffs, N.Y.
Trotter, L. E. (1973). Solution Characteristics and Algorithms for the Vertex Packing Problem. *Technical report No. 168, Cornell University.*
Trotter, L. E. (1974). A Class of Facet Producing Graphs for Vertex Packing

Polyhedra. *Technical Report No. 78, Yale University*, February 1974.
Trubin, V. A. (1969). On a Method of Solution of Integer Linear Programming Problems of A Special Kind. *Soviet Math. Dokl.*, **10**, 1544–1596.
Weyl, H. (1935). Elementare Theorie der konvexen Polyeder. *Comm. Math. Helv.*, **7**, 290–306. (Translated in *Contributions to the Theory of Games*, Vol. I, pp. 3–18, Annals of Mathematics Studies, No. 24, Princeton.)
Wolsey, L. A. (1975). Faces for Linear Inequalities in Zero–One Variables. *Mathematical Programming*, **8**, 165–178.
Wolsey, L. A. (1974). Facets and Strong Valid Inequalities for Integer Programming. *Core, Heverlee, Belgium April* 1974.
Zemel, E. (1974). Lifting the facets of 0–1 polytopes. *MSRR 354, Carnegie-Mellon University, December* 1974.
Zwart, P. B. (1972). Intersection Cuts for Separable Programming. *Sch. of Eng. Appl. Sci., Washington University, St. Louis, Missouri, January* 1972.

Appendix. A Bibliography on Applications

This bibliography lists articles, research reports, and a few books on applications of the set partitioning (and in some cases, the set covering) model, by area of applications, in chronological order within each area.

Crew Scheduling (Airline, Railroad, etc.)

1. Charnes, A. and M. H. Miller (1956). A Model for Optimal Programming of Railway Freight Train Movements. *Man. Sci.*, **3**, 74–92.
2. McCloskey, J. F. and F. Hanssman (1957). An Analysis of Stewardess Requirements and Scheduling for a Major Airline. *Naval Res. Log. Quart.*, **4**, 183–192.
3. Evers, G. H. E. (1965). Relevant Factors Around Crew-Utilization. *AGIFORS Symposium, KLM.*
4. Steiger, F. (1965). Optimization of Swissair's Crew Scheduling by an Integer Linear Programming Model. *Swissair O.R. SDK* 3.3.911.
5. Kolner, T. N. (1966). Some Highlights of a Scheduling Matrix Generator System. *United Airlines.*
6. Niederer, M. (1966). Optimization of Swissair's Crew Scheduling by Heuristic Methods Using Integer Linear Programming Models. *AGIFORS Symposium, September* 1966.
7. Agard, J. (1966). Monthly Assignment of Stewards. *Air France, AGIFORS Symposium, Killarney.*
8. Arabeyre, J. P. (1966). Methods of Crew Scheduling. *AGIFORS, Air France.*
9. Moreland, J. A. (1966). Scheduling of Airline Flight Crews. *M.S. Thesis, M.I.T.*
10. Agard, J. J. P. Arabeyre, and J. Vautier (1967). Génération Automatique de Rotations d'équipages. *RIRO I-6*, 107–117.
11. Steiger, F. and M. Niederer (1968). Scheduling Air Crews by Integer Programming. Presented at *IFIP Congress, Edinburg.*
12. Thiriez, H. M. (1968). Implicit Enumeration Applied to the Crew Scheduling Algorithm. *Department of Aeronautics, M.I.T.*
13. Azabeyte, J. P. J. Feaznley, F. Steiger, and W. Teather (1969). The Air Crew Scheduling Problem: A Survey. *Trans. Sci.*, **3**, 140–163.
14. Thiriez, H. (1969). Airline Crew Scheduling—A Group Theoretic Approach. *Ph.D. Dissertation, M.I.T.*
15. Spitzer, M. (1971). Solutions to the Crew Scheduling Problem. *AGIFORS Symposium, October* 1971.

Airline Fleet Scheduling

16. Levin, A. (1969). Fleet Routing and Scheduling Problems for Air Transportations Systems. *Ph.D. Dissertation, M.I.T.*

Truck Delivery

17. Dantzig, G. B. and J. H. Ramser (1960). The Truck Dispatching Problem. *Man. Sci.*, **6**, 80–91.
18. Balinski, M. L. and M. H. Quandt (1964). On an Integer Program for a Delivery Problem. *Opns. Res.*, **12**, 300–304.
19. Clarke, G. and S. W. Wright (1964). Scheduling of Vehicles from a Central Depot to a Number of Delivery Points. *Opns. Res.*, **12**, 4, 568–581.
20. Pierce, J. F. (1968). Application of Combinatorial Programming Algorithms for a Class of All Zero–One Integer Programming Problems. *Man. Sci.*, **15**, 191–209.

Stock Cutting

21. Pierce, J. F. (1970). Pattern Sequencing and Matching in Stock Cutting Operations. *Tappi*, **53**, 4, 668–678.

Line and Capacity Balancing

22. Salveson, M. E. (1955). The Assembly Line Balancing Problem. *Jour. of Indus. Eng.*, **6**, 3, 519–526.
23. Freeman, D. R. and J. V. Jucher (1967). The Line Balancing Problem. *Journal of Industrial Engineering*, **18**, 361–364.
24. Steinman, H. and R. Schwinn (1969). Computational Experience with a Zero–One Programming Problem. *Opns. Res.*, **17**, 5, 917–920.

Facility Location

25. Revelle, C., D. Marks, and J. C. Liebman (1970). An Analysis of Private and Public Sector Location Models. *Man. Sci.*, **16**, 12, 692–707.
26. Toregas, C., R. Swain, C. Revelle, and L. Bergman (1971). The Location of Emergency Service Facilities. *Opns. Res.*, **19**, 1363–1373.
26a. Garfinkel, R. S., A. W. Neebe, and M. R. Rao (1974). The m-Center Problem: Bottleneck Facility Location. *Working paper NO. 7414, Graduate School of Management, University of Rochester, Rochester, N.Y.*

Capital Investment

27. Valenta, J. R. (1969). Capital Equipment Decisions: A Model for Optimal Systems Interfacing. *M.S. Thesis, M.I.T.*

Switching Current Design and Symbolic Logic

28. Roth, J. P. (1950). Algebraic Topological Methods for the Synthesis of Switching Systems—I. *Trans. Amer. Math. Soc.*, **88**, 301–326.
29. Quine, W. V. (1955). A Way to Simplify Truth Functions. *Am. Math. Mon.*, **62**, 627–631.
30. McCluskey, E. J., Jr. (1956). Minimization of Boolean Functions. *Bell System Tech. Journal*, **35**, 1412–1444.

31. Petrick, S. R. (1956). A Direct Determination of the Redundant Forms of a Boolean Function from the Set of Prime Implicants. *AFCRC-TR*-56-110, *Air Force Cambridge Research Centre.*
32. Paul, M. C. and S. H. Unger (1959). Minimizing the Number of States in Incompletely Specified Sequential Functions. *IRE Trans. on Electronic Computers*, Ec-8, 356–367.
33. Pyne, I. B. and E. J. McCluskey, Jr. (1961). An Essay on Prime Implicant Tables. *Siam J.*, **9,** 604–631.
34. Cobham, A., R. Fridshal, and J. H. North (1961). An Application of Linear Programming to the Minimization of Boolean Functions. *Res. Rep. RC-472, IBM.*
35. Cobham, A., R. Fridshal, and J. H. North (1962). A Statistical Study of the Minimization of Boolean Functions Using Integer Programming. *Res. Rep. R.C.-756, IBM.*
36. Cobham, A. and J. H. North (1963). Extension of the Integer Programming Approach to the Minimization of Boolean Functions. *Res. Rep. R.C.-915, IBM.*
37. Root, J. C. (1964). An Application of Symbolic Logic to a Selection Problem. *Opns. Res.*, **12,** 4, 519–526.
38. Balinski, M. L. (1965). Integer Programming: Methods, Uses, Computation. *Man. Sci.*, **12,** 3, 253–313.
39. Gimpel, J. F. (1965). A Reduction Technique for Prime Implicant Tables. *IEEE Trans. on Electronic Computers*, **EC-14,** 535–541.

Information Retrieval

40. Day, R. H. (1965). On Optimal Extracting from a Multiple File Data Storage System: An Application of Integer Programming. *Opns. Res.*, **13,** 3, 489–494.

Marketing

41. Shanker, R. J., R. K. Turner, and A. A. Zoitners (1972). Integrating the Criteria for Sales Force Allocation: A Set-Partitioning Approach. *Working paper #48-72-3, CSIA, C,M.U.*

Political Districting

42. Garfinkel, R. S. (1968). Optimal Political Districting. *Ph.D. Dissertation, Johns Hopkins University.*
43. Wagner, W. H. (1968). An Application of Integer Programming to Legislative Redistricting. Presented at *34th National Meeting of ORSA.*
44. Garfinkel, R. S. and G. L. Nemhauser (1970). Optimal Political Districting by Implicit Enumeration Techniques. *Man. Sci.*, **16,** B495–B508.

CHAPTER 8

The Graph-Colouring Problem

SAMUEL M. KORMAN
Operational Research Unit, Department of Industry, London

8.1 Introductory Aspects

8.1.1 Introduction to Graph Colouring

This chapter deals with the problem of colouring the vertices of a general graph using the fewest possible number of colours, subject to the condition that no two adjacent vertices are to be of the same colour. This problem is known as the graph-colouring problem (GCP), and although various heuristic procedures exist for its approximate 'solution' (e.g. Matula, Marble, and Issacson, 1974; Wood 1969) we consider here only exact algorithms which guarantee optimal solution.

8.1.1.1 *Applications of the Graph-Colouring Problem*

As a means of introducing the GCP and as a simple but typical example of its application, consider a situation where it is desired to schedule school examinations into time periods, where the particular courses chosen by any one student throughout the year involve a wide range of subjects. Then it is necessary that no two examinations be scheduled into the same time period if there is any student taking both of the corresponding courses. Although one could, for example, schedule each examination into a different time period, it is often desired to determine a schedule involving the fewest time periods into which the examinations can take place. (Neufield and Tartar, 1974; Welsh and Powell, 1967; Williams, 1968; Wood, 1969).

For example, suppose seven subjects are offered, call them $x_1, x_2, x_3, x_4, x_5, x_6, x_7$, of which two or three must be chosen, and that, of the 35, say, students, eight take x_2, x_4, and x_7, four take x_6 and x_7, one takes x_1 and x_2, two take x_1, x_2, and x_6, eleven take x_1, x_5, and x_6, eight take x_1, x_3, and x_5, and one takes x_3 and x_4; where it is desired to establish a minimum-period examination timetable.

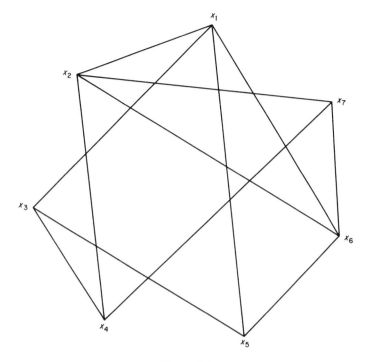

Figure 8.1

The above information can be represented in the form of a graph G by letting the vertices correspond to subjects, and joining by an arc any two vertices representing subjects which are jointly taken by one or more students. The graph G thus obtained can be represented as in Figure 8.1.

The problem of scheduling the examinations into the fewest time periods can then be solved by partitioning the set of vertices $\{x_1, x_2, x_3, x_4, x_5, x_6, x_7\}$ into the smallest number of disjoint subsets such that no two vertices within any subset are joined by an arc, i.e. by solving the colouring problem for the graph G, where a 'colour' corresponds to a time period. In the above example, the minimum number of time periods is in fact four, with many different optimal schedules, a typical one being $\{x_1, x_4\}$; $\{x_2, x_5\}$; $\{x_3, x_6\}$; $\{x_7\}$.

Other applications of the GCP include resource allocation (Christofides, 1975), loading problems (Eilon and Christofides, 1971), and wiring of printed circuits (Liu, 1968).

Although the reader is assumed to have knowledge of the basic graph-theoretic definitions, we next define some colour-related concepts which will be used in this chapter.

The Graph-Colouring Problem

8.1.1.2 Colouring-related Definitions and Elementary Consequences

An *independent set* of a graph $G = (X, \Gamma)$ is a set of vertices of G, no two of which are adjacent. An independent set $X' \subset X$ is a *maximal independent set*, denoted MIS, if there is no independent set X'' with $X' \subset X'' \subseteq X$.

A *completely connected set* of $G = (X, \Gamma)$ is a set of vertices of G, each pair of which is adjacent. A maximal completely connected set is called a *clique*. It is seen that a clique of G is an MIS of $\bar{G}$, its complement graph, and vice versa.

A *colouring* of a graph G is a manner of assigning colours to all the vertices of G so that adjacent vertices are not of the same colour. This is done by defining a *colouring function* $C(\cdot)$ on the vertices of G, where $C(x_i) = j$ if and only if vertex x_i is coloured with colour j.

A graph is said to be *r-colourable* if it can be coloured with r (or less) colours. The *chromatic number* of a graph G, denoted by $\gamma(G)$, is the minimum number of colours required to colour G. If $\gamma(G) = q$ then G is said to be *q-chromatic*. (Thus if G is q-chromatic, it is r-colourable $\forall r \geq q$).

The *independence number of G*, denoted $\alpha(G)$, is the number of vertices in the largest MIS of G. The *clique-number of G*, denoted $\omega(G)$, is the number of vertices in the largest clique of G. (Obviously $\alpha(G) = \omega(\bar{G})$ and vice versa.)

The following results are then immediate consequences of the above definitions.

(i) Since, in any colouring of graph $G = (X, \Gamma)$, no identically coloured vertices can be adjacent, all vertices of any clique must be assigned different colours. Thus $\gamma(G) \geq \omega(G)$. Note that this bound can be a very poor bound as can be seen from the following theorem.

Theorem 8.1 (Mycielski, 1955) *For any $q \geq 0$, there exists a graph G with $\omega(G) = 2$ for which $\gamma(G) > q$.*

(ii) Any set of identically coloured vertices must be an independent set, hence $\gamma(G) \geq n/\alpha(G)$, where $n = |X|$ is the number of vertices of G.

(iii) No clique, Q, of G can have more than one vertex in common with any MIS, M, of G; i.e. $|Q \cap M| \leq 1$.

8.1.1.3 Some Known Results on Colouring

In addition to the two simple bounds above, there have been other bounds established for the chromatic number of graph $G = (X, \Gamma)$ with n vertices and m arcs, which we list below without proof (Brooks, 1941; Berge, 1973)

(i) $\gamma(G) \geq n^2/(n^2 - 2m)$
(ii) $\gamma(G) \leq 1 + (2m(n-1)/n)^{\frac{1}{2}}$
(iii) $\gamma(G) \leq \max_{x_i \in X}\{d(x_i)\} + 1$, where $d(x_i)$ is the degree of vertex x_i.

Finally, we quote two well-known results on graph colouring.

Theorem 8.2 (*Konig*, 1950) *A graph is 2-colourable if and only if it contains no cycles of odd length.*

Theorem 8.3 (*Heawood*, 1890) *Any planar graph is 5-colourable.*

Indeed, it has recently been shown (Appel and Haken, 1976) that the famous Four-Colour Conjecture, which states that all planar graphs are 4-colourable, is also true.

8.1.2 Various Formulations of the Graph-Colouring Problem

Like many problems in computational graph theory, the GCP can be formulated in a number of ways as a general (0, 1) integer programming (IP) problem (e.g. Christofides, 1975). However, such a formulation does not lead to a practicable method for solving the GCP since the number of variables necessary for even moderately sized graphs is of the order nq (where n is the number of vertices and q the chromatic number of the graph) and so is too large to be efficiently solved as a general (0, 1) IP. As a result, other formulations as (0, 1) IP's of a more specific type have been proposed for the GCP.

8.1.2.1 *Formulation as a Set Partitioning/Covering Problem*

As was seen above, in any feasible colouring of G, the set of vertices which are coloured identically must be an independent set of G. This leads naturally to the following formulation of the GCP as a set partitioning problem.

Suppose all independent sets $S_1, S_2, \ldots, S_t$ of G have been enumerated and define the 0–1 integer variables w_j, $j = 1, \ldots, t$, and e_{ij}, $i = 1, \ldots, n$, $j = 1, \ldots, t$, respectively as

$w_j = 1$ if set S_j is part of a feasible colouring of G, i.e.
    if S_j represents all the vertices of a certain colour
= 0 otherwise
$e_{ij} = 1$ if vertex x_i is in set S_j
= 0 otherwise.

The Graph-Colouring Problem

Then it is easily seen that the GCP is equivalent to the IP:

$$\min \quad z = \sum_{j=1}^{t} w_j$$

$$\text{s.t.} \quad \sum_{j=1}^{t} e_{ij} w_j = 1 \quad i = 1, \ldots, n$$

$$w_j = 0, 1 \quad j = 1, \ldots, t.$$

In this formulation, too, the number of variables (being equal to the number of independent sets of the graph G) is again quite large, although, due to its specific structure, it does lead to a more efficient solution than the general IP above. However, by using a result mentioned in a later section, the size of the set partitioning problem above can be reduced by considering only MIS of G. Although in this case some vertices may be 'over-coloured' by being contained in more than one MIS chosen in the colouring of G, in practice one would then make an (arbitrary) choice of one of these possible colours for any such vertex.

More precisely, this formulation of the GCP as a unicost version of the well-known set covering problem (SCP) (see Chapter 7 of this book) is as follows.

Suppose that $M_1, M_2, \ldots, M_p$ represent the family of MIS of G, and for each M_j define the 0–1 variable y_j as:

$y_j = 1$ if M_j is present in the optimal colouring of G

$= 0$ otherwise.

Again with

$e_{ij} = 1$ if $x_i \in M_j$

$= 0$ otherwise

we have the following program:

$$\min \quad z = \sum_{j=1}^{p} y_j$$

$$\text{s.t.} \quad \sum e_{ij} y_j \geq 1 \quad i = 1, \ldots, n$$

$$y_j = 0, 1 \quad j = 1, \ldots, p$$

where the inequalities correspond to possible over-colourings.

Thus for the example of Figure 8.1, there are 9 MIS, viz. $M_1 = \{x_1, x_4\}$, $M_2 = \{x_1, x_7\}$, $M_3 = \{x_2, x_3\}$, $M_4 = \{x_2, x_5\}$, $M_5 = \{x_3, x_6\}$, $M_6 = \{x_3, x_7\}$, $M_7 = \{x_4, x_5\}$, $M_8 = \{x_4, x_6\}$, $M_9 = \{x_5, x_7\}$, so that $p = 9$ with the matrix $[e_{ij}]$ being as shown in Figure 8.2.

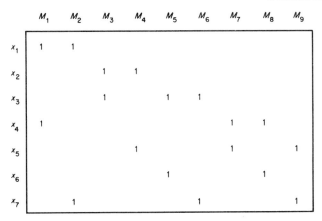

Figure 8.2

We note that the optimum solution given earlier can be derived from any one of the following three solutions to the SCP.
(a) $y_1 = y_2 = y_4 = y_5 = 1$; $y_3 = y_6 = y_7 = y_8 = y_9 = 0$
(b) $y_1 = y_4 = y_5 = y_6 = 1$; $y_2 = y_3 = y_7 = y_8 = y_9 = 0$
(c) $y_1 = y_4 = y_5 = y_9 = 1$; $y_2 = y_3 = y_6 = y_7 = y_8 = 0$

these corresponding respectively to the over-colouring of vertices x_1, x_3, and x_5.

This latter formulation is considerably better than those mentioned earlier, and although it involves the additional problem of determining all the MIS of a graph, it is a viable method of solution for the GCP, as efficient algorithms exist (e.g. Bron and Kerbosch, 1973) for MIS enumeration.

Before going on to consider more specific methods for solving the GCP, we conclude this section by describing some simple and well-known results (Berge, 1973; Roschke and Furtado, 1973) which often allow a reduction in the amount of work involved in solving the GCP.

8.1.3 Elementary Reductions in Problem Size

(i) Suppose x_s and x_t are two vertices of the graph $G = (X, \Gamma)$ with the property that $\Gamma x_s \subseteq \Gamma x_t$. (Note that this implies, in particular, that x_s is not adjacent to x_t.) It is then readily seen that any colour which is feasible for vertex x_t (in the course of any algorithm) would also be feasible for x_s, so that we need only in fact use the algorithm to optimally colour the (sub)graph $G' = (X', \Gamma)$ where $X' = X - \{x_s\}$, and then assign the colour $C(x_t)$ to x_s.

The Graph-Colouring Problem

(ii) Again, suppose that the graph G contains an articulating set which is a clique, Q, of cardinality p, say. That is, removing Q from the graph will divide the graph into two, day, disjoint components, $G_1 = (X_1, \Gamma_1)$ and $G_2 = (X_2, \Gamma_2)$, where $X_1 \cup X_2 = X - Q$ so that the colouring of X_1 is unaffected by the colouring of X_2, and vice versa. Then it is readily seen that the graphs $G' = (X_1 \cup Q, \Gamma)$ and $G'' = (X_2 \cup Q, \Gamma)$—where Γ is restricted in each case to the appropriate vertex set $(X_i \cup Q)$—can each be coloured separately (using at least p colours), and then combined (overlapping at Q) to obtain an optimal colouring of G, with $\gamma(G) = \max [\gamma(G'), \gamma(G'')]$.

Thus, in this case, the problem of optimally colouring G is reduced to the simpler problems of colouring G' and G''; and obviously an analogous reasoning applies should Q divide G into more than two components.

We note here that two results most often used for graph reduction (Roschke and Furtado, 1973) correspond to the cases $p = 1$ and $p = 2$.

8.2 Vertex-Sequential Graph-Colouring Algorithms

In this and the following sections we develop various tree search methods for solving the GCP directly.

8.2.1 The Basic Algorithm and Some Variations

8.2.1.1 A Simple Vertex-by-Vertex Colouring Algorithm

In what follows we shall assume that the vertices of $G = (X, \Gamma)$ have an associated sequential ordering (at this stage arbitrary) $x_1, x_2, \ldots, x_n$. An elementary and well-known algorithm (Brown, 1972) for finding an optimal colouring of G is then to assign colours to the vertices of G in ascending order of their index as follows.

Initially we assign colour 1 to x_1, and then colour the remaining vertices in sequence by assigning to vertex x_i, $2 \le i \le n$, the colour $C(x_i)$ represented by the lowest integer which has not been assigned to any of the vertices of index lower than i which are adjacent to x_i. That is, each vertex in order is assigned the lowest feasible colour, so that $C(x_i) \le \max_{j<i} \{C(x_j)\} + 1$, $i = 1, 2, \ldots, n$.

When x_n has been coloured in this manner, a feasible colouring of G will have been achieved using q_1, say, colours where $q_1 = \max_{1 \le i \le n} \{C(x_i)\}$. Though this colouring will rarely be optimal (after the first pass), q_1 will obviously serve as an upper bound for the next phase of the algorithm, whereby we attempt to generate a feasible colouring using $q_2 \le q_1 - 1$ colours. To accomplish this we would of course have to recolour (at least) those vertices to which we had assigned colour q_1. Hence we backtrack to vertex x_k, where x_{k+1} is the vertex of lowest index which has been assigned colour q_1.

We then attempt to colour x_k with its smallest (feasible) alternative colour greater than its current colour $C(x_k)$. If no such alternative colour, c, exists which is smaller then q_1 we backtrack to x_{k-1}. On the other hand if $c < q_1$, we recolour x_k with c, and proceed forward as before, colouring x_{k+1} with its smallest feasible colour, etc., until such time as either x_n is coloured or some vertex x_l is reached which requires colour q_1.

In the former case a feasible colouring using $q_2 < q_1$ colours has been obtained, and by replacing q_1 by q_2 in the above procedure we backtrack again in an attempt to find a colouring using $q_3 \leq q_2 - 1$ colours; while in the latter case we backtrack again from x_l and proceed forward as before. When we eventually seek to backtrack past x_1, the algorithm terminates with q as the chromatic number of G, where q is the least integer for which a feasible colouring has been obtained, with the actual colouring being given by the $C(x_i)$ in force at that time.

Example Consider the 7-vertex graph G in Figure 8.3. The sequential

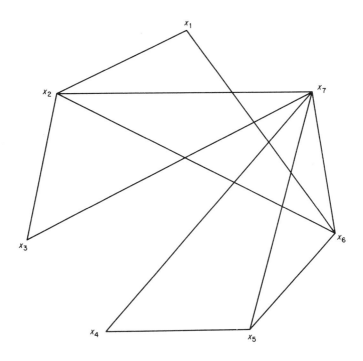

Figure 8.3

The Graph-Colouring Problem

values of the $C(x_i)$ assigned to the vertex of G will then be as follows:

$$C(x_1)=1 \; C(x_2)=2 \; C(x_3)=1 \; C(x_4)=1 \; C(x_5)=2 \; C(x_6)=3 \; C(x_7)=4.$$

(Thus the first feasible colouring obtained necessitates the use of 4 colours.)

$$C(x_5)=3$$
$$C(x_4)=2 \; C(x_5)=1 \; C(x_6)=3$$
$$C(x_5)=3$$
$$C(x_4)=3 \; C(x_5)=1 \; C(x_6)=3$$
$$C(x_5)=2 \; C(x_6)=3$$
$$C(x_3)=3 \; C(x_4)=1 \; C(x_5)=2 \; C(x_6)=3$$
$$C(x_5)=3$$
$$C(x_4)=2 \; C(x_5)=1 \; C(x_6)=3$$
$$C(x_5)=3$$
$$C(x_4)=3 \; C(x_5)=1 \; C(x_6)=3$$
$$C(x_5)=2 \; C(x_6)=3 \; C(x_7)=1.$$

Thus a colouring has been found involving only 3 colours, viz. $\{(x_1, x_7), (x_2, x_5), (x_3, x_4, x_6)\}$, and since no further backtracking possibilities exist, the algorithm terminates, with $\gamma(G)=3$.

8.2.1.2 Improvements to the Simple Algorithm

In the presentation given above, we have implicitly made use of the fact that if a vertex x_i cannot be coloured with any of the colours used for the previous $(i-1)$ vertices, then one need only consider the assignment of colour $q+1$ to x_i, where $q = \max_{j<i} \{C(x_j)\}$ as the sole alternative for forward branching at this stage. This result, which is formally proved in (Brown, 1972), leads naturally to some limitation of the size of the search tree.

In particular, we know then that the only colour which will be assigned to vertex x_1 throughout the course of the algorithm will be colour 1. Similarly, the only alternatives which will be considered for vertex x_2 will be colours 1 and 2, depending on whether x_1 and x_2 are to have the same or different colours. Thus, if x_2 is adjacent to x_1 (so that they cannot be coloured the same) then the only colour considered for x_2 would be colour 2. Repeating this argument, we see that if the vertices $x_1, x_2, \ldots, x_p$ constitute a clique, there will be but one colour assigned to each throughout the course of the algorithm; therefore the algorithm can be terminated when all backtrackings from vertex x_{p+1} have been completed, thus again reducing the search.

Recalling that the ordering of the vertices prior to the algorithm was arbitrary, it is then obvious that arranging the vertices so that the first $\omega(G)$

vertices constitute a clique of the graph G will improve the efficiency of the algorithm. A further enhancement is to order the remaining vertices so that, for any i, vertex x_i is adjacent to more of the vertices $x_1, x_2, \ldots, x_{i-1}$ than are any of the vertices $x_{i+1}, \ldots, x_n$ (ties being broken by choosing the vertex of greater degree). Indeed, as this procedure will in general lead to a near-maximum clique anyway, it is reasonable to use this criterion to order all the vertices, with x_1 then being a vertex of maximum degree of G.

It is also possible to improve the simple algorithm by using a look-ahead procedure during the course of the algorithm in an effort to reduce the number of backtracking steps, and also to direct the search (Brown, 1972). However, the basic algorithm can be greatly improved by incorporating the concept of a dynamic re-ordering of the as yet uncoloured vertices as the algorithm progresses, so that the vertex to be coloured next (at any stage) is that one, amongst all uncoloured vertices, which has the smallest number of feasible possible colour assignments available to it. This heuristic strategy will then naturally result in a reduction of the size of the total search-tree for the problem, by terminating those forward branchings which do not lead to the optimal solutions at as early a stage as possible (Korman, 1975).

The above procedure is relatively easy to incorporate into the basic algorithm, and in fact, as will be seen from the computational results of Section 8.4, gives rise to a marked reduction in computing times. Moreover, using the vertex re-ordering procedure at each stage will serve a similar purpose as the backtracking criterion of the look-ahead procedure in Brown (1972). Thus, if at any stage of the algorithm there exists some vertex x_k which cannot be assigned any colour which is less than q_h, where q_h is the number of colours in the best colouring to date (so that x_k has no possible feasible assignments open to it) then x_k will be chosen as the next vertex to be coloured and a backtracking step will immediately ensue.

Table 8.1

Stage	Activity	$P(x_1)$	$P(x_2)$	$P(x_3)$	$P(x_4)$	$P(x_5)$	$P(x_6)$	$P(x_7)$
0	Initialize $P(i) = n\ \forall i$	7	7	7	7	7	7	7
1	$C(x_1) = 1$	—	6	7	7	7	6	7
2	$C(x_2) = 2$	—	—	6	7	7	5	6
3	$C(x_6) = 3$	—	—	6	7	6	—	5
4	$C(x_7) = 1$	—	—	5	6	5	—	—
5	$C(x_3) = 3$	—	—	—	6	5	—	—
6	$C(x_5) = 2$	—	—	—	5	—	—	—
7	$C(x_4) = 3$	—	—	—	—	—	—	—
8	Stop—as no backtracking possibilities exist.							

The Graph-Colouring Problem   221

Example Consider again the graph G in Figure 8.3. Let us define a 'Possibility Table' to give the number of feasible possibilities $P(x_i)$ open to each vertex x_i at each stage of the algorithm. The computation then proceeds as Table 8.1 (where ties for minimization are taken in favour of the least index).

It can thus be seen that by dynamically re-ordering the vertices much backtracking can be saved relative to the original algorithm (although, of course, there will in general be some backtracking).

8.2.2 A Dichotomous Search Algorithm

Although the methods just described lead to effective algorithms for graph colouring, they do not really involve any graph-theoretic principles relating to the problem itself. It is natural then to seek methods of solution which are based on some underlying graph-theoretic concepts, and this we do in the remaining parts of this chapter. Thus, we now discuss a method of a different type (Zykov, 1952) which can also be classed as a vertex-by-vertex approach.

8.2.2.1 *Zykov's Original Method*

Let us consider any two non-adjacent vertices, x_1 and x_2, say, of the n-vertex graph $G = (X, \Gamma)$ which is to be coloured. Now any feasible colouring of G must fall into one of two distinct mutually exclusive classes, consisting respectively of those colourings wherein x_1 and x_2 are assigned the same colour (whichever it might be) and those wherein x_1 and x_2 are assigned different colours. In the former case, G is colour-equivalent† to the $(n-1)$-vertex pseudo-graph, $\tilde{G}$, obtained by coalescing x_1 and x_2 into a single 'pseudo-vertex;, $\tilde{x}_1$, which is joined to all vertices of $\Gamma x_1 \cup \Gamma x_2$. In the latter case, as x_1 and x_2 are to be coloured differently, it is then immaterial whether x_1 is or is not adjacent to x_2, so that G is colour-equivalent to the n-vertex graph $G^+ = (X, \Gamma^+)$ obtained from G by adding the arc (x_1, x_2). Also, as the two classes above exhaust all the possibilities, it follows that $\gamma(G) = \min(\gamma(\tilde{G}), \gamma(G^+))$. (It is to be noted that both $\tilde{G}$ and G^+ (called the reduced graphs of G) are nearer to being complete graphs than is G, since $\Gamma\tilde{x}_1 \supseteq \Gamma x_1$, and $\Gamma^+ x_1 \supset \Gamma x_1$, $\Gamma^+ x_2 \supset \Gamma x_2$, respectively.)

Next, we consider, in turn, $\tilde{G}$ and G^+ and, taking any two non-adjacent vertices of that graph, we re-apply the above procedure to each of $\tilde{G}$ and

---

† A graph G and a pseudo-graph $\tilde{G}$, derived from G, are said to be *colour-equivalent* if $\gamma(\tilde{G}) = \gamma(G)$.

G^+, thus obtaining

$$\gamma(G) = \min \{\gamma(\tilde{\tilde{G}}), \gamma(\tilde{G}^+), \gamma(\widetilde{G^+}), \gamma(G^{++})\}.$$

The same procedure is then applied to each of these new (reduced) graphs, etc., with the entire process continuing until there is no remaining reduced graph which contains any pair of non-adjacent vertices, in which case $\gamma(G)$ will be the minimum chromatic number of any of these irreducible, completely connected graphs.

Thus the optimum colouring of G is that one associated with a completely connected graph of least verticcs obtained by the above dichotomous method, the actual colouring being determined by colouring with the same colour those vertices which have been coalesced with one another during the course of the algorithm.

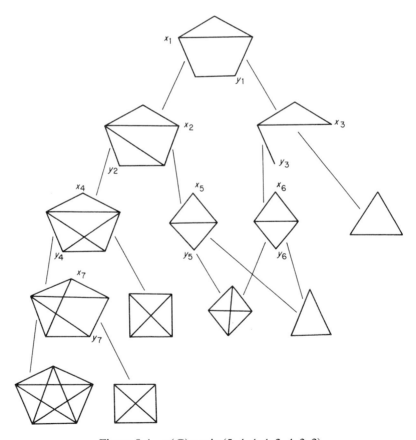

Figure 8.4 $\gamma(G) = \min (5, 4, 4, 4, 3, 4, 3, 3)$

Example We illustrate this method by showing the tree generated by the application of the algorithm to the 5-vertex graph in Figure 8.4, where the x_r and y_r represent the two vertices under consideration at each stage r.

8.2.2.2 Improvements to the Basic Algorithm

It is readily seen that the above algorithm is relatively cumbersome in both computing time and storage in that (a) many reduced graphs will be generated which will in the final reckoning have no bearing on the determination of the chromatic number; and moreover, (b) there are likely to be a large number of reduced graphs which are isomorphic, or nearly isomorphic, to one another, thus leading to great replications when each of these graphs is then (individually) reduced.

With a view to improving the efficiency of this algorithm, a depth-first tree-search variant of this method has been proposed (Corneil and Graham, 1973). This algorithm makes use of the fact that the number of vertices, q, in any generated completely connected reduced graph is an obvious upper bound on $\gamma(G)$, so that it would not be necessary to branch from any reduced graph (at any stage) containing a clique of cardinality q. Also, by suitably choosing the two non-adjacent vertices to be considered at any branch—as being elements of a subset of 'densely-connected' vertices whenever possible—one can attempt to direct heuristically the search with the aim of producing completely connected reduced graphs as early as possible.

However, as is indicated by the computational results in Corneil and Graham (1973), these modifications, although greatly improving the algorithm, cannot completely overcome the innate disadvantages of the main method. Therefore, in the following sections we consider a completely different class of graph-colouring algorithms which are also based on graph-theoretic aspects of the GCP.

8.3 Colour-Sequential Graph-Colouring Algorithms

It was seen above that the graph-colouring problem (GCP) could be formulated, and solved, as a more general set covering problem (SCP) by using the maximal independent sets (MIS) of the given graph as the covering sets. Theorem 8.4 below, on which the SCP approach is based, can in fact be used to derive a tree-search directly. This method involves, at any stage of the search, the simultaneous colouring of *all* the vertices for which it is proposed to assign a particular colour. Thus, one may equivalently treat this approach as the assignment of sets of vertices to the colours, in sequential order of colours.

This section describes such a strategy, from its inception as a simple breadth-first tree-search algorithm, through a sequence of modifications, leading finally to a depth-first tree-search procedure which makes use of further graph-theoretic results.

8.3.1 Preliminaries

The basis for all the colour-sequential algorithms of this section is the following theorem (Christofides, 1971).

Theorem 8.4 Any graph $G = (X, \Gamma)$ can be optimally coloured by colouring with the first colour some maximal independent set, M_1, of G then colouring with the second colour some maximal independent set, M_2, of $G_1 \equiv G - M_1$ and so on until all vertices have been coloured, where $G_k = (X_k, \Gamma)$ with $X_k = X - \bigcup_{j=1}^{k} M_j$, and $G_0 \equiv G$.

A complete breadth-first tree-search algorithm based directly on this theorem is given in Christofides (1971), the method itself being more easily described by introducing the concept of a *maximal r-colourable subset*, $M_{(r)}$, of a graph G as any maximal subset of vertices of G which requires no more than r colours for a feasible colouring. (Thus, an independent set of G is 1-colourable, and only independent sets are 1-colourable, while $\gamma(G)$ is obviously the least integer r for which X, the vertex set of G, is r-colourable; the family of MIS of G represents the maximal 1-colourable subsets of G, while it is easily seen that X itself is the unique maximal γ-colourable subset). Theorem 8.4 then implies that only maximal r-colourable subsets need be considered for potential colourings. Furthermore, as can be easily proved by contradiction the family, $\mathbf{M}_{(r)}$ (where $\mathbf{M}$ denotes the family of sets M, etc.) of maximal r-colourable subsets, $M_{(r)}^{j}$ of G can be determined from the family $\mathbf{M}_{(r-1)}$ by means of the following recurrence relation:

$$M_{(r)}^{j} = M_{(r-1)}^{i} \cup M_{(1)}^{k}(G_{r-1}') \quad (\text{R1})$$

for some ith set $M_{(r-1)}^{i} \in \mathbf{M}_{(r-1)}$ and kth set $M_{(1)}^{k} \in \mathbf{M}_{(1)}(G_{r-1}')$, where $G_{r-1}' \equiv (X - M_{(r-1)}^{i}, \Gamma)$. (For convenience of presentation, we shall henceforth denote by $G - M$ the subgraph $G' \equiv (X - M, \Gamma)$.)

Since $\mathbf{M}_{(1)} = \mathbf{M}(G)$, the family of MIS of G, the algorithm proceeds as follows.

Step 1 Determine $\mathbf{M}_{(1)} = \mathbf{M}(G)$, by using a standard MIS/clique determination algorithm (e.g. Bron and Kerbosch, 1973). Set $r = 1$.

Step 2 Set $r = r + 1$. Calculate $\mathbf{M}_{(r)}^{+}$ as the family of all sets $M_{(r)}^{j}$ obtained from relation (R1) by using all possible values of i and k.

The Graph-Colouring Problem

Step 3 Derive the family $\mathbf{M}_{(r)}$ by removing from $\mathbf{M}_{(r)}^+$ any sets which are subsets of other sets of $\mathbf{M}_{(r)}^+$ (leaving one copy of each maximal set so obtained).

Step 4 If $X \in \mathbf{M}_{(r)}$, terminate with $\gamma(G) = r$ and the corresponding colouring associated with the build-up of X; otherwise return to Step 2.

Although the method just described is relatively costly in terms of both computation time and storage, a number of simple improvements can be made which make the algorithm more efficient.

8.3.2 Improvements to the Simple Algorithm

Instead of determining directly all the MIS of the graph $G' = G - M_{(r)}^j$ at each stage, one can utilize the fact that all MIS of G' must be of the form $M - M_{(r)}^j$, where M is an MIS of the original graph G, to determine $M(G - M_{(r)}^j)$ directly from $M(G)$ (Roschke and Furtado, 1973).

Further improvement can be achieved by pruning the search-tree as follows. Given any optimal colouring of a graph G for which vertex x_i is

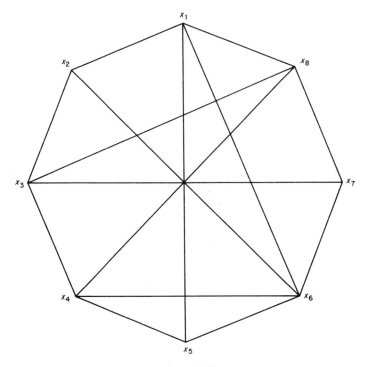

Figure 8.5

assigned colour $C(x_i)$, it is readily apparent that one can obtain an equivalent optimal colouring for which some specific vertex, x_0, say, is assigned a specific colour, c, say. This simple idea can be applied to the tree-search algorithm of the previous section, in order to reduce the size of the search-tree, by predetermining (at any stage) some specific vertex which is to be assigned the next colour in the sequence of colours.

More specifically, at the first stage of the algorithm one need only consider those MIS, $M_{(1)}^i$, which contain some specific vertex x_1^* of X. Similarly, at any level r we need only consider for further branching those MIS of $G - M_{(r)}^i$ which contain some specific vertex $x_r^* \in X - M_{(r-1)}^i$, an obvious choice for the x_r^* being that vertex which belongs to the fewest MIS of $G - M_{(r-1)}^i$ ($M_0^i = \varnothing$).

As is reported in Roschke and Furtado (1973) and Wang (1974), the above modifications, when incorporated into the basic tree-search algorithm, lead to a significant increase in efficiency.

Example We illustrate the modified algorithm by finding an optimum colouring for the graph G (taken from Wang, 1974) depicted in Figure 8.5, with the resultant search-tree being shown in Figure 8.6. In Figure 8.6, G_r^j

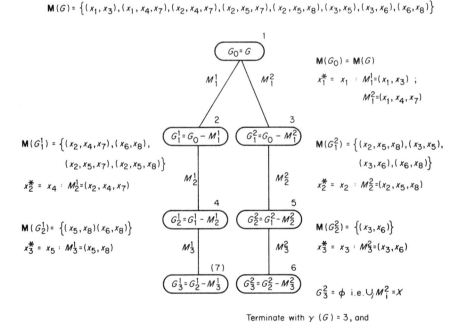

Figure 8.6

The Graph-Colouring Problem

represents the jth subgraph of the as yet uncoloured vertices to be considered at any level r, M_r^j represents the set of vertices to be assigned colour r (so leading to subgraph G_r^j), while the order in which the nodes are considered for branching purposes is indicated by the index to the upper right of each node.

8.3.3 An Equivalent Depth-First Approach

The breadth-first tree search just described can easily be adapted to a depth-first procedure in the usual fashion. We detail below the exact description of a depth-first graph-colouring algorithm (Wang, 1974) which corresponds to the method described in the previous section.

8.3.3.1 A Depth-First Graph-Colouring Algorithm

Step 0 Calculate $\mathbf{M}(G)$, the family of MIS of G. Set $\bar{q} = |X|$ and initialize $C(j) = j, j = 1, \ldots, \bar{q}$.

Step 1 Set $G' = G$, $r = 1$.

Step 2 If $r \geq \bar{q}$ go to 8, otherwise continue to 3.

Step 3 Derive $\mathbf{M}(G')$ (from $\mathbf{M}(G)$) and establish x_r^* as that vertex contained in the least number of sets in $\mathbf{M}(G')$, say these are $M_r^1, M_r^2, \ldots, M_r^m$.

Step 4 Set $p_r = m$, $h_r = 0$.

Step 5 If $h_r = p_r$ go to 8, otherwise increment $h_r = h_r + 1$ and continue to Step 6.

Step 6 Set $M' = m_r^{h_r}$, $G' = G' - M'$. If $G' = \emptyset$ go to 7, otherwise increment $r = r + 1$ and return to Step 2.

Step 7 A feasible colouring using r colours has been attained, viz. $M_1^{h_1}$, $M_2^{h_2}, \ldots, M_r^{h_r}$; set $\bar{q} = r$ and define $C(j) = \{x_i \mid x_i \in M_j^{h_j}\}$ for $j = 1, \ldots, r$.

Step 8 If $r = 1$, STOP with $\gamma(G) = \bar{q}$ and associated colouring function $C(\cdot)$, otherwise set $r = r - 1$, $G' = G' \cup M_r^{h_r}$, and return to Step 5.

Example The search-tree obtained when applying the above algorithm to the graph G in Figure 8.5 is essentially similar to that depicted in Figure 8.6, but with the nodes considered in a different order.

Thus we would first proceed down the left-most path of the tree considering in turn nodes 1, 2, 4, and 7, and so generate a further node (emanating from 7) by assigning colour 4 to the vertex set $M_4^1 = \{x_6\}$. At this stage a feasible colouring using 4 colours is achieved and we would backtrack to node 1 (via nodes 7, 4, and 2 as no branching alternatives exist for these nodes) and then proceed down the alternative path in an attempt to find a colouring involving 3 colours or less.

As such a colouring is found we use this latter to update the feasible solution found earlier and backtrack once more, but as no further branching alternatives exist, the algorithm would terminate with the identical optimal solution found in the previous example.

(It is to be noted that if we would have considered M_1^2 for branching purposes before M_1^1, so that the first feasible colouring generated would have been (M_1^2, M_2^2, M_3^2), only as many nodes would have been considered as were used in the breadth-first method of Figure 8.6, since node 7 would not have been considered due to the bound cut-off.)

We consider next a tree-search algorithm for graph colouring which involves an alternative strategy in an attempt to reduce further the size of the search-tree.

8.3.4 A Multiple Algorithm for Graph Colouring

In this section we describe a depth-first tree-search algorithm, based on Theorem 8.5 below, which (relative to the algorithms based on Theorem 8.4) further restricts the choice of sets to be considered at each stage of the tree-search.

8.3.4.1 An Intersection Theorem and Some Consequences

Let us denote the family of cliques of G of maximum size (hereafter termed the maximum cliques of G) as:

$$\hat{\mathbf{Q}}(G) = \{Q_i \mid Q_i \in \mathbf{Q}(G), |Q_i| = \omega(G)\}$$

where $\mathbf{Q}(G)$ represents the family of cliques of G (so that $\hat{\mathbf{Q}}(G) \subseteq \mathbf{Q}(G)$).

Theorem 8.5 (Intersection Theorem) (Korman, 1975) If for a graph G, $\gamma(G) = \omega(G)$, then an optimum colouring of G can be obtained by first colouring with colour 1 an MIS *of G, say M_1, such that*

$$|M_1 \cap Q_i| = 1 \quad \forall Q_i \in \hat{\mathbf{Q}}(G)$$

then colouring with colour 2 an MIS *of the graph G_1, say M_2, such that*

$$|M_2 \cap Q_i| = 1 \quad \forall Q_i \in \hat{\mathbf{Q}}(G_1), \quad \text{etc.}$$

until all the vertices of G are coloured; where, as before,

$$G_k = (X_k, \Gamma) \quad \text{with} \quad X_k = X - \bigcup_{j=1}^{k} M_j \quad (\text{and } G_0 = G).$$

The Graph-Colouring Problem

The implementation of the actual algorithms can be simplified by using instead the graphs $\tilde{G}_k$ obtained by successively joining, at every stage k, a pseudo-vertex $\tilde{x}_k$, representing M_k, to each vertex of $(G_k + \{\tilde{x}_1, \tilde{x}_2, \ldots, \tilde{x}_{k-1}\})$.

Thus, consider any arbitrary sequence of sets $\{M_1, M_2, \ldots, M_p\}$ where M_k is an MIS of the graph G_{k-1} in the sequence of graphs $G_0, G_1, \ldots, G_p$ defined as above, where for notational convenience we now write G_k for $\tilde{G}_k$.

Lemma The graph sequence $G_1, G_2, \ldots, G_p$ satisfies the property

$$\omega(G_{k-1}) \leq \omega(G_k) \leq \omega(G_{k-1}) + 1 \quad \text{for} \quad k = 1, \ldots, p.$$

Moreover, $\omega(G_k) = \omega(G_{k-1})$ if and only if

$$|M_k \cap Q_i| = 1 \quad \forall Q_i \in \hat{\mathbf{Q}}(G_{k-1}).$$

Corollary *If $M_1, M_2, \ldots, M_q$ represent the coloured sets in an optimal'colouring of G, so that $q = \gamma(G)$, then in the chain of inequalities*

$$\omega(G_0) \leq \omega(G_1) \leq \cdots \leq \omega(G_q) \quad (G_0 = G)$$

there are $\omega(G)$ equalities and $(\gamma(G) - \omega(G))$ strict inequalities.

8.3.4.2 An Overview of the Multiple Algorithm

The basic strategy of the Multiple Algorithm can now be informally described as follows. We first assume that $\gamma(G) = \omega(G)$, so that Theorem 8.5 would apply to graph G, and embark on a depth-first search which is basically similar to that described in the previous section but in which the number of branchings at any stage is considerably reduced by the application of Theorem 8.5 to the (sub)graph G_u corresponding to those vertices of G not yet coloured at that stage. If at any stage there exists an uncoloured vertex, x_0, such that none of the MIS of G_u containing x_0 intersect all maximum cliques of G_u, then a backtracking step is taken. Hence two possibilities can finally occur.

(a) A feasible colouring is eventually obtained (which necessarily satisfies the intersection condition at every stage); in which case, by Theorem 8.5, the colouring is optimal and the procedure can be terminated at this point.

(b) We backtrack to level 0 without attaining a feasible colouring, in which case (again by Theorem 8.5) the assumption $\gamma(G) = \omega(G)$ is shown to be false and $\gamma(G) \geq \omega(G) + 1$. It is then supposed that $\gamma(G) = \omega(G) + 1$, or, equivalently, the graph $G^1 = (X^1, \Gamma^1)$ is constructed from the graph G by adjoining a 'dummy' vertex to a maximum clique, Q_0, of G to form a clique of cardinality $(\omega(G) + 1)$.† For this graph G^1 one then has by assumption

---

† It is obvious that $\omega(G^1) = \omega(G) + 1$, while it can be easily seen that $\gamma(G^1) = \gamma(G)$ (assuming $\gamma(G) \geq \omega(G^1)$).

$\gamma(G^1) = \omega(G^1)$ so that the whole procedure can be repeated using graph G^1 instead of graph G, until either

(i) a feasible (and so optimal) colouring of G^1 is obtained which induces an optimal colouring of G in $\omega(G)+1$ colours (by assigning to the vertices of G the colours assigned to them by the optimal colouring of G^1 just found), or

(ii) no feasible colouring can be obtained in which case $\gamma(G^1) > \omega(G^1)$ (i.e. $\gamma(G) > \omega(G)+1$), so that the graph $G^2 = (X^2, \Gamma^2)$ is constructed by adjoining a further dummy vertex to the (unique) maximum clique of G^1. Then by assuming $\gamma(G^2) = \omega(G^2)$ (i.e. $\gamma(G) = \omega(G)+2$) the whole procedure is repeated on G^2, etc.

Eventually, since $\gamma(G) \leqslant n$ where $n = |X|$, we must reach the point where $\gamma(G^h) = \omega(G^h)$ for some $h \geqslant 0$ ($G^0 \equiv G$), in which case an optimal colouring (using $(\omega(G)+h$ colours) is obtained in the obvious manner.

We shall now discuss the implementation of this algorithm in somewhat greater detail.

8.3.4.3 Some Comments on Implementation

As mentioned in the previous section, there are two separate phases to the algorithm:

(a) the transition from a graph G^{i-1} to a graph G^i, and
(b) the main depth-first strategy applied to the graph G^i.

We now consider in more detail the practical implementation of these phases.

(a) It is shown in Korman (1975) that it is not necessary actually to add any dummy vertices. The conceptual equivalence of G^i is merely to allow exactly i nodes on every particular path from the root of the search tree for which we do *not* require the satisfaction of the full intersection condition of Theorem 8.5. Instead, at each one of these i nodes we require the satisfaction of the intersection condition only for all but one maximum clique, the exception being that clique of G_u deriving from Q_0. The transition from G^{i-1} to G^i is then simply to allow one further node of this type on every path from the root of the search tree.

(b) Although the algorithm as described would require the computation of maximum cliques of a different graph at each stage of the algorithm, it is readily apparent that the family $\mathbf{Q}(G)$ itself can be used for the intersection test at each stage, provided only that there exists some means of establishing at any level r exactly which cliques, Q_k, of G would generate the maximum cliques $(Q_k - M_r^j)$ of subgraph $G - M_r^j$.

The Graph-Colouring Problem

This can, however, be simply achieved by the adoption of a counter variable QSIZE(k) for each clique Q_k of G, initialized as the cardinality of Q_k and reduced (increased) by unity every time a vertex of Q_k is coloured (uncoloured) during the search.

Again, when a transition phase arises, all that is necessary is to increase QSIZE(1)† by unity to simulate the augmentation of Q_1 with a dummy vertex so that QSIZE(1) = $\omega(G^i) + 1 = \omega(G^{i+1})$ and, exactly as before, at any level r, the cliques which are used for the intersection check will all satisfy QSIZE(i) = QSIZE(1) at that stage. Recalling that during the depth-first search applied to the conceptional graph G^i there will be exactly i stages on every forward path from the root for which the intersection condition will be relaxed, we can now formally describe the algorithm

8.3.4.4 Description of the Multiple Algorithm

Step 0 Calculate $\mathbf{M}(G)$, $\mathbf{Q}(G)$, and set QSIZE(j) = $|Q_j|$ $\forall j$; set $s = 0$.

Step 1 Set $G' = G$, $t = 0$, $r = 1$.

Step 2 Determine $\mathbf{M}(G')$ and $\hat{\mathbf{Q}}(G') = \{Q_j \in \mathbf{Q}(G') \mid \text{QSIZE } (j) = \text{QSIZE}(1)\}$. Establish $\hat{\mathbf{M}}(G') = \{M_i \in \mathbf{M}(G') \mid M_i \cap Q_j \neq \emptyset \forall Q_j \in \hat{\mathbf{Q}}(G')\}$. If $\mathbf{M}(G') = \emptyset$, go to Step 7; otherwise continue to Step 3.

Step 3 Determine x_r^* as the vertex of G' contained in the least number of sets in $\hat{\mathbf{M}}(G')$, say $M_r^1, M_r^2, \ldots, M_r^m$.

Step 4 Set $p_r = m$, $h_r = 0$.

Step 5 If $h_r = p_r$ go to Step 9; otherwise increment $h_r = h_r + 1$, and continue to Step 6.

Step 6 Set $M' = M_r^{h_r}$, $G' = G' - M'$. If $G' = \emptyset$ go to Step 11; otherwise update QSIZE(1) = QSIZE(1) − 1, QSIZE(j) = QSIZE(j) − $|Q_j \cap M'|$, $\forall j > 1$, increment $r = r + 1$ and return to Step 2.

Step 7 If $t \geq s$ go to Step 9; otherwise re-establish

$$\hat{\mathbf{M}}(G') = \{M_i \in \mathbf{M}(G') \mid M_i \cap Q_j \neq \emptyset \ \forall Q_j \in \hat{\mathbf{Q}}(G'), j \neq 1\}$$

and continue to Step 8.

Step 8 If $\hat{\mathbf{M}}(G') = \emptyset$ go to 9; otherwise increment $t = t + 1$ and return to Step 3.

---

† For convenience we assume that the cliques of G have been ordered such that the maximum clique to which any dummy vertices would be adjoined is the first clique, Q_1.

232 Samuel M. Korman

Step 9 If $r=1$, increment $s=s+1$, set QSIZE (1) = QSIZE $(1)+1$, and return to Step 2; otherwise update $r=r-1$ and continue to Step 10.

Step 10 Set $G' = G' \cup M_r^{h_r}$, $t = t + |Q_1 \cap M_r^{h_r}| - 1$, update

$$\text{QSIZE}(1) = \text{QSIZE}(1) + 1, \text{QSIZE}(j) = \text{QSIZE}(j) + |Q_j \cap M_r^{h_r}|, \forall j > 1$$

and return to Step 5.

Step 11 STOP with $\gamma(G) = r$, and corresponding optimal solution $M_1^{h_1}, M_2^{h_2}, \ldots, M_r^{h_r}$.

$\mathbf{M}(G) = \{(x_1, x_3), (x_1, x_4, x_7), (x_2, x_4, x_7), (x_2, x_5, x_8), (x_3, x_5), (x_3, x_6), (x_6, x_8)\}$

$\mathbf{Q}(G) = \{(x_1, x_2, x_6), (x_1, x_5, x_6), (x_1, x_8), (x_2, x_3), (x_3, x_4, x_8), (x_3, x_7, x_8), (x_4, x_5, x_6), (x_6, x_7)\}$

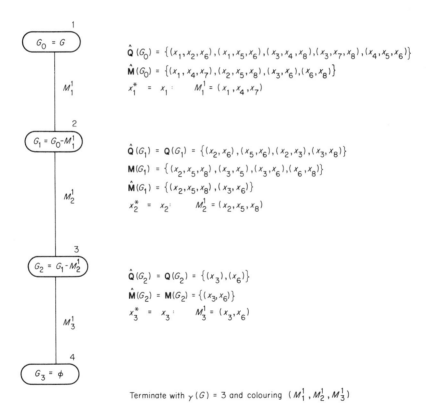

Figure 8.7

The Graph-Colouring Problem

Example Figure 8.7 depicts the search-tree obtained by applying the Multiple Algorithm to the graph G of Figure 8.5 above. Again G_r represents the uncoloured subgraph to be considered at level r, and M_r represents the set of vertices to be assigned colour r, while $\hat{Q}(G_r)$ represents the family of maximum cliques of G_r and $\hat{M}(G_r)$ represents the family of MIS of G_r which satisfy the intersection condition involving $\hat{Q}(G_r)$.

It is to be noted that although an optimal colouring was obtained very quickly in the above example, indeed with no backtracking being necessary, such an occurrence is not of course general, the good performance in the above case being due to the fact that G satisfies $\gamma(G) = \omega(G)$ (which as mentioned earlier, is the optimum situation for an algorithm based on the Intersection Theorem).

8.4 Computational Results and Conclusions

In this final section, we present computational results for the algorithms described in the previous sections. The random graph problems which were used for testing the various algorithms, were generated as follows.

For each order (number of vertices) of a graph beginning at 25 and increasing in units of 5, ten graphs were generated for each of the densities 0.1, 0.3, 0.5, 0.7, 0.9, (where the density of a graph with n vertices and m arcs is defined as $2m/n(n-1)$, $n(n-1)/2$ being the maximum number of possible arcs in a (complete) graph). The algorithms were then run on all ten graphs in any category and the results averaged. If not all the graphs were tested before the termination of the computer run, that particular category of (partial) results was ignored. Also only nontrivial results are recorded.

It is to be noted that as all the algorithms discussed in this chapter involve subsetting operations of some type, it is important to use boolean operations, if these are available; as indeed they are on the CDC-6000 Computer series. Thus all the computational results presented in Table 8.2 are FORTRAN programs incorporating boolean operations on a CDC-6400.

8.4.1 Summary

In summary we can enumerate the main conclusions as follows:

(i) Vertex-sequential methods would currently seem to represent the most effective algorithms for the optimal colouring of graphs.

(ii) Dynamic vertex re-ordering produces a great improvement when incorporated into the above methods.

(iii) The effectiveness of this class of algorithms decreases as the density (and hence γ) of the graph increases, except for the extremely dense graphs which are nearly complete.

(iv) The Intersection Theorem, when introduced into algorithms of the

Table 8.2 Computational results (Korman, 1975)

Ve	De	Times (CDC 6400 seconds)			
		A	B	C	D
25	0.1	—	—	13.9	0.6
	0.3	—	—	11.6	0.3
	0.5	—	—	7.8	0.3
	0.7	—	—	2.0	0.2
	0.9	—	—	0.1	0.3
30	0.1	0.2	0.2		1.8
	0.3	1.2	0.6		0.9
	0.5	2.9	0.6	155.6	1.0
	0.7	4.5	1.4	28.3	0.6
	0.9	0.6	0.3	0.8	0.7
35	0.1	0.4	0.5		
	0.3	3.5	1.0		18.8
	0.5	10.4	3.1		4.9
	0.7	31.2	6.8		1.7
	0.9	0.4	0.6		1.2
40	0.1	0.5	0.3		
	0.3	27.5	1.5		
	0.5	50.8	14.3		30.1
	0.7	71.4	20.6		26.2
	0.9	0.7	1.6		
45	0.1	0.9	0.5		
	0.3	87.1	3.0		
	0.5	120.5	37.8		
	0.7	180.8	74.2		
	0.9	0.8	7.9		
50	0.1		0.7		
	0.3		21.3		
	0.5				
	0.7				
	0.9				

A   Basic vertex-by-vertex colouring algorithm
B   As A, but incorporating dynamic vertex re-ordering
C   Depth-first method Wang (1974)
D   The multiple algorithm

colour-sequential type, establishes a great improvement in computational efficiency.

(v) Unlike the vertex-sequential schemes, these methods improve in effectiveness as the density of the graph increases, since $\gamma(G)$ gets closer to $\omega(G)$ and so the number of branchings decreases.

References

Appel, K. and W. Haken (1976). Every Planar Map is 4-Colourable. *Bulletin American Mathematical Society*, **82,** 711.
Berge, C. (1973). *Graphs and Hypergraphs*, North Holland Publishing Company, Amsterdam.
Brooks, R. L. (1941). On Colouring the Nodes of a Network. *Proc. Cambridge Philosophical Soc.*, **37,** 456.
Brown, J. R. (1972). Chromatic Scheduling and the Chromatic Number Problem. *Management Science*, **19,** 456.
Bron, C. and J. Kerbosch, (1973). Finding all Cliques of an Undirected Graph. *Comm. A.C.M.*, **16,** 575.
Christofides, N. (1971). An Algorithm for the Chromatic Number of a Graph. *Computer Journal*, **14,** 38.
Christofides, N. (1975). *Graph Theory*, Academic Press, London.
Corneil, D. G. and B. Graham (1973). An Algorithm for Determining the Chromatic Number of a Graph. *SIAM Journal of Computing*, **2,** 311.
Eilon, S. and N. Christofides (1971). The Loading Problem. *Management Science*, **17,** 259.
Heawood, P. J. (1890). Map-Colour Theorems. *Quarterly Journal of Mathematics*, **24,** 332.
Konig, D. (1950). *Theorie der Endlichen un Urendlichen Graphen*, Chelsea, N.Y.
Korman, S. M. (1975). Graph Colouring and Related Problems in Operations Research. Ph.D. Thesis, Imperial College, London.
Liu, C. L. (1968). *Introduction to Combinatorial Mathematics*, McGraw-Hill, N.Y.
Matula, D. W., G. Marbele, and J. D. Issacson (1974). Graph Colouring Algorithms. In Read (Ed.), *Graph Theory and Computing*, Academic Press, N.Y.
Mycielski, J. (1955). Sur le Coloriage des Graphes. *Colloq. Math*, **3,** 161.
Neufield, B. A. and J. Tartar (1974). Graph Colouring Conditions for the Existence of Solutions to the Time-Table Problem. *Communications of the A.C.M.*, **17,** 450.
Roschke, S. I. and A. L. Furtado (1973). An Algorithm for Obtaining the Chromatic Number and an Optimal Colouring of a Graph. *Information Processing Letters*, **2,** 34.
Wang, C. C. (1974). An Algorithm for the Chromatic Number of a Graph. *Journal of A.C.M.*, **21,** 385.
Welsh, D. J. A. and M. B. Powell (1967). An Upper Bound on the Chromatic Number of a Graph and its Applications to Time-Tabling Problems. *Computer Journal*, **10,** 85.
Williams, M. R. (1968). A Graph Theory Model for the Solution of Timetables. Ph.D. Thesis, University of Glasgow.
Wood, D. C. (1968). A Technique for Colouring a Graph Applicable to Large Scale Timetabling Problems. *Computer Journal*, **12,** 317.
Zykov, A. A. (1952). On Some Properties of Linear Complexes. *Amer. Math Soc., Translation*, **79,** p. 81.

CHAPTER 9

The 0–1 Knapsack Problem

SILVANO MARTELLO and PAOLO TOTH
Istituto di Automatica, University of Bologna

9.1 Introduction

9.1.1 The Knapsack Problem

Suppose a knapsack has to be filled without exceeding a prescribed total weight. If different types of objects, each of which having a value and a weight, are available, the following problem arises: find a feasible combination of objects so that the total value of the objects put in the knapsack is maximized.

Let W be the total weight limitation; let the integers $1, 2, \ldots, n$ denote the n available types of objects, p_j and w_j the value (or profit) and the weight of the jth type: then the *Knapsack Problem* can be mathematically expressed as:

$$\max \sum_{j=1}^{n} x_j p_j \qquad (9.1)$$

subject to

$$\sum_{j=1}^{n} x_j w_j \leq W \qquad (9.2)$$

$$x_j \geq 0 \text{ and integer } (j = 1, \ldots, n) \qquad (9.3)$$

where x_j represents the number of objects of type j which are selected.
Without loss of generality, the following assumptions can be made:

$$p_j, w_j \text{ positive}\dagger \text{ integers } (j = 1, \ldots, n) \qquad (9.4)$$

$$w_j \leq W \ (j = 1, \ldots, n). \qquad (9.5)$$

The Knapsack problem is intensively studied because it may represent many industrial situations—as will be shown in the next subsection—and because it arises as a subproblem in various integer programming problems.

† Only in the particular case where in (9.2) the equality constraint holds can nonpositive profits p_j also be considered.

The most commonly utilized methods for solving knapsack problems are basically of four types:

(a) Implicit enumeration;
(b) Dynamic programming;
(c) Network approaches;
(d) Generalized lagrangian methods.

Dynamic programming algorithms are computationally efficient when the value of W is small; when it is quite large they generally tend to be very inefficient because of the excessive storage required, whereas the enumeration techniques are much less affected by this disadvantage. Network approaches, discussed in Shapiro (1968), Shapiro and Wagner (1967), and in Frieze (1976), formulate the knapsack problem as a shortest route problem: they are usually inefficient because of the enormous size of the resulting networks, so they will not be discussed here. Also, we will not analyse the methods of type (d), whose performance is computationally satisfactory only when approximate solutions are required; classical studies on the solution of discrete programming problems through lagrangian multipliers can be found in Everett (1963), Brooks and Geoffrion (1966), Nemhauser and Ullman (1968), Shapiro (1971); applications to the knapsack problem are in Gulley, Swanson, and Woolsey (1972).

This survey considers the 0–1 form of the above problem, where it is supposed that only one object of each type is available. The *Zero–One Knapsack Problem* is then defined by (9.1), (9.2), and

$$x_j = 0 \text{ or } 1 \quad (j = 1, \ldots, n) \tag{9.6}$$

and, for obvious reasons, the following additional assumption is introduced:

$$W < \sum_{j=1}^{n} w_j. \tag{9.7}$$

The next sections review the most efficient algorithms of type (a) and (b) for the solution of the Zero–One Knapsack Problem and illustrate methods useful for improving the algorithms' performance (upper bounds, reduction procedures, dominance criteria); the relative efficiencies of the presented algorithms are analysed through computational comparisons. An Appendix reviews other kinds of knapsack problems and some related problems.

9.1.2 Applications of the Knapsack Problem

The present subsection briefly illustrates two classical applications of the knapsack problem: industrial production and capital budgeting. More

The 0-1 Knapsack Problem

detailed reviews can be found in Salkin (1975) and Salkin and de Kluyver (1975).

Industrial Production

The *Cargo Loading Problem* directly arises from the theoretical formulation of the knapsack problem: the loading of a ship with the most valuable cargo is discussed, for example, in the excellent book of Bellman and Dreyfus (1962).

Alternatively, the *Cutting Stock Problem* consists in finding the most valuable way of cutting a portion of space into pieces of different values: one-dimensional and two-dimensional knapsack problems arising from this formulation are discussed and solved through dynamic programming algorithms in the fundamental papers of Gilmore and Gomory (1961, 1963, 1965, 1966).

Capital Budgeting

The problem of selecting among various investment possibilities so as to maximize the total payoff without exceeding the available funds can be directly expressed as a zero–one knapsack problem (each investment possibility is either accepted or rejected). When the situation is broken into periods, each having available funds and investment costs, we have the *Multi-Period Capital Budgeting Model*: this zero–one multi-dimensional knapsack problem is discussed and solved through dynamic programming algorithms in Weingartner (1963, 1968) and in Weingartner and Ness (1967); Cord (1964) and Kaplan (1966) utilize lagrangian multipliers to obtain approximate solutions by supposing that the constraints may be violated by an acceptable amount.

9.2 Upper Bounds

Assume that the objects have been ordered so that

$$p_1/w_1 \geq p_2/w_2 \geq \ldots \geq p_n/w_n \qquad (9.8)$$

and let

$$s = \text{largest integer for which } \sum_{j=1}^{s} w_j \leq W. \qquad (9.9)$$

Dantzig (1957) has proved the following theorem.

Theorem 9.1 *The optimal solution to the associated continuous problem, defined by* (9.1), (9.2), *and*

$$0 \leq x_j \leq 1 \quad (j=1,\ldots,n)$$

is

$$x_j = 1 \quad (j=1,\ldots,s)$$
$$x_j = 0 \quad (j=s+2,\ldots,n)$$
$$x_{s+1} = \left(W - \sum_{j=1}^{s} w_j\right) / w_{s+1}.$$

Corollary 9.1 *The value*

$$\mathrm{UB}_1 = \sum_{j=1}^{s} p_j + \left\lfloor \left(W - \sum_{j=1}^{s} w_j\right) p_{s+1}/w_{s+1} \right\rfloor^{\dagger} \tag{9.10}$$

is an upper bound of the solution to the Zero–One Knapsack Problem.

An improvement on Dantzig's upper bound has been found by Martello and Toth (1977a):

Theorem 9.2 *Let*

$$B_1 = \sum_{j=1}^{s} p_j + \left\lfloor \left(W - \sum_{j=1}^{s} w_j\right) p_{s+2}/w_{s+2} \right\rfloor$$

$$B_2 = \sum_{j=1}^{s} p_j + \left\lfloor p_{s+1} - \left(w_{s+1} - \left(W - \sum_{j=1}^{s} w_j\right)\right) p_s/w_s \right\rfloor.$$

Then

$$\mathrm{UB}_2 = \max\{B_1, B_2\} \tag{9.11}$$

is an upper bound of the solution to the Zero–One Knapsack Problem.

Proof Because of definitions (9.8) and (9.9), the optimal solution to the problem can be obtained in two ways: without inserting the $(s+1)$th object or by inserting it. In the former case the solution obviously cannot exceed the value of B_1. In the latter case, since it is necessary to remove at least one of the first s objects, the best possible solution is given by B_2, where we have supposed that the object to be removed has exactly the minimum necessary value of w (i.e. $w_{s+1} - (W - \sum_{j=1}^{s} w_j)$) and the worst value of p/w (i.e. p_s/w_s). The greater one out of B_1 and B_2 is then a valid upper bound.

† $\lfloor z \rfloor$ = greatest integer less than or equal to z; $\lceil z \rceil$ = smallest integer greater than or equal to z.

The 0–1 Knapsack Problem

Theorem 9.3 *The value UB_2 given by (9.11) is less than or equal to the value UB_1 given by (9.10).*

Proof We have to prove that

(i) $$\text{UB}_1 \geq B_1$$

and that

(ii) $$\text{UB}_1 \geq B_2.$$

(i) is obvious, since $p_{s+1}/w_{s+1} \geq p_{s+2}/w_{s+2}$. To prove (ii), it is sufficient to prove that

$$\left(W - \sum_{j=1}^{s} w_j\right) p_{s+1}/w_{s+1} \geq p_{s+1} - \left(w_{s+1} - \left(W - \sum_{j=1}^{s} w_j\right)\right) p_s/w_s.$$

Transforming

$$\left(W - \sum_{j=1}^{s} w_j\right)(p_{s+1}/w_{s+1} - p_s/w_s) \geq w_{s+1}(p_{s+1}/w_{s+1} - p_s/w_s)$$

we have

(iii) $$\left(W - \sum_{j=1}^{s} w_j - w_{s+1}\right)(p_{s+1}/w_{s+1} - p_s/w_s) \geq 0.$$

Because of (9.9), $W - \sum_{j=1}^{s} w_j - w_{s+1} < 0$; because of (9.8), $p_{s+1}/w_{s+1} - p_s/w_s \leq 0$. It follows that (iii), and hence (ii), is always true.

Example Let $n=6$, $W=60$, and suppose that the objects have been arranged as in (9.8):

$$(p_j) = (15, 14, 14, 18, 17, 12)$$
$$(w_j) = (12, 14, 15, 24, 24, 17).$$

From (9.9) we have $s=3$, $\sum_{j=1}^{3} w_j = 41$, $\sum_{j=1}^{3} p_j = 43$.
Dantzig's bound: Corollary 9.1 gives

$$\text{UB}_1 = 43 + \lfloor 19 \times 18/24 \rfloor = 57.$$

Martello-Toth's bound: Theorem 9.2 gives

$$B_1 = 43 + \lfloor 19 \times 17/24 \rfloor = 56$$
$$B_2 = 43 + \lfloor 18 - 5 \times 14/15 \rfloor = 56$$

hence $\text{UB}_2 = 56 < \text{UB}_1$.

The optimal solution is here given by $(x_j) = (1, 1, 1, 0, 0, 1)$ with a total value of 55.

The following improvement on Martello-Toth's upper bound has been proposed by Hudson (1977).

Theorem 9.4 *Let*

$$s^* = \text{largest integer for which } \sum_{j=1}^{s^*} w_j \leq W - w_{s+1}$$

$$B_3 = p_{s+1} + \sum_{j=1}^{s^*} p_j + \left\lfloor \left(W - w_{s+1} - \sum_{j=1}^{s^*} w_j \right) p_{s^*+1} / w_{s^*+1} \right\rfloor.$$

Then

$$\text{UB}_3 = \max \{B_1, B_3\}$$

is an upper bound for the zero–one knapsack problem and

$$\text{UB}_3 \leq \text{UB}_2.$$

Proof. If we set $x_{s+1} = 1$ then we are left with another knapsack problem for the other objects with a weight limit $W - w_{s+1}$; using Dantzig's bound for this smaller problem, we find that an upper bound to all solutions to the original problem which have $x_{s+1} = 1$ is B_3. Similarly we find that an upper bound to all solutions to the original problem which have $x_{s+1} = 0$ is B_1 of Theorem 9.2. Thus UB_3 is an upper bound for the zero–one knapsack problem.

In order to prove that $\text{UB}_3 \leq \text{UB}_2$, consider that obviously $s^* < s$. Now if $w_{s+1} - w_s \leq W - \sum_{j=1}^{s} w_j$, then $s^* = s-1$, $B_3 = B_2$ and $\text{UB}_3 = \text{UB}_2$; otherwise,

$$B_2 - B_3 = \sum_{j=s^*+2}^{s} \left(p_j - \frac{w_j p_s}{w_s} \right) - \left(W - w_{s+1} - \sum_{j=1}^{s^*+1} w_j \right) \left(\frac{p_{s^*+1}}{w_{s^*+1}} - \frac{p_s}{w_s} \right) \geq 0.$$

Another improvement on UB_2 has been presented by Müller-Merbach (1978) through a different approach:

Theorem 9.5. *Let p_j^* be the dual variables of the optimal continuous solution of Theorem 1, that is*

$$p_j^* = p_j - w_j p_{s+1}/w_{s+1} \quad \text{for } j = 1, \ldots, s$$

$$p_j^* = -p_j + w_j p_{s+1}/w_{s+1} \quad \text{for } j = s+2, \ldots, n.$$

Then

$$\text{UB}_4 = \sum_{j=1}^{s} p_j + \left\lfloor \left(W - \sum_{j=1}^{s} w_j \right) p_{s+1}/w_{s+1} - \min \left\{ \left(W - \sum_{j=1}^{s} w_j \right) p_{s+1}/w_{s+1}, \right. \right.$$

$$\left. \left. \min_j \{p_j^* | j \neq s+1\} \right\} \right\rfloor$$

is an upper bound for the Zero–One Knapsack Problem.

The 0-1 Knapsack Problem

Proof In order to obtain an integer solution from the continuous solution, either only the fractional variable x_{s+1} has to be set to 0 (without any change of the other variables), or at least one of the other variables, say x_j, has to change its value (from 1 to 0 or from 0 to 1): if any of the latter changes take place, the value of the optimal continuous solution will be reduced by at least the corresponding p_j^* if the first change takes place by at least $p_{s+1}(W - \sum_{j=1}^{s} w_j)/w_{s+1}$. This leads to UB$_4$.

Note that UB$_4$ is not greater than UB$_1$, but it can be less, equal, or greater than both UB$_2$ and UB$_3$; an improved bound can be obtained by taking the minimum between UB$_3$ and UB$_4$.

Also note that UB$_4$ requires a computational effort much greater than those required for UB$_3$ and UB$_2$.

Example For the problem previously considered the new upper bounds will be:

Hudson's bound: Theorem 9.4 gives

$$s^* = 2$$
$$B_3 = 18 + 29 + \lfloor 10 \times 14/15 \rfloor = 56$$
$$\text{UB}_3 = 56 = \text{UB}_2.$$

Müller-Merbach's bound: Theorem 9.5 gives

$$p_1^* = 15 - 12 \times 18/24 = 6.00$$
$$p_2^* = 14 - 14 \times 18/24 = 3.50$$
$$p_3^* = 14 - 15 \times 18/24 = 2.75$$
$$p_5^* = -17 + 24 \times 18/24 = 1.00$$
$$p_6^* = -12 + 17 \times 18/24 = 0.75$$

$$\text{UB}_4 = 43 + \lfloor 19 \times 18/24 - \min\{19 \times 18/24, 0.75\} \rfloor = 56 = \text{UB}_2 = \text{UB}_3.$$

9.3 Implicit Enumeration Algorithms

The first enumeration method for the solution of the zero-one knapsack problem was the breadth-first branch and bound procedure presented by Kolesar (1967); the large computer memory and time requirements of Kolesar's algorithm were greatly reduced by the depth-first branch and bound technique of Greenberg and Hegerich (1970). Recently, Horowitz and Sahni (1974) proposed a highly efficient branch and bound procedure, based on Greenberg-Hegerich's scheme; the same algorithm was independently obtained by Ahrens and Finke (1975). Further improvements have been presented by Barr and Ross (1975), Fayard and Plateau (1975), Zoltners (1978), Nauss (1976), Suhl (1977), Martello and Toth (1977a).

9.3.1 The Algorithm of Horowitz and Sahni

We will use the following notation:

$p = \text{profit}\left(\sum_{j=1}^{n} p_j x_j\right)$ associated with the current solution

$(x_j) = (x_1, x_2, \ldots, x_n)$

$P = \text{profit}\left(\sum_{j=1}^{n} p_j X_j\right)$ associated with the currently best

solution $(X_j) = (X_1, X_2, \ldots, X_n)$.

Horowitz and Sahni (1974) have proposed the following procedure for the solution of the zero–one knapsack problem:

Step 1 (*Initialize*) Order the objects in decreasing order of p_j/w_j; set $p_{n+1} = 0$, $w_{n+1} = \infty$. Set $P = p = 0$, $(X_j = x_j = 0, j = 1, \ldots, n)$, $i = 1$.

Step 2 (*Test heuristic*) Find $z = \sum_{j=i}^{s} p_j + (W - \sum_{j=i}^{s} w_j) p_{s+1}/w_{s+1}$, with $s = $ largest index for which $\sum_{j=i}^{s} w_j \leq W$ (if $w_i > W$, set $s = i - 1$). If $P \geq \lfloor z \rfloor + p$, go to Step 5.

Step 3. (*Build a new current solution*) If $w_i \leq W$ and $i \leq n$, set $W = W - w_i$, $p = p + p_i$, $x_i = 1$, $i = i + 1$ and repeat 3. Otherwise, if $i \leq n$, set $x_i = 0$, $i = i + 1$. In any case, if $i < n$, go to Step 2; if $i = n$, repeat Step 3.

Step 4 (*Update the current optimal solution*) If $P < p$, set $P = p$ $(X_j = x_j, j = 1, \ldots, n)$. In any case, set $i = n$ and, if $x_n = 1$, set $W = W + w_n$, $p = p - p_n$, $x_n = 0$.

Step 5 (*Backtrack*) Find the largest $k < i$ for which $x_k = 1$. If no such k exists, the maximum total profit is P and the corresponding solution is given by (X_j). Otherwise, set $W = W + w_k$, $p = p - p_k$, $x_k = 0$, $i = k + 1$ and go to Step 2.

The algorithm starts by building, through iterated executions of Step 3, the first current solution (coinciding with the integer solution given by Theorem 9.1 if x_{s+1} is set to 0). A depth-first branch and bound search is then performed.

A *forward move* consists of inserting (through Step 3) the largest set of new consecutive objects into the current solution under condition (9.2). A *backtracking move* consists of removing (at Step 5) from the current solution the object with the largest index.

Whenever a forward move is exhausted, Step 2 tests whether further forward moves could lead to a current solution improving on the current optimal solution: if so, a new forward move is performed; otherwise a backtracking follows. When, during the execution of Step 3, the last object

The 0-1 Knapsack Problem

has been considered, Step 4 tests whether the current solution improves on the current optimal solution and, if so, the current optimal solution is updated. The algorithm stops when no further backtracking is possible.

Example Let $n=7$, $W=50$, and suppose that the objects have been arranged as required by Step 1:

$$(p_i) = (70, 20, 39, 37, 7, 5, 10)$$

$$(w_i) = (31, 10, 20, 19, 4, 3, 6).$$

Step 1. $P = p = 0$, $(X_j) = (x_j) = (0, 0, 0, 0, 0, 0, 0)$, $i = 1$, $p_8 = 0$, $w_8 = \infty$.
Step 2 $s = 2$, $z = 90 + 9 \times 39/20 = 107.55$. $P < 107 + 0$.
Step 3 $W = 50 - 31 = 19$, $p = 0 + 70 = 70$, $x_1 = 1$, $i = 2$.
Step 3 $W = 19 - 10 = 9$, $p = 70 + 20 = 90$, $x_2 = 1$, $i = 3$.
Step 3 $w_3 > 9$, $i < n$: $x_3 = 0$, $i = 4 < n$.
Step 2 $s = 2$, $z = 0 + 9 \times 39/20 = 17.55$. $P < 17 + 90$.
Step 3 $w_4 > 9$, $i < n$: $x_4 = 0$, $i = 5 < n$.
Step 2 $s = 6$, $z = 12 + 2 \times 10/6 = 15.\bar{3}$. $P < 15 + 90$.
Step 3 $W = 9 - 4 = 5$, $p = 90 + 7 = 97$, $x_5 = 1$, $i = 6$.
Step 3 $W = 5 - 3 = 2$, $p = 97 + 5 = 102$, $x_6 = 1$, $i = 7$.
Step 3 $w_7 > 2$, $i = n$: $x_7 = 0$, $i = 8$.
Step 4 $P = 102$, $(X_j) = (1, 1, 0, 0, 1, 1, 0)$; $i = 7$.
Step 5 $k = 6$, $W = 2 + 3 = 5$, $p = 102 - 5 = 97$, $x_6 = 0$, $i = 7$.
Step 2 $s = 6$, $z = 0 + 5 \times 10/6 = 8.\bar{3}$. $P < 97 + 8$.
Step 3 $w_7 > 5$, $i = n$: $x_7 = 0$, $i = 8 > n$.
Step 4 $P > p : i = 7$.
Step 5 $k = 5$, $W = 5 + 4 = 9$, $p = 97 - 7 = 90$, $x_5 = 0$, $i = 6$.
Step 2 $s = 7$, $z = 15 + 0 = 15$. $P < 15 + 90$.
Step 3 $W = 9 - 3 = 6$, $p = 90 + 5 = 95$, $x_6 = 1$, $i = 7$.
Step 3 $W = 6 - 6 = 0$, $p = 95 + 10 = 105$, $x_7 = 1$, $i = 8$.
Step 3 $i > n$.
Step 4 $P = 105$, $(X_j) = 1, 1, 0, 0, 0, 1, 1)$; $i = 7$, $W = 0 + 6 = 6$, $p = 105 - 10 = 95$, $x_7 = 0$.
Step 5 $k = 6$, $W = 6 + 3 = 9$, $p = 95 - 5 = 90$, $x_6 = 0$, $i = 7$.
Step 2 $s = 7$, $z = 10 + 0$. $P > 90 + 10$.
Step 5 $k = 2$, $W = 9 + 10 = 19$, $p = 90 - 20 = 70$, $x_2 = 0$, $i = 3$.
Step 2 $s = 2$, $z = 0 + 19 \times 39/20 = 37.05$. $P < 37 + 70$.
Step 3 $w_3 > W$, $i < n$: $x_3 = 0$, $i = 4 < n$.
Step 2 $s = 4$, $z = 37 + 0 = 37$. $P < 37 + 70$.
Step 3 $W = 19 - 19 = 0$, $p = 70 + 37 = 107$, $x_4 = 1$, $i = 5$.
Step 3 $w_5 > W$, $i < n : x_5 = 0$, $i = 6 < n$.
Step 2 $s = 5$, $z = 0$. $P < 0 + 107$.
Step 3 $w_6 > W$, $i < n : x_6 = 0$, $i = 7 = n$.
Step 3 $w_7 > W$, $i = n : x_7 = 0$, $i = 8$.

Step 4 $P = 107$, $(X_j) = (1, 0, 0, 1, 0, 0, 0)$; $i = 7$.
Step 5 $k = 4$, $W = 0 + 19 = 19$, $p = 107 - 37 = 70$, $x_4 = 0$, $i = 5$.
Step 2 $s = 7$, $z = 22 + 0 = 22$. $P > 22 + 70$.
Step 5 $k = 1$, $W = 19 + 31 = 50$, $p = 70 - 70 = 0$, $x_1 = 0$, $i = 2$.
Step 2 $s = 4$, $z = 96 + 1 \times 7/4 = 97.75$. $P > 97 + 0$.
Step 5 $\not\exists k$: stop.

Figure 9.1 shows the decision-tree of the example.

Variants of the above method have been proposed by Nauss (1976) and by Martello and Toth (1977a); the following subsection presents this last algorithm, which is the one involving the least computation time.

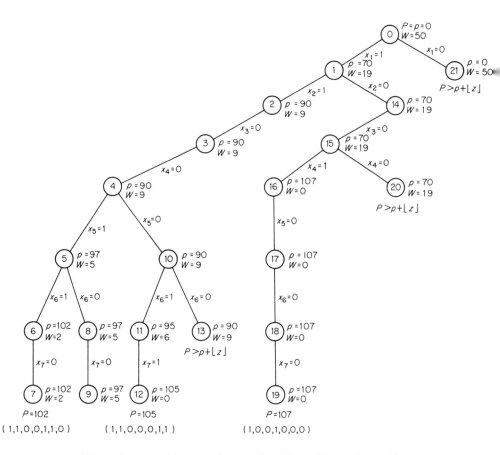

Figure 9.1 Decision tree for the algorithm of Horowitz and Sahni

The 0–1 Knapsack Problem

9.3.2 The Algorithm of Martello and Toth

Step 1 (Initialize) Order the objects in decreasing order of p_j/w_j. Compute $p^* = \sum_{j=1}^{s} p_j$, with s = largest index for which

$$w^* = \sum_{j=1}^{s} w_j \leq W.$$

If $w^* = W$, the optimal solution is given by $P = p^*$, $(X_j = 1, j = 1, \ldots, s)$ $(X_j = 0, j = s+1, \ldots, n)$. Stop.

Otherwise, compute $(M_j = \min \{w_k | j < k \leq n\}$, $j = 1, \ldots, n-1)$, $M_n = \infty$. Set $U = \mathrm{UB}_2$ (see (9.11), Theorem 9.2), $p = P = 0$, $(x_j = 0, j = 1, \ldots, n)$, $i = 1$, $\bar{s} = n$. Go to Step 4.

Step 2 (*Try to insert the i-th object into the current solution*) If $w_i \leq W$, go to Step 3. Otherwise, if $P \geq p + \lfloor Wp_{i+1}/w_{i+1} \rfloor$, go to Step 5; if not, set $i = i+1$ and repeat Step 2.

Step 3 (*Build a new current solution*) Compute $p^* = \bar{p}_i + \sum_{j=z_i}^{s} p_j$, with s = largest index for which $w^* = \bar{w}_i + \sum_{j=z_i}^{s} w_j \leq W$ and $s \leq n$ (if $\bar{w}_i + w_{z_i} > W$, set $s = \bar{z}_i - 1$). Two possibilities exist:

(a) $w^* = W$ and $s < n$: if $P \geq p + p^* + \lfloor (W - w^*)p_{s+1}/w_{s+1} \rfloor$, go to Step 6. Otherwise, go to Step 4.
(b) $w^* = W$ or $s = n$: if $P \geq p + p^*$, go to Step 6. Otherwise, set $P = p + p^*$, $(X_j = x_j, j = 1, \ldots, i-1)$, $(X_j = 1, j = i, \ldots, s)$, $(X_j = 0, j = s+1, \ldots, n)$: if $P = U$, stop; if not, go to Step 6.

Step 4 (*Save the current solution*) Set $W = W - w^*$, $p = p + p^*$, $(x_j = 1, j = i, \ldots, s)$. Compute $\bar{w}_i = w^*$, $\bar{p}_i = p^*$, $\bar{z}_i = s+1$, $(\bar{w}_j = \bar{w}_{j-1} - w_{j-1}, \bar{p}_j = \bar{p}_{j-1} - p_{j-1}, \bar{z}_j = s+1$ for $j = i+1, \ldots, s)$, $(\bar{w}_j = \bar{p}_j = 0, \bar{z}_j = j$ for $j = s+1, \ldots, \bar{s})$; set $\bar{s} = s$. Three possibilities exist:

(a) $s < n-2$: set $i = s+2$. If $W < M_{i-1}$, go to Step 5; otherwise, go to Step 2.
(b) $s = n-2$: if $W \geq w_n$, set $W = W - w_n$, $p = p + p_n$, $x_n = 1$. In any case, set $i = n-1$ and go to Step 5.
(c) $s = n-1$: set $i = n$ and go to Step 5.

Step 5 (*Save the current optimal solution*) If $P < p$, set $P = p$, $(X_j = x_j, j = 1, \ldots, n)$; if $P = U$, stop. Otherwise $(P \geq p$ or $P \neq U)$, if $x_n = 1$, set $W = W + w_n$, $p = p - p_n$, $x_n = 0$; in any case, go to Step 6.

Step 6 (*Backtrack*) Find the largest $k < i$ for which $x_k = 1$. If no such k exists, stop. Otherwise, set $R = W$, $W = W + w_k$, $p = p - p_k$, $x_k = 0$. If $R \geq M_k$, set $i = k+1$ and go to 2. Otherwise, set $i = k$, $h = k+1$ and go to Step 7.

Step 7 (*Try to substitute the h-th object for the k-th*) If $h > n$ or $P \geq$

$p + \lfloor Wp_h/w_h \rfloor$, go to Step 6. Otherwise, set $D = w_h - w_k$; three possibilities exist:

(a) $D = 0$: set $h = h + 1$ and repeat Step 7.
(b) $D > 0$: if $D > R$ or $P \geq p + p_h$, set $h = h + 1$ and repeat Step 7. Otherwise, set $P = p + p_h$, $(X_j = x_j, j = 1, \ldots, k)$, $(X_j = 0, j = k+1, \ldots, n, j \neq h)$, $X_h = 1$. If $P = U$, stop; if not, set $R = R - D$, $k = h$, $h = h + 1$ and repeat Step 7.
(c) $D < 0$: if $R - D < M_h$, set $h = h + 1$ and repeat Step 7. Otherwise, if $P \geq p + p_h + \lfloor (R-D)p_h/w_h \rfloor$, go to Step 6; if not, set $W = W - w_h$, $p = p + p_h$, $x_h = 1$, $i = h + 1$, $\bar{w}_h = w_h$, $\bar{p}_h = p_h$, $\bar{z}_h = h + 1$, $(\bar{w}_j = \bar{p}_j = 0, \bar{z}_j = j$ for $j = h+1, \ldots, \bar{s})$, $\bar{s} = h$ and go to Step 2.

Vector (M_j), defined at Step 1, enables one to know whether, given the current value of W, at least one more object can be introduced into the solution, that is whether Steps 2 and 3 have to be performed. The aim of vectors $(\bar{p}_j)$, $(\bar{w}_j)$, $(\bar{z}_j)$ is to save parts of the current solution which could be reutilized: suppose in fact that a current solution has been built by introducing the objects from the ith to the sth: then, when trying to insert objects from an $\bar{i}$th $(i < \bar{i} \leq s)$, if nothing has changed in the current solution before the ith object, it is certainly possible to introduce into the new current solution the sequence of elements from the $\bar{i}$th to the sth.

A *forward move* is here split into Steps 2, 3, and 4. Step 2 is only a preliminary to the effective forward Step 3, where the new current solutions are built. In case 3(b), p^* cannot grow any more with the present value of i: so, if it is worthwhile, a new optimal solution is saved, but vector (x_j) is not updated, so that needless backtrackings on values $(x_i, \ldots, x_s)$ are avoided. In case 3(a), if the current solution found can—through subsequent forward moves—improve on the current optimal solution, then Step 4 is performed; otherwise the backtracking Step 6 immediately follows.

At step 6 the backtracking on the kth object is followed by a normal forward move only if R (value of W preceding the backtracking) is large enough to allow the introduction into the solution of at least one of the objects following the kth. Otherwise, a particular forward procedure (Step 7) is utilized, based on the following consideration: the current solution could be improved only if the kth object is replaced by an object (say the hth) having a greater p_h and a w_h small enough to allow its introduction, or by at least two objects (say the hth and the $(h+r)$th) with w_h and w_{h+r} less than w_k.

So, Step 7 considers the objects following the kth until the last one which can be of any use. In some cases (7(a); 7(b) with $D > R$ or $P \geq p + p_h$; 7(c) with $R - D < M_h$) the object considered is rejected, since it cannot improve on the solution or since it is too large. In case 7(b) with $D \leq R$ and $P < p + p_h$, if the replacement improves on the current optimal solution, P

The 0–1 Knapsack Problem

and vector (X_j) are updated and the search starts again from the new situation. In case 7(c) with $R - D \geq M_h$, the current solution is updated and a normal forward search then follows.

The local upper bound of Step 3(a) has been computed through the classical Dantzig's way, but it could also be computed similarly to the upper bound of the problem, that is the test could be: $P \geq p + p^* + \max\{\lfloor(W-w^*)p_{s+2}/w_{s+2}\rfloor, \lfloor p_{s+1}-(w_{s+1}-(W-w^*))p_s/w_s\rfloor\}$. In this case the new bound can also be proved to be better than Dantzig's, but the last one requires significantly fewer operations, which can make up for the larger number of forward and backtracking moves involved; so, an absolute strategy cannot be defined and only empirical studies on the particular data sets can suggest a choice between the two possibilities. A mixed strategy is also possible;

(i) If $P \geq p + p^* + \lfloor(W-w^*)p_{s+1}/w_{s+1}\rfloor$, go to Step 6.
(ii) If $P < p + p^* + \lfloor(W-w^*)p_{s+2}/w_{s+2}\rfloor$, go to Step 4.
(iii) If $P \geq p + p^* + \lfloor p_{s+1}-(w_{s+1}-(W-w^*))p_s/w_s\rfloor$, go to Step 6; otherwise, go to Step 4.

Example We will solve the same problem utilized to illustrate Horowitz–Sahni's algorithm: $n = 7$, $W = 50$,

$$(p_j) = (70, 20, 39, 37, 7, 5, 10)$$

$$(w_j) = (31, 10, 20, 19, 4, 3, 6)$$

Step 1 $s = 2$: $p^* = 90$, $w^* = 41 \neq W$. $(M_i) = (3, 3, 3, 3, 3, 6, \infty)$; $U = 90 + \max\{\lfloor 9 \times 37/19\rfloor, \lfloor 39 - 11 \times 20/10\rfloor\} = 107$, $P = p = 0$, $(x_j) = (0, 0, 0, 0, 0, 0, 0)$, $i = 1$, $\bar{s} = 7$.

Step 4 $W = 50 - 41 = 9$, $p = 0 + 90 = 90$, $x_1 = x_2 = 1$; $(\bar{w}_j) = (41, 10, 0, 0, 0, 0, 0)$, $(\bar{p}_j) = (90, 20, 0, 0, 0, 0, 0)$, $(\bar{z}_j) = (3, 3, 3, 4, 5, 6, 7)$, $\bar{s} = 2$. $s < n - 2 \rightarrow (a)$. $i = 4$.

Step 2 $w_4 > 9$, $P < 90 + \lfloor 9 \times 7/4\rfloor = 105$: $i = 5$.

Step 2 $w_5 < 9$.

Step 3 $s = 6$, $p^* = 0 + 12 = 12$, $w^* = 0 + 7 = 7 < W \rightarrow (a)$. $P < 90 + 12 + \lfloor 2 \times 10/6\rfloor = 105$.

Step 4 $W = 9 - 7 = 2$, $p = 90 + 12 = 102$, $x_5 = x_6 = 1$; $\bar{w}_5 = 7$, $\bar{p}_5 = 12$, $\bar{z}_5 = 7$, $\bar{w}_6 = 3$, $\bar{p}_6 = 5$, $\bar{z}_6 = 7$, $\bar{s} = 6$. $s = n - 1 \rightarrow (c)$. $i = 7$.

Step 5 $P = 102$, $(X_j) = (1, 1, 0, 0, 1, 1, 0)$; $P < U$.

Step 6 $k = 6$, $R = 2$, $W = 2 + 3 = 5$, $p = 102 - 5 = 97$, $x_6 = 0$. $R < M_6$: $i = 6$, $h = 7$.

Step 7 $P < 97 + \lfloor 5 \times 10/6\rfloor = 105$: $D = 9 - 6 = 3 > 0 \rightarrow (b)$. $D > R$: $h = 8$.

Step 7 $h > n$.

Step 6 $k = 5$, $R = 5$, $W = 5 + 4 = 9$, $p = 97 - 7 = 90$, $x_5 = 0$. $R > M_5$: $i = 6$.

Step 2 $w_6 < 9$.

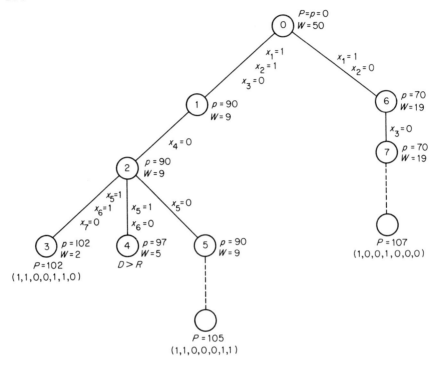

Figure 9.2 Decision tree for the algorithm of Martello and Toth

Step 3 $s = 7$, $p^* = 5 + 10 = 15$, $w^* = 3 + 6 = 9 = W \to (b)$. $P < 90 + 15$: $P = 105$, $(X_j) = (1, 1, 0, 0, 0, 1, 1)$; $P < U$.
Step 6 $k = 2$, $R = 9$, $W = 9 + 10 = 19$, $p = 90 - 20 = 70$, $x_2 = 0$. $R > M_2$: $i = 3$.
Step 2 $w_3 > 19$; $P < 70 + \lfloor 19 \times 37/19 \rfloor = 107$: $i = 4$.
Step 2 $w_4 = 19$.
Step 3 $s = 4$, $p^* = 0 + 37 = 37$, $w^* = 0 + 19 = 19 = W \to (b)$. $P < 70 + 37$: $P = 107$, $(X_j) = (1, 0, 0, 1, 0, 0, 0)$; $P = U$: stop.

Figure 9.2 shows the decision tree of the example.

9.4 Dynamic Programming Algorithms

Let us define for each integer $m (1 \leq m \leq n)$ and for each integer z $(0 \leq z \leq W)$:

$$f_m(z) = \max \left\{ \sum_{j=1}^m p_j x_j \, \Big| \, \sum_{j=1}^m w_j x_j \leq z, \, x_j = 0, 1 \text{ for } j = 1, \ldots, m \right\}. \quad (9.12)$$

The 0–1 Knapsack Problem

From (9.12) it follows that

$$f_1(z) = 0 \quad \text{for} \quad 0 \leq z < w_1$$
$$f_1(z) = p_1 \quad \text{for} \quad w_1 \leq z \leq W.$$

The recursive equations for the mth stage ($m = 2, \ldots, n$) are given by:

$$f_m(z) = f_{m-1}(z) \quad \text{for} \quad 0 \leq z < w_m$$
$$f_m(z) = \max \{f_{m-1}(z), f_{m-1}(z - w_m) + p_m\}, \quad \text{for} \quad w_m \leq z \leq W.$$

Toth (1977) presented a procedure derived directly from the above recursive equations; the following variables are assumed to be defined before the execution of the procedure at the mth stage:

$$v = \min \left\{ \sum_{j=1}^{m-1} w_j, W \right\}$$

$$b = 2^{m-1}$$

$$F_z = f_{m-1}(z) \quad \text{for} \quad z = 0, 1, \ldots, v$$
$$X_z = \{x_{m-1}, x_{m-2}, \ldots, x_1\}, \quad \text{for} \quad z = 0, 1, \ldots, v,$$

where x_j defines the value of the jth variable of the partial optimal solution corresponding to $f_{m-1}(z)$, i.e.

$$z = \sum_{j=1}^{m-1} w_j x_j \quad \text{and} \quad f_{m-1}(z) = \sum_{j=1}^{m-1} p_j x_j.$$

From a computational point of view, it is worthwhile to express each set X_z as a bit string, so this notation will be used in the following. After the execution of the procedure, v and the vectors (F_z) and (X_z) are relative to the mth stage.

Procedure P1

Step 1 If $v = W$, go to Step 4; otherwise, set $u = v$, $v = \min \{v + w_m, W\}$, $F_v = F_u$, $X_v = X_u$.

Step 2 Set $z = v - 1$.

Step 3 If $z \leq u$, go to Step 4; otherwise, set $F_z = F_u$, $X_z = X_u$, $z = z - 1$ and repeat Step 3.

Step 4 Set $y = v - w_m$, $g = F_y + p_m$. If $F_v < g$, set $F_v = g$, $X_v = X_y + b$.

Step 5 Set $z = v - 1$.

Step 6 If $z < w_m$, return.

Step 7 Set $y = z - w_m$, $g = F_y + p_m$. If $F_z < g$, set $F_z = g$, $X_z = X_y + b$; in any case, set $z = z - 1$ and go to Step 6.

In many cases it is possible to reduce greatly the number of states considered at a given stage eliminating all the *dominated states*, that is the states (F_z, X_z) for which there exists at least one state (F_y, X_y) having $F_y \geq F_z$ and $y < z$. This technique has been utilized by Horowitz and Sahni (1974) and Ahrens and Finke (1975).

The undominated states at the mth stage can be computed through the procedure proposed by Toth (1977). The following variables are assumed to be defined before the execution of the procedure at the mth stage:

s_{m-1} = number of states at stage $(m-1)$

$b = 2^{m-1}$

$L1_i$ = total weight of the ith state $(i = 1, \ldots, s_{m-1})$

$F1_i$ = total profit of the ith state $(i = 1, \ldots, s_{m-1})$

$X1_i = \{x_{m-1}, x_{m-2}, \ldots, x_1\}$, for $i = 1, \ldots, s_{m-1}$,

where values x_j represent the partial solution of the ith state, that is,

$$L1_i = \sum_{j=1}^{m-1} w_j x_j \quad \text{and} \quad F1_i = \sum_{j=1}^{m-1} p_j x_j.$$

After the execution of the procedure, the total weight, the total profit, and the set of the partial solution relative to state k of the mth stage are represented, respectively, by $L2_k$, $F2_k$, and $X2_k$.

The sets $X1_i$ and $X2_k$ are also expressed as bit strings. The vectors $(L1_i)$, $(F1_i)$, $(L2_k)$, and $(F2_k)$ are ordered according to increasing values.

Procedure P2

Step 1 Set $L1_0 = F1_0 = X1_0 = F2_0 = 0$, $h = 1$, $k = 0$, $i = 0$, $y = w_m$.
Step 2 Three possibilities exist:
  (a) $L1_h < y$: if $F1_h > F2_k$, set $k = k + 1$, $L2_k = L1_h$, $F2_k = F1_h$, $X2_k = X1_h$. In any case, if $h = s_{m-1}$, go to Step 3; otherwise, set $h = h + 1$ and repeat Step 2.
  (b) $L1_h > y$: set $g = F1_i + p_m$. If $g > F2_k$, set $k = k + 1$, $L2_k = y$, $F2_k = g$, $X2_k = X1_i + b$; in any case, set $i = i + 1$, $y = L1_i + W_m$ and repeat Step 2.
  (c) $L1_h = y$: set $g = F1_i + p_m$, $x = X1_i + b$, $i = i + 1$, $y = L1_i + w_m$. If $F1_h > g$, set $g = F1_h$, $x = X1_h$. If $g > F2_k$, set $k = k + 1$, $L2_k = y$, $F2_k = g$,

The 0–1 Knapsack Problem

$X2_k = x$; in any case, if $h = s_{m-1}$, go to Step 3; otherwise, set $h = h + 1$ and repeat Step 2.

Step 3 If $y > W$, go to Step 4; otherwise, set $g = F1_i + p_m$. If $g > F2_k$, set $k = k + 1$, $L2_k = y$, $F2_k = g$, $X2_k = X1_i + b$; in any case, if $i = s_{m-1}$, go to Step 4; otherwise, set $i = i + 1$, $y = L1_i + w_m$ and repeat Step 3.

Step 4 Set $s_m = k$; return.

It must be noted that the maximum value of s_m is given by min $\{2^m - 1, W\}$.

Procedure P2 requires no specific ordering of the variables $x_1, x_2, \ldots, x_m$. However, its efficiency greatly increases if the variables are ordered according to decreasing values of the ratios p_j/w_j, because then the number of undominated states at each stage is reduced; therefore in the following this ordering will be assumed.

Example Let $n = 6$, $W = 190$,

$$(p_j) = (50, 50, 64, 46, 50, 5)$$

$$(w_j) = (56, 59, 80, 64, 75, 17)$$

Table 9.1 gives the values of the total weights L_i and the total profits F_i of the undominated states at each stage. The optimal solution of the problem is $(x_j) = (1, 1, 0, 0, 1, 0)$ with a maximum profit of 150. The total number of the undominated states is 29. Applying procedure P1 the number of states should be 866.

Table 9.1

i	$m=1$		$m=2$		$m=3$		$m=4$		$m=5$		$m=6$	
	L_i	F_i	L_i	F_i	L_i	F_i	L_i	F_i	L_i	F_i	L_i	F_i
1	56	50	56	50	56	50	56	50	56	50	17	5
2			115	100	80	64	80	64	80	64	56	50
3					115	100	115	100	115	100	73	55
4					136	114	136	114	136	114	80	64
5							179	146	179	146	97	69
6									190	150	115	100
7											132	105
8											136	114
9											153	119
10											179	146
11											190	150

9.4.1 The Algorithm of Horowitz and Sahni

Horowitz and Sahni (1974) presented an algorithm based on the subdivision of the original problem of n variables into two subproblems respectively of $q = \lceil n/2 \rceil$ and $r = n - q$ variables. For each subproblem a list containing all the undominated states relative to the last (respectively the qth and the rth) stage is computed; then the two lists are merged in order to find the optimal solution to the original problem.

The main feature of Horowitz–Sahni's algorithm is given by the property of having, for the worst cases, two lists each of $(2^q - 1)$ states, instead of a list of $(2^n - 1)$ states as required by the original problem. So, for the worst cases, the algorithm reduces computing times and storage requirements by a square root factor, with respect to a direct application or procedure P2 to the original problem. However, in almost the totality of problems, the number of the undominated states is much less than the corresponding maximum number, both because many states are dominated and because generally the value of W is much less than such maximum number,† so the improvement given by Horowitz–Sahni's algorithm is greatly reduced.

Ahrens and Finke (1975) proposed an algorithm where the technique utilized by Horowitz and Sahni is combined with a branch and bound procedure in order to reduce the storage requirements. This algorithm works very well for hard problems having low values of n and very high values of w_i and W, but has the disadvantage that the branch and bound procedure is always executed, even if the storage requirements are not excessive and therefore its execution could be avoided.

Example Let us consider the previous example in order to illustrate Horowitz–Sahni's algorithm. Table 9.2 gives the undominated states at each

Table 9.2

i	$m=1$		$m=2$		$m=3$		$m=6$		$m=5$		$m=4$	
	L_i	F_i	L_i	F_i	L_i	F_i	L_i	F_i	L_i	F_i	L_i	F_i
1	56	50	56	50	56	50	17	5	64	46	64	46
2			115	100	80	64	64	46	75	50		
3					115	100	75	50	139	96		
4					136	114	81	51				
5							92	55				
6							139	96				
7							156	101				

† Consider that for $n = 60$, average value of $w_i = 100$ and $W = 0.5 \sum_{j=1}^{n} w_j$, we have $2^q - 1 = 2^{30} - 1 \simeq 10^9$ and $W \simeq 3000$.

The 0–1 Knapsack Problem

stage of the two subproblems ($q = 3$). The number of undominated states is now 18.

9.4.2 The Algorithm of Toth

Toth (1977) presented a dynamic programming algorithm based on the elimination of the *unutilized states* and on the combination of procedures P1 and P2.

Elimination of the Unutilized States

Several states defined through the application, at a given stage, of procedures P1 and P2 are never utilized in the following stages.

The following rule can be applied for the elimination of the unutilized states.

If a state, defined at the mth stage, has a total weight L such that at least one of the conditions

(i) $$L < A_m = \max\left\{1, W - \sum_{j=m+1}^{n} w_j\right\}$$

(ii) $$B_m = W - \min_{m < j \leq n}\{w_j\} < L < W$$

holds, then the state will be never utilized in the stages following the mth one.

The following changes can be made to procedure P1. Replace Steps 2 and 5 with
  Set $z = \min\{v - 1, B_m\}$.
Replace Step 6 with
  If $z < \max\{w_m, A_m\}$, return.

After the execution of Procedure P2, a state i satisfying one of the conditions

$$L2_i < A_m \text{ and } L2_{i+1} \leq A_m$$
$$B_m < L2_i < W$$

can be eliminated. A remaining state k is therefore:
$$k = s_m \text{ or } r_m \leq k \leq q_m$$

where
  r_m such that $L2_{r_m - 1} < A_m$, $L2_{r_m} \geq A_m$†
  $q_m = \max\{i \mid L2_i \leq B_m, i < s_m\}$.

† It is assumed $L2_0 = 0$.

Table 9.3

i	$m=1$ $A_1=1$ $B_1=173$		$m=2$ $A_2=1$ $B_2=173$		$m=3$ $A_3=34$ $B_3=173$		$m=4$ $A_4=98$ $B_4=173$		$m=5$ $A_5=173$ $B_5=173$		$m=6$ $A_6=190$ $B_6=190$	
	L_i	F_i	L_i	F_i	L_i	F_i	L_i	F_i	L_i	F_i	L_i	F_i
1	56	50	56	50	56	50	80	64	136	114	190	150
2			115	100	80	64	115	100	190	150		
3					115	100	136	114				
4					136	114	179	146				

In the following, the modified versions of procedures P1 and P2 will be called, respectively, procedure P1(a) and procedure P2(a).

The above rules (i) and (ii) cannot obviously be applied to Horowitz–Sahni's algorithm, since the merging procedure requires all the undominated states relative to the two subproblems.

Example For the example previously considered, Table 9.3 gives the results obtained by means of the elimination of the unutilized states. The total number of states is 14.

Combination of Procedures P1 and P2

Generally, the utilization of the algorithms derived from procedure P2 (i.e. Horowitz–Sahni's algorithm) and procedure P2(a) greatly reduces the number of the undominated states and, consequently, the computation times and the storage requirements. It must be noted however that, for the calculation of a single state, the computation times and the storage requirements of procedure P2 are greater than those corresponding to the utilization of procedure P1(a). In fact, with regard to the storage requirements, procedure P2 needs 3 words (total weight, total profit, and partial solution) for each state of the current stage and of the previous one, while on the other hand procedure P1(a) needs only 2 words (total profit and partial solution) for each state of the current stage. Besides, with regard to the computation times, it clearly appears from the detailed steps of the procedures that the time required for the calculation of a single state in procedure P2 is greater (by a factor of 4) than the corresponding one in procedure P1(a). From these considerations it follows that the utilization of procedure P2 is worthwhile only in those cases—quite common—where the number of undominated states generated at each stage is much less than the number of states generated by procedure P1(a).

However, when 'hard' problems are considered, the number of states generated by Horowitz–Sahni's algorithm or by procedure P2(a) tends to be almost equal to the number of states generated by procedure P1(a); therefore, for such problems, the latter procedure must be preferred both from the computation times and storage requirements points of view.

A dynamic programming algorithm which efficiently solves both easy and hard problems can be obtained combining the best characteristics of the two approaches. This can be achieved by utilizing procedure P2(a) as long as the number of the generated states is low, and then by utilizing procedure P1(a). A simple heuristic rule to determine automatically, during the execution of the algorithm, the stage when it is worthwhile to change the procedure can be obtained in the following way.

For the current stage, say m ($m < n$), let us define:

$$\text{NS1}_m = \min \left\{ \sum_{j=1}^{m} w_j, B_m + 1 \right\} - A_m + 1$$

= number of states at stage m when procedure P1(a) is utilized

NS2_m = current value of s_m = number of states at stage m when procedure P2(a) is utilized

R = (average computation time of one state utilizing procedure P2(a))/ (average computation time of one state utilizing procedure P1(a)).

If condition

$$\text{NS2}_m > \text{NS1}_m / R$$

holds, then at stage m it is worthwhile to pass from procedure P2(a) to procedure P1(a).

It must be noted that the computation of NS1_m requires no extra time because it is independent of the states currently defined at stage m. In addition, it is worthwhile to point out that the combination of procedures P2 and P1 cannot obviously be applied with good results to Horowitz–Sahni's algorithm.

Example A case showing the bad performance of procedure P2(a), when applied to a hard problem, is now considered. Let $n = 7$, $W = 31$, $(p_j) = (5, 6, 7, 9, 11, 16, 16)$, $(w_j) = (4, 5, 6, 8, 10, 15, 15)$. Table 9.4 gives the undominated states and the value of NS1_m and NS2_m at each stage m.

Reduction Procedure

The number of variables and the value of W of the zero–one knapsack problem can be reduced through the use of a *reduction procedure* presented by Ingargiola and Korsh (1973) and improved by Toth (1976).

Table 9.4

	$m=1$ $A_1=1$ $B_1=26$ $\text{NS}1_1=4$ $\text{NS}2_1=1$		$m=2$ $A_2=1$ $B_2=25$ $\text{NS}1_2=9$ $\text{NS}2_2=3$		$m=3$ $A_3=1$ $B_3=23$ $\text{NS}1_3=15$ $\text{NS}2_3=7$		$m=4$ $A_4=1$ $B_4=21$ $\text{NS}1_4=22$ $\text{NS}2_4=15$		$m=5$ $A_5=1$ $B_5=16$ $\text{NS}1_5=17$ $\text{NS}2_5=12$		$m=6$ $A_6=16$ $B_6=16$ $\text{NS}1_6=2$ $\text{NS}2_6=2$		$m=7$ $A_7=31$ $B_7=31$	
i	L_i	F_i	L_i	F_i	L_i	F_i	L_i	F_i	L_i	F_i	L_i	F_i	L_i	F_i
1	4	5	4	5	4	5	4	5	4	5	16	18	31	34
2			5	6	5	6	5	6	5	6	31	34		
3			9	11	6	7	6	7	6	7				
4					9	11	8	9	8	9				
5					10	12	9	11	9	11				
6					11	13	10	12	10	12				
7					15	18	11	13	11	13				
8							12	14	12	14				
9							13	15	13	15				
10							14	16	14	16				
11							15	18	15	18				
12							17	20	31	33				
13							18	21						
14							19	22						
15							23	27						

The procedure partitions the set $N = \{1, 2, \ldots, n\}$ into three subsets I_0, I_1, and R such that

$$x_j = 0 \forall j \in I_0, \quad x_j = 1 \forall j \in I_1, \quad R = N - I_1 - I_2.$$

The original knapsack problem can so be transformed into the reduced form

$$\max \sum_{j \in R} p_j x_j + \sum_{j \in I_1} p_j$$

subject to $\sum_{j \in R} w_j x_j \leq W - \sum_{j \in I_1} w_j, \quad x_j = 0, 1 \forall j \in R.$

Let LB be a lower bound of the solution to the problem defined by (9.1), (9.2), and (9.6); subsets I_0 and I_1 can then be obtained in the following way.

Let us compute, for any $i \in N$, the upper bounds UB_i and $\overline{\text{UB}}_i$ of the solutions to the problems defined, respectively, by (9.1), (9.2), (9.6), $x_i = 1$ and (9.1), (9.2), (9.6), $x_i = 0$; we then have

$$I_0 = \{i \in N | \text{UB}_i < \text{LB}\},$$
$$I_1 = \{i \in N | \overline{\text{UB}}_i < \text{LB}\}.$$

The 0–1 Knapsack Problem

The lower bound LB is obtained through a heuristic solution of the original problem; the upper bounds UB_i and $\overline{\text{UB}}_i$ can be computed efficiently through theorem 9.2. The detailed steps of the reduction procedure follow.

Procedure R

Step 1 (*Initialize*) Order the objects according to decreasing values of p_j/w_j. Compute $P = \sum_{j=1}^{s} p_j$, with $s = \max \{k \in N | \sum_{j=1}^{k} w_j < W\}$. Set $\overline{W} = W - \sum_{j=1}^{s} w_j$, $r = s+1$. If $\overline{W} = w_r$, set $I_1 = \{j \in N | j \leq r\}$, $I_0 = \{j \in N | j > r\}$ and stop. Otherwise, set $w_{n+1} = w_{n+2} = \infty$, $p_{n+1} = p_{n+2} = 0$, $w_0 = 1$, $p_0 = p_1$, $\text{LB} = P$, $i = 1$.

Step 2 (*Compute $\overline{\text{UB}}_i$ for $i \leq r$*) Compute $P^* = P - p_i + \sum_{j=r}^{s} p_j$, with $s = \max \{k \in N | \sum_{j=r}^{k} w_j \leq \overline{W} + w_i, \ k \geq r-1\}$. Set $C = \overline{W} + w_i - \sum_{j=r}^{s} w_j$, $\overline{\text{UB}}_i = P^* + \max \{\lfloor Cp_{s+2}/w_{s+2} \rfloor, \lfloor p_{s+1} - (w_{s+1}-C)p_s/w_s \rfloor\}$, $\text{LB} = \max \{\text{LB}, P^*\}$. If $i < r$, set $i = i+1$ and repeat Step 2.

Step 3 (*Compute UB_i for $i \geq r$*) Compute $P^* = P + p_i - \sum_{j=s}^{r-1} p_j$, with $s = \max \{k \in N | \sum_{j=k}^{r-1} w_j + \overline{W} \geq w_i, \ k \leq r\}$. Set $C = \overline{W} + \sum_{j=s}^{r-1} w_j - w_i$, $\text{UB}_i = P^* + \max \{\lfloor Cp_{s+1}/w_{s+1} \rfloor, \lfloor p_s - (w_s - C)p_{s-1}/w_{s-1} \rfloor\}$, $\text{LB} = \max \{\text{LB}, P^*\}$. If $i < n$, set $i = i+1$ and repeat Step 3.

Step 4 (*Build subsets I_1 and I_0*) Set $I_1 = \{i \in N | \overline{\text{UB}}_i < \text{LB}, \ i \leq r\}$. Compute $W^* = W - \sum_{i \in I_1} w_i$. Set $I_0 = \{i \in N | \text{UB}_i < \text{LB}, \ i \geq r\} \cup \{i \in N - I_1 | w_i > W^*\}$, $R = N - I_1 - I_0$. If $W^* = \sum_{i \in R} w_i$, set $I_1 = I_1 \cup R$, $R = \emptyset$; in any case, stop.

It must be pointed out that it is useless to compute UB_i for $i < r$ or $\overline{\text{UB}}_i$ for $i > r$, because these values coincide with the upper bound of the solution to the original problem, and are therefore greater than or equal to the final value of LB.

In Step 4, a further reduction of the number of elements of subset R has been obtained by imposing constraints (9.5) and (9.7) for the reduced problem.

Example Let us consider the example used in Sections 9.3.1 and 9.3.2: $n = 7$, $W = 50$,

$$(p_j) = (70, 20, 39, 37, 7, 5, 10)$$

$$(w_j) = (31, 10, 20, 19, 4, 3, 6).$$

Step 1 $s = 2$, $P = 90$, $\overline{W} = 9$, $r = 3$. $w_8 = w_9 = \infty$, $p_8 = p_9 = 0$, $w_0 = 1$, $p_0 = 70$, $\text{LB} = 90$, $i = 1$.
Step 2 $s = 4$, $P^* = 90 - 70 + 76 = 96$, $C = 9 + 31 - 39 = 1$, $\overline{\text{UB}}_1 = 96 + \max \{\lfloor 1 \times 5/3 \rfloor, \lfloor 7 - 3 \times 37/19 \rfloor\} = 97$, $\text{LB} = 96$, $i = 2$.
Step 2 $s = 2$, $P^* = 90 - 20 + 0 = 70$, $C = 9 + 10 - 0 = 19$, $\overline{\text{UB}}_2 = 70 + \max \{\lfloor 19 \times 37/19 \rfloor, \lfloor 39 - 1 \times 20/10 \rfloor\} = 107$, $\text{LB} = 96$, $i = 3$.

Step 2 $s=3$, $P^*=90-39+39=90$, $C=9+20-20=9$, $\overline{\text{UB}}_3=90+\max\{\lfloor 9\times 7/4\rfloor, \lfloor 37-10\times 39/20\rfloor\}=107$, LB $=96$.
Step 3 $s=1$, $P^*=90+39-90=39$, $C=9+41-20=30$, $\overline{\text{UB}}_3=39+\max\{\lfloor 30\times 20/10\rfloor, \lfloor 70-1\times 70/1\rfloor\}=99$, LB $=96$, $i=4$.
Step 3 $s=2$, $P^*=90+37-20=107$, $C=9+10-19=0$, $\text{UB}_4=107+\max\{0, \lfloor 20-10\times 70/31\rfloor\}=107$, LB $=107$, $i=5$.
Step 3 $s=3$, $P^*=90+7-0=97$, $C=9+0-4=5$, $\text{UB}_5=97+\max\{\lfloor 5\times 37/19\rfloor, \lfloor 39-15\times 20/10\rfloor\}=106$, LB $=107$, $i=6$.
Step 3 $s=3$, $P^*=90+5-0=95$, $C=9+0-3=6$, $\text{UB}_6=95+\max\{\lfloor 6\times 37/19\rfloor, \lfloor 39-14\times 20/10\rfloor\}=106$, LB $=107$, $i=7$.
Step 3 $s=3$, $P^*=90+10-0=100$, $C=9+0-6=3$, $\text{UB}_7=100+\max\{\lfloor 3\times 37/19\rfloor, \lfloor 39-17\times 20/10\rfloor\}=105$, LB $=107$.
Step 4 $I_1=\{1\}$, $W^*=50-31=19$; $I_0=\{3,5,6,7\}\cup\{3\}=\{3,5,6,7\}$; $R=\{2,4\}$.

A faster, but less efficient, reduction can be obtained through the following procedure of Dembo and Hammer (1975):

Order the objects according to decreasing p_i/w_i ratios. Compute LB $=\sum_{j=1}^{s} p_j$ with $s=\max\{k\in N|\sum_{j=1}^{k} w_j \leq W\}$ and set $\bar{W}=W-\sum_{j=1}^{s} w_j$, UB $=$ LB $+\bar{W}p_{s+1}/w_{s+1}$; for $j=s+2,\ldots,n$, if $w_j\leq\bar{W}$, set $\bar{W}=\bar{W}-w_j$ and LB $=$ LB $+p_j$ (this heuristic solution is generally called *greedy solution*).

Set $I_1=\{i\in N|p_i-w_ip_{s+1}/w_{s+1}>\text{UB}-\text{LB}, i\leq s\}$. Set $I_0=\{i\in N|-p_i+w_ip_{s+1}/w_{s+1}>\text{UB}-\text{LB}, i>s+1\}$.

The procedure can be easily justified by considering that upper bounds $\overline{\text{UB}}_i$ and UB_i (worse than the previously defined ones) can be obtained by setting

$$\overline{\text{UB}}_i = \sum_{j=1}^{s} p_j - p_i + \left(w_i + W - \sum_{j=1}^{s} w_j\right)p_{s+1}/w_{s+1}$$
$$= \text{UB} - p_i + w_ip_{s+1}/w_{s+1}$$

$$\text{UN}_i = \sum_{j=1}^{s} p_j + p_i - \left(w_i - W + \sum_{j=1}^{s} w_j\right)p_{s+1}/w_{s+1}$$
$$= \text{UB} + p_i - w_ip_{s+1}/w_{s+1}.$$

Dembo–Hammer's procedure can also be employed as a preprocessing of procedure R.

9.6 Dominance and Fathoming Criteria

As shown in Sections 9.3 and 9.4, the zero–one knapsack problem can be efficiently solved by means of branch and bound or dynamic programming algorithms. This consideration leads to the attempt to combine the two

The 0–1 Knapsack Problem 261

approaches in order to improve on the performance of the algorithms; so, in the present section, the introduction of dominance criteria in a branch and bound algorithm and the utilization of upper bounds in a dynamic programming procedure will be analysed.

9.6.1 Dominance Criteria in the Branch and Bound Algorithms

Each node k of the decision tree generated during the execution of the branch and bound algorithms given in Section 9.3 and corresponding to the current solution $(x_1, x_2, \ldots, x_i)$, can be described by the state (L_k, F_k, I_k), where

$$L_k = \sum_{j=1}^{i} w_j x_j \text{ is the total weight}$$

$$F_k = \sum_{j=1}^{i} p_j x_j \text{ is the total profit}$$

$$I_k = i+1 \quad \text{is the next object to be examined.}$$

It is now possible to give the following *dominance criterion*:

Given two nodes k and h, if the relations

$$L_k \leq L_h, \quad F_k \geq F_h, \quad I_k \leq I_h$$

hold, then node h is dominated by node k and can be eliminated from the decision tree.

The values L_k and I_k of any node k can be increased by means of the following considerations:

(i) Set $I_k = m = \min \{j | w_j \leq W - L_k, i < j \leq n\}$; if no such m exists, set $I_k = n+1$, $L_k = W$.

(ii) If $W - L_k > \sum_{j=I_k}^{n} w_j$, set $L_k = W - \sum_{j=I_k}^{n} w_j$.

The dominance criterion can improve on the performance of the branch and bound algorithms when hard problems are to be solved, since for such problems many nodes of the decision tree having near values of L_k, F_k and I_k are generated. With respect to the dynamic programming procedures, the introduction of the dominance criterion in a branch and bound algorithm leads to a smaller number of states to be generated and stored, but needs the simultaneous storage of all the undominated nodes generated during the execution of the algorithm, and not only of the states corresponding to the current stage and to the previous one. This last disadvantage can be overcome by storing only a subset of all the undominated nodes; this obviously implies a reduction in the efficiency of the dominance criterion.

9.6.2 Fathoming Criteria in the Dynamic Programming Procedures

The computation of upper bounds can be introduced in the dynamic programming procedures in order to *fathom* the states not leading to optimal solutions. This approach has been firstly proposed by Morin and Marsten (1976) and applied to the solutions of the travelling salesman problem and of the nonlinear knapsack problem. Here the application to the zero–one knapsack problem presented by Toth (1977) is described.

At each stage m $(1 < m < n)$ a lower bound LB_m of the optimal solution to the original problem can be given by

$$\text{LB}_m = \max \{F^* + \beta(m, L^*), \text{LB}_{m-1}\}$$

where F^* and L^* are the total profit and the total weight corresponding to the last state of stage m and $\beta(m, L^*)$ is a lower bound of the solution to the subproblem defined by

$$\max \sum_{j=m+1}^{n} p_j x_j$$

$$\text{subject to} \sum_{j=m+1}^{n} w_j x_j \leq W - L^*; \; x_j = 0, 1, \text{ for } j = m+1, \ldots, n.$$

The value $\beta(m, L^*)$ can be obtained by means of a heuristic procedure, for example by inserting in the knapsack as many objects having the highest value of the ratio p_j/w_j as possible (*greedy solution*).

For each state k defined at the mth stage and having total weight L_k and total profit F_k, an upper bound UB_{m,L_k} to the solution of the subproblem defined by

$$\max \sum_{j=m+1}^{n} p_j x_j$$

$$\text{subject to} \sum_{j=m+1}^{n} w_j x_j \leq W - L_k; \; x_j = 0, 1, \text{ for } j = m+1, \ldots, n$$

can be computed through the application of Theorem 9.2.

If the condition

$$F_k + \text{UB}_{m,L_k} \leq \text{LB}_m$$

holds, then state k can be fathomed.

It is to be noted that the fathoming criteria cannot be inserted in procedure P1(a) because such procedure needs, at each stage, all the states

Table 9.5

i	m = 1, LB$_1$ = 146		m = 2, LB$_2$ = 146			m = 3, LB$_3$ = 146			m = 4, LB$_4$ = 146			m = 5, LB$_5$ = 150			m = 6	
	L$_i$	F$_i$	L$_i$	F$_i$	UB$_i$	L$_i$	F$_i$	UB$_i$	L$_i$	F$_i$	UB$_i$	L$_i$	F$_i$	UB$_i$	L$_i$	F$_i$
1	56	50	56	50	102	56	50	92*	115	100	50	136	114	5*	190	150
2			115	100	—	80	64	75*	136	114	34	190	150	—		
3						115	100	50	179	146	—					
4						136	114	—								

having total weights in a given interval. For the computation of the upper bounds corresponding to all the states defined at the mth stage ($1 < m < n$), the following steps can be added to procedure P2(a):

Step 1 Set $k = q_m$, $j = m + 1$, $S = P = 0$, $w_{n+1} = w_{n+2} = \infty$, $p_{n+1} = p_{n+2} = 0$.

Step 2 Set $c = W - L2_k$.

Step 3 If $c \leq S + w_j$, go to Step 4; otherwise, set $S = S + w_j$, $P = P + p_j$, $j = j + 1$ and repeat Step 3.

Step 4 Set $\text{UB} = P + \max \{\lfloor (c - S)p_{j+1}/w_{j+1} \rfloor, \lfloor p_j - (w_j - (c - S))p_{j-1}/w_{j-1} \rfloor\}$. If $F2_k + \text{UB} \leq \text{LB}_m$, fathom state k. In any case, if $k = r_m$, return; otherwise, set $k = k - 1$ and go to Step 2.

The efficiency of the fathoming criteria tends to increase when high values of m are considered, since, as m grows, both the lower bound LB$_m$ increases, and because of the ordering assumed for the objects, the upper bounds decrease. Besides, with regard to the Horowitz–Sahni algorithm, it is worthwhile to note that for the most part of the stages of the subproblem corresponding to the last $(n - \lceil n/2 \rceil)$ variables, the efficiency of the fathoming criteria decreases greatly, both because the lower bounds LB$_m$ are low and because only states having low values of the total weight are considered.

The application of the fathoming criteria to the example utilized in Sections 9.4, 9.4.1, and 9.4.2 gives the results shown in Table 9.5 (the total number of states is 10); an asterisk indicates the fathomed states.

9.7 Computational Results

The present section analyses the computational performance of the algorithms described in the previous sections. Since the computation times greatly depend on the correlation between profits and weights, three uniformly randomly generated data sets, with different degrees of correlation, will be considered:

Data Set UCR (Uncorrelated): $w_{min} \leq w_j \leq w_{max}$ $(j = 1, \ldots, n)$
$p_{min} \leq p_j \leq p_{max}$ $(j = 1, \ldots, n)$.

Data Set WCR (Weakly Correlated): $w_{min} \leq w_j \leq w_{max}$ $(j = 1, \ldots, n)$
$w_j - r \leq p_j \leq w_j + s$ $(j = 1, \ldots, n)$.

Data Set SCR (Strongly Correlated) $w_{min} \leq w_j \leq w_{max}$ $(j = 1, \ldots, n)$
$p_j = w_j + c$ $(j = 1, \ldots, n)$.

All the considered algorithms require that the objects are arranged according to decreasing values of the ratio p_j/w_j: the average sorting times depend mainly on the value of n and are shown in Table 9.6. This table also gives the average times required to reduce the problems through procedure R of Section 9.5 (such times are also roughly independent of data sets); the percentage of remaining objects was approximately 5–10 per cent for data set UCR, 15–35 per cent for data set WCR, and 50 per cent for data set SCR.

Table 9.6 Sorting and reduction times. Entries are the average times for 200 problems

n	50	100	200	500	1000	2000	5000	10000
Sorting time	2	5	12	32	72	156	437	941
Reduction time	4	7	15	37	72	139	298	551

The times of Table 9.6, as well as the times of all the tables of the present section, are expressed in milliseconds, are relative to a CDC-6600 computer, and have been obtained from sets of 200 problems each.

Sorting was obtained through the COMPASS Subroutine SORTZV of the CERN library; the reduction algorithm, as well as all the algorithms analysed in the following tables, was coded in FORTRAN IV.

Tables 9.7, 9.8, and 9.9 compare, on medium-size problems, the branch and bound algorithms of Horowitz–Sahni (BBHS) and of Martello–Toth (BBMT) and the dynamic programming algorithms of Horowitz–Sahni (DPHS) and of Toth (DPT1 and DPT2, relative, respectively, to Sections 9.4.2 and 9.6.2). The branch and bound algorithms have been run both directly and by previous application of reduction procedure R; the dynamic programming algorithms, being greatly affected by the value of n and of W,

The 0–1 Knapsack Problem

Table 9.7 Uncorrelated data sets; $w_{\min} = p_{\min} = 1$, $w_{\max} = p_{\max} = 100$. Average times for 200 problems, sorting time excluded

W	n	BBHS	BBMT	R+BBHS	R+BBMT	R+DPHS	R+DPT1	R+DPT2
200	50	8	4	4	4	5	6	5
	100	16	5	8	7	10	9	9
	200	29	7	14	13	16	14	15
$0.5 \sum_{j=1}^{n} w_j$	50	13	6	6	5	8	8	8
	100	35	12	10	9	20	21	17
	200	109	21	18	15	56	41	32
$0.8 \sum_{j=1}^{n} w_j$	50	15	6	4	4	5	5	5
	100	47	11	8	7	10	10	10
	200	167	20	15	14	20	19	18

Table 9.8 Weakly correlated data sets; $w_{\min} = 1$, $w_{\max} = 100$, $r = s = 10$. Average times for 200 problems, sorting time excluded

W	n	BBHS	BBMT	R+BBHS	R+BBMT	R+DPHS	R+DPT1	R+DPT2
200	50	13	7	7	5	16	13	12
	100	21	10	9	8	25	16	16
	200	35	13	14	13	32	20	20
$0.5 \sum_{j=1}^{n} w_j$	50	17	10	11	8	79	46	34
	100	35	15	16	11	303	93	72
	200	84	19	23	16	902	163	117
$0.8 \sum_{j=1}^{n} w_j$	50	19	9	11	9	73	48	39
	100	46	14	16	11	214	87	65
	200	137	17	23	16	664	166	110

only by previous application of R. The average times given in the tables are comprehensive of the reduction times but not of the sorting times.

The data sets have been obtained by using the following values:

$$w_{\min} = p_{\min} = 1$$

$$w_{\max} = p_{\max} = 100$$

$$r = s = 10$$

$$c = 10.$$

Three different values of W have been considered.

Table 9.9 Strongly correlated data sets; $w_{min} = 1$, $w_{max} = 100$, $c = 10$. Average times for 200 problems; sorting time excluded

W	n	BBHS	BBHT	R+BBHS	R+BBMT	R+DPHS	R+DPT1	R+DPT2
	50	130	17	51	15	27	16	19
200	100	510	39	189	35	49	24	29
	200	1796*	210	1373*	182	79	36	44
	50	12 probl.	1210	12 probl.	1025	364	148	179
$0.5 \sum_{j=1}^{n} w_j$	100	—	3 probl.	—	8 probl.	1673*	548	702
	200	—	—	—	—	—	1506*	1957*
	50	1507*	1225*	1409*	1094	332	151	192
$0.8 \sum_{j=1}^{n} w_j$	100	—	—	—	3 probl.	1326*	487	683
	200	—	—	—	—	—	1476*	2117*

A time-limit of 250 seconds was assigned to each algorithm for the solution of the 600 problems relative to each value of W. Whenever this time was not large enough to solve the whole set of problems, the average times were given only if the number of problems solved was significant (such situations are indicated by an asterisk); otherwise, only this number was given.

For uncorrelated data sets (Table 9.7), all the algorithms, with the exception of BBHS, have about the same computational performance and a linear dependence on n; the best method seems to be BBMT. It is to be noted that BBMT is only slightly improved by the previous application of R, contrary to what happens for BBHS.

Branch and bound algorithms clearly perform better than dynamic programming algorithms for weakly correlated data sets (Table 9.8). Again the best method is BBMT, mainly for large values of W; among the dynamic programming procedures the best one is DPT2, mainly for high values of n.

The general trend of Table 9.8 reverses itself in Table 9.9 (strongly correlated data sets), where the best performances are obtained by dynamic programming algorithms. In this case the best method was DPT1; among the branch and bound methods, BBMT confirms its superiority over BBHS.

9.8 Large Size Problems

In many practical applications knapsack problems with a high number of variables have to be solved: enumerative methods are quite efficient for such

The 0–1 Knapsack Problem

cases, while dynamic programming algorithms cannot generally be applied because of their high storage requirements.

When solving large size knapsack problems through the branch and bound methods of Section 9.3, the problem arises of the relatively high computing time required to sort the objects. A comparison between Tables 9.6 and 9.7 shows, for example, that the percentage of such time over the total computational effort—relative to uncorrelated data sets solved through reduction procedure R and Martello-Toth's algorithm—is approximately 32% when $n = 50$, 39% when $n = 100$, 46% when $n = 200$. Computational experiences with large-size problems (Martello and Toth, 1977b) have shown a further increase in this percentage.

9.8.1 The Algorithm of Balas and Zemel

In order to reduce the computational effort, Balas and Zemel (1977) have proposed an algorithm to solve zero–one knapsack problems without sorting the objects.

The method is based on the definition of what the authors call the core problem. Suppose the objects ordered according to decreasing p_j/w_j ratios: for an optimal solution $(\hat{x}_j)$ to the zero–one knapsack problem, let

$$j_* = \min\{j | \hat{x}_j = 0\}$$
$$j_{**} = \max\{j | \hat{x}_j = 1\}$$
$$j_1 = \min\{j_*, j_{**}\}$$
$$j_2 = \max\{j_*, j_{**}\}.$$

Then the *Core Problem* is defined as

$$\max \sum_{j=j_1}^{j_2} x_j p_j$$

$$\text{subject to } \sum_{j=j_1}^{j_2} x_j w_j \leq W - \sum_{j=1}^{j_1-1} w_j$$

$$x_j = 0 \text{ or } 1 \quad (j = j_1, \ldots, j_2).$$

While the core problem cannot be identified without solving the given zero–one knapsack problem, a satisfactory approximation can be found by solving the associated continuous problem. To this end, instead of employing Dantzig's method (see Theorem 9.1) to determine the *critical ratio* $\lambda_f = p_{s+1}/w_{s+1}$, Balas and Zemel have developed a binary-search-type method (procedure 1 below) which solves the continuous problem without any ordering of the objects.

For any positive scalar λ, denote

$$S_1(\lambda) = \sum_{i \in \{j | p_j/w_j > \lambda\}} w_i, \quad S_2(\lambda) = S_1(\lambda) + \sum_{i \in \{j | p_j/w_j = \lambda\}} w_i.$$

Then λ_f is obviously the unique solution to the pair of inequalities

$$S_1(\lambda) < W \leq S_2(\lambda).$$

Procedure 1

Step 1 Set $N_0 = N_1 = \emptyset$, $N_F = \{1, 2, \ldots, n\}$.

Step 2 Choose λ to be the median of the first 3 ratios p_j/w_j in N_F; partition N_F into:

$$N^> = \{j \in N_F | p_j/w_j > \lambda\}$$
$$N^< = \{j \in N_F | p_j/w_j < \lambda\}$$
$$N^= = \{j \in N_F | p_j/w_j = \lambda\}$$

and calculate

$$S_1(\lambda) = \sum_{i \in N^>} w_i, \quad S_2(\lambda) = S_1(\lambda) + \sum_{i \in N^=} w_i.$$

Step 3 If $S_1(\lambda) < W \leq S_2(\lambda)$, an optimal solution $(\bar{X}_j)$ to the associated continuous problem is obtained by setting ($\bar{X}_j = 1$ for $j \in N_1 \cup N^>$), ($\bar{X}_j = 0$ for $j \in N_0 \cup N^<$) and then 'filling the knapsack' with variables in $N^=$ (any, possibly including one at a fractional value): stop.

If $S_1(\lambda) \geq W$, set $N_0 = N_0 \cup N^< \cup N^=$, $N_F = N^>$. If $S_2(\lambda) < W$, set $N_1 = N_1 \cup N^> \cup N^=$, $N_F = N^<$.

Step 4 If $|N_F| > \theta$ (θ is a fixed threshold value: $\theta = 25$ is considered an adequate value), go to Step 2. Order N_F according to decreasing p_j/w_j ratios. Solve the continuous problem defined on N_F through Dantzig's method and extend the solution by setting ($\bar{X}_j = 1$ for $j \in N_1$), ($\bar{X}_j = 0$ for $j \in N_0$); stop.

As a byproduct, procedure 1 yields an approximate core problem. Suppose $N = \{1, 2, \ldots, n\}$ ordered according to decreasing p_j/w_j ratios: any set $I \subset N$ containing the index f of the fractional component of an optimal solution to the continuous problem can be viewed as an approximation to the core problem; so, it is natural to assume

$$I = N_F \quad \text{if } |N_F| \leq \theta$$
$$I = N_F^\theta \quad \text{otherwise}$$

where N_F is the set ordered at step 4, while N_F^θ is an (ordered) subset of N_F

The 0–1 Knapsack Problem

such that $|N_F^\theta| = \theta$ and the index f of the fractional component of $(\bar{X}_j)$ is contained in the middle third of N_F^θ; this last configuration must be obtained even if it is set $I = N_F$, possibly by enlarging the interval defined by N_F in the desired direction.

A heuristic (procedure 2 below) is now employed to find a 'good' integer solution to the approximate core problem.

Procedure 2

Step 1 Build a first integer solution $(\tilde{X}_j)$ as follows: set $\tilde{X}_j = 1$ if $\bar{X}_j = 1$, $\tilde{X}_f = 0$, then examine each remaining variable in turn and, if possible, set it to 1 (*greedy solution*). Find that $i_t \in I = \{i_1, \ldots, i_t\}$ such that $i_1 = f$. Compute $\text{MIN} = \min_{i \in I} \{w_i\}$. Set $\hat{x}_f = 0$, $q = 1$.

Step 2 Set $i = i_q$, $\hat{x}_i = 0$, $u = w_i + w_f \bar{X}_f$, $h = l$.

Step 3 Set $k = i_h$. If $u = w_k$ or $u - w_k \geq \text{MIN}$, set $\hat{x}_k = 1$, $u = u - w_k$; otherwise, set $\hat{x}_k = 0$. If $u = 0$, go to Step 5. If $h = t$, set $h = l$ and go to Step 4. Set $h = h + 1$ and repeat Step 3.

Step 4 Set $k = i_h$. If $w_k \leq u$ and $\hat{x}_k = 0$, set $\hat{x}_k = 1$ and go to Step 5. If $h < t$, set $h = h + 1$ and repeat Step 4.

Step 5 Complete the solution $(\hat{x}_j)$ by setting $\hat{x}_j = \bar{X}_j$ for all components whose value has not yet been assigned, compare $(\hat{x}_j)$ with (X_j) and store in $(\tilde{X}_j)$ the better of the two. If $q = l - 1$, stop; otherwise, set $q = q + 1$ and go to Step 2.

A slightly improved version of procedure 2 makes use of the following dominance relation: if objects i, j are such that

$$p_i \geq p_j \text{ and } w_i \leq w_j$$

it is said that object i *dominates* object j; as a result, if $\hat{x}_i$ is set to 0, then $\hat{x}_j$ can also be set to 0.

Consider now the difference between the total profits of the continuous and of the heuristic solutions:

$$\text{Gap} = \sum_{j=1}^{n} p_j \bar{X}_j - \sum_{j=1}^{n} p_j \tilde{X}_j.$$

If Gap < 1, then $(\tilde{X}_j)$ is an optimal solution to the zero–one knapsack problem; if Gap ≥ 1, slightly modified versions of the reduction procedures of Section 9.5 are applied to define a new approximate core problem:

Procedure 3 consists in fixing as many variables as possible, firstly through the method of Dembo and Hammer (1975), then through procedure R; in both the algorithms, the lower bound LB is given by the value of the heuristic solution found by procedure 2. All the variables that were not fixed are then introduced into the approximate core problem.

If this redefined approximate core problem has more than s variables (s a parameter, say $s = 50$), then procedure 2 is again applied to it, followed by procedure 3; otherwise one of the enumerative algorithms from the literature is applied.

We have found, through extensive computational experience, that iterations on procedures 2 and 3 above are generally inefficient. So, in the next subsection, we always apply the enumerative method after procedure 3.

9.8.2 Computational Results for Large-Size Problems

Balas–Zemel's algorithm, here referred as BZ, was coded in FORTRAN IV and compared, on a CDC-6600, with the branch and bound algorithms of Horowitz–Sahni (BBHS) and of Martello–Toth (BBMT), both preceded by reduction procedure R. The enumerative method applied in BZ after procedure 3 was always BBMT.

Table 9.10 is relative to the same strongly correlated data sets considered for Table 9.9, but here the average times of R + BBHS and of R + BBMT are comprehensive of the time needed to sort the objects.

In spite of what is stated by Balas and Zemel (1977), the table shows that strongly correlated data sets are 'hard problems' for algorithm BZ as for other algorithms. The computational performance of BZ was slightly worse

Table 9.10 Strongly correlated data sets; $w_{min} = 1$, $w_{max} = 100$, $c = 10$. Average times for 200 problems; sorting time included

W	n	R + BBHS	R + BBMT	BZ
	50	53	17	29
200	100	194	40	53
	200	1385*	194	213
	50	12 probl.	1027	1058
$0.5 \sum_{j=1}^{n} w_j$	100	—	8 probl.	8 probl.
	200	—	—	—
	50	1411*	1096	1120
$0.8 \sum_{j=1}^{n} w_j$	100	—	3 probl.	3 probl.
	200	—	—	—

The 0-1 Knapsack Problem

than the one of R+BBMT, since the number of variables left after procedure 3 was always very near to n, so that most of the computation time was taken by BBMT.

In order to compare the algorithms' performance on large size problems, all the next tables of the present section are relative to uncorrelated and weakly correlated data sets. Each entry is the average time, expressed in CDC-6600 milliseconds, for 50 problems; for R+BBHS and R+BBMT the entries are comprehensive of the reduction and sorting times, which are shown separately in Table 9.6.

The algorithms' performance with varying n up to 10 000 are analysed in Tables 9.11 and 9.12, relative, respectively, to uncorrelated and weakly

Table 9.11 Uncorrelated data sets; $w_{min} = p_{min} = 1$, $w_{max} = p_{max} = 1000$. Average times for 50 problems; sorting time included. $W = 0.5 \sum_{j=1}^{n} w_j$

n	R+BBHS	R+BBMT	BZ
50	9	8	19
100	17	15	28
200	31	29	40
500	84	73	80
1000	183	158	159
2000	339	308	257
5000	846	766	572
10000	1604	1510	1042

Table 9.12 Weakly correlated data sets; $w_{min} = 1$, $w_{max} = 1000$, $r = s = 100$. Average times for 50 problems; sorting time included. $W = 0.5 \sum_{j=1}^{n} w_j$

n	R+BBHS	R+BBMT	BZ
50	37	26	34
100	55	38	56
200	89	59	79
500	161	115	143
1000	264	202	225
2000	418	333	361
5000	895	767	668
10000	1712	1516	1144

correlated data sets with

$$W = 0.5 \sum_{j=1}^{n} w_j,$$

$$w_{\min} = p_{\min} = 1$$

$$w_{\max} = p_{\max} = 1000$$

$$r = s = 100.$$

The fastest methods are R+BBMT, for problems with n less than a certain n^*, and BZ, for problems with $n > n^*$; the tables seem to show $n^* \cong 1000$ for uncorrelated data sets, $2000 < n^* < 5000$ for weakly correlated data sets. The average times of the three algorithms grow linearly with the value of n.

In order to analyse the dependence of the average execution times on the value of W and on the range of the problem coefficients, we employed the uncorrelated data sets above with n fixed to 5000. (We will not compare here the algorithms' performances since for such values of n, BZ will obviously be the fastest method)

Table 9.13 shows that R+BBMT is practically independent of the value of W, while the times of R+BBHS and mainly of BZ seem to decrease for low and high values of W.

Table 9.13 Uncorrelated data sets; $w_{\min} = p_{\min} = 1$, $w_{\max} = p_{\max} = 1000$. Average times for 50 problems; sorting time included. $n = 5000$

W	R+BBHS	R+BBMT	BZ
$0.01 \sum_{j=1}^{n} w_j$	813	760	448
$0.2 \sum_{j=1}^{n} w_j$	849	767	593
$0.5 \sum_{j=1}^{n} w_j$	846	766	572
$0.8 \sum_{j=1}^{n} w_j$	797	758	504

Table 9.14 shows that the three algorithms have average computing times growing almost logarithmically with the range of the coefficients; the difference between the times of the branch and bound algorithms (mainly of BBMT) and the ones of BZ decreases when the range of the coefficients grows.

The 0–1 Knapsack Problem

Table 9.14 Uncorrelated data sets; $w_{min} = p_{min} = 1$. Average times for 50 problems; sorting time included. $n = 5000$; $W = 0.5 \sum_{j=1}^{n} w_j$

Range	R+BBHS	R+BBMT	BZ
$w_{max} = p_{max} = 50$	730	710	434
$w_{max} = p_{max} = 100$	743	714	443
$w_{max} = p_{max} = 1000$	846	766	572
$w_{max} = p_{max} = 10000$	1024	883	719

Balas and Zemel have found that problems 'hard' for their method are those for which the gap between the optimum profit of the continuous and of the integer problem is high and the average size of $|p_j - w_j \lambda_f|$ is low. This situation arises when no optimal integer solution satisfies constraint (9.2) with equality and the p_j/w_j ratios are close to each other; both such conditions are generally verified for the strongly correlated data sets which, as shown above, are really 'hard' not only for BZ but for all the enumerative algorithms.

In order to impose that no optimal integer solution satisfies constraint (9.2) with equality, a new data set was employed by randomly generating the w_j multiple of 5 in the range (5, 100) and by setting $W \cong 0.05 \sum_{j=1}^{n} w_j$ not a multiple of 5; the p_j were randomly generated in (5, 100), so the p_j/w_j ratios were not so close to each other as above. Table 9.15 shows that such problems are also hard for all the considered algorithms, though the times are less than the corresponding ones for strongly correlated data sets.

We can conclude that, for large size knapsack problems, the best algorithms are R+BBMT and BZ. The choice depends on the value of n when, given the p_j and the w_j, only one knapsack problem has to be solved.

Table 9.15 Uncorrelated data sets; $w_{min} = p_{min} = 5$, $w_{max} = p_{max} = 100$, w_j multiple of 5 $\forall j$. Average times for 50 problems; sorting time included. $W = \lfloor 0.5 \sum_{j=1}^{n} w_j - 0.5 \rfloor$

n	R+BBHS	R+BBMT	BZ
50	8	7	20
100	17	15	29
200	34	30	43
500	338	209	221
1000	2371	1205	1266

If, for the same set of objects, several values of W have to be considered—as often happens in practice and when the knapsack problem arises as a subproblem—R+BBMT is clearly to be preferred, since the object sorting needs to be executed only once.

Appendix

A1 The Unbounded Knapsack Problem

The problem initially considered in Section 9.1.1:

$$\max \sum_{j=1}^{n} x_j p_j \qquad (9.1)$$

$$\text{subject to } \sum_{j=1}^{n} x_j w_j \leq W \qquad (9.2)$$

$$x_j \geq 0 \text{ and integer } (j = 1, \ldots, n) \qquad (9.3)$$

where no bound exists on the availability of each object, is generally referred to as the *Unbounded Knapsack Problem*.

The most efficient dynamic programming algorithm for the solution of this problem is probably that of Gilmore and Gomory (1966), which can be further improved through a property presented by Hu (1969).

Enumeration methods have been proposed by Gilmore and Gomory (1963) and by Cabot (1970); a more efficient branch and bound algorithm has been proposed by Martello and Toth (1977d).

The unbounded knapsack problem can also be indirectly solved through conversion into an equivalent zero–one problem; this can be obtained by means of the following steps:

Step 1 Set $k = 0$, $j = 1$.

Step 2 Set $c = \lfloor W/w_j \rfloor$, $e = 1$.

Step 3 Set $k = k+1$, $\bar{p}_k = e p_j$, $\bar{w}_k = e w_j$, $e = 2e$. If $e \leq c$, repeat Step 3.

Step 4 If $j < m$, set $j = j+1$ and go to Step 2. Otherwise, stop: the zero–one knapsack problem to be solved is

$$\max \sum_{j=1}^{k} \bar{x}_j \bar{p}_j$$

$$\text{subject to } \sum_{j=1}^{k} \bar{x}_j \bar{w}_j \leq W, \bar{x}_j = 0 \text{ or } 1 \ (j = 1, \ldots, k)$$

Computational comparisons (Martello and Toth, 1977b, 1977d) show that the fastest method is the one of Martello and Toth, which solves problems

The 0–1 Knapsack Problem

up to 10000 variables (with p_j, w_j uniformly random in the range (1, 1000) and $W = 0.5 \sum_{j=1}^{n} w_j$) in average time of 0.02 seconds (+0.94 seconds for sorting the objects) on a CDC-6600. Among the other methods, the most efficient is the improved version of the Gilmore–Gomory dynamic programming approach, which however cannot solve large-size problems because of the high storage requirements. Also the conversion into zero–one form is quite efficient only for small values of W and n, since the transformed problem involves

$$k = \sum_{j=1}^{n} \lceil \log_2 \lfloor W/w_j \rfloor \rceil \text{ variables.}$$

A2 The Bounded Knapsack Problem

When positive integer bounds $b_j (j = 1, \ldots, n)$ exist on the availability of each object, the problem, mathematically expressed by (9.1), (9.2), (9.3), and

$$x_j \leq b_j \quad (j = 1, \ldots, n)$$

with the obvious assumptions

$$\sum_{j=1}^{n} b_j w_j > W \text{ and } b_j w_j \leq W \quad (j = 1, \ldots, n)$$

is called the *Bounded Knapsack Problem*. The most efficient method for its solution is the branch and bound algorithm of Martello and Toth (1977d); in Martello and Toth (1977b) it has been shown that this method solves randomly generated problems up to 1000 variables (with p_j, w_j in the range (1, 1000), b_j in (1, 10) and $W = 0.5 \sum_{j=1}^{n} w_j$) in average time of 4.1 seconds (+0.07 seconds for sorting the objects) on a CDC-6600.

The dynamic programming approach of Nemhauser and Ullman (1969) has high computing times even for small values of n and W (about 80 times the corresponding computing times of the Martello–Toth algorithm).

The Bounded Knapsack Problem too can be transformed into zero–one form:

Step 1 Set $k = 0$, $j = 1$.

Step 2 Set $c = b_j$.

Step 3 Set $e = \lceil c/2 \rceil$, $k = k+1$, $\bar{p}_k = ep_j$, $\bar{w}_k = ew_j$, $c = \lfloor c/2 \rfloor$. If $c > 0$, repeat Step 3.

Step 4 If $j < n$, set $j = j+1$ and go to Step 2. Otherwise, stop: the zero–one knapsack problem to be solved is:

$$\max \sum_{j=1}^{k} \bar{x}_j \bar{p}_j$$

$$\text{subject to } \sum_{j=1}^{k} \bar{x}_j \bar{w}_j \leq W, \ \bar{x}_j = 0 \text{ or } 1 \ (j = 1, \ldots, k)$$

Since the transformed problem has $k = \sum_{j=1}^{n} \lceil \log_2 b_j \rceil$ variables, this method is efficient only for small values of n and of the bounds b_j.

A3 The Value-Independent Knapsack Problem

The extreme case of correlation between profits and weights—$p_j = w_j$, $j = 1, \ldots, n$—leads to the so-called (*Bounded*) *Value-Independent Knapsack Problem*:

$$\max \sum_{j=1}^{n} x_j h_j$$

$$\text{subject to } \sum_{j=1}^{n} x_j h_j \leq W$$

$$0 \leq x_j \leq b_j \qquad (j = 1, \ldots, n)$$

$$x_j \text{ integer} \qquad (j = 1, \ldots, n)$$

The need for a special definition of this problem arises from the fact that very high computing times are generally necessary to solve it through the usual algorithms for knapsack problems when no solution exactly 'fills the knapsack'. This is mainly because no object can here be considered 'better' than another ($p_j/w_j = 1$, $j = 1, \ldots, n$), so no help can be given from sorting the objects: such a handicap particularly affects implicit enumeration methods, where no result can be obtained by the bounding procedures.

On the other hand, it must be noted that, for randomly generated problems, the optimal solution generally 'fills the knapsack', so the performance of implicit enumeration methods is generally satisfactory.

Faaland (1973) presented a dynamic programming algorithm for the value-independent knapsack problem; its storage requirements and computing times depend highly on the value of W. Ahrens and Finke (1975) studied the zero–one case—they call the resulting problem 0–1 *Stickstacking Problem*—and proposed a method based on a listing and a merging of the objects. Their procedure is again of the dynamic programming type but allows the solution of problems with high values of W; a single problem of 40 variables, having h_j randomly generated in the range $(1, 10^{12})$ and $W = 10^{13}$ has been solved in 648.073 seconds on a CDC-6400.

A4 The Change-Making Problem

Consider an unbounded knapsack problem where all the profits have the same negative value and where, in condition (9.2), only the equality constraint holds. The resulting problem can be expressed as

$$\min \sum_{j=1}^{n} x_j$$

$$\text{subject to } \sum_{j=1}^{n} x_j w_j = W, \; x_j \geq 0 \text{ integer } (j=1,\ldots,n)$$

and is generally referred to as the *Unbounded Change-Making Problem:* it can be viewed, in fact, as the problem of exactly assembling a given change W using the least number of coins of specified values w_j, in the case where the available number of coins is unlimited.

Note that, because of the equality constraint, a solution to the problem does not necessarily exist unless we impose that one among the w_j has value 1; under this condition, a recursive solving procedure was developed by Chang and Gill (1970).

The general case was solved through a dynamic programming algorithm by Wright (1975) and through branch and bound algorithms by Martello and Toth (1977c, 1979). These last methods have been shown to involve computing times that are very much smaller than those of Wright's and Chang and Gill's algorithms: problems up to 1000 variables (with w_j uniformly random in the range (1, 2000) and $W = \sum_{j=1}^{n} w_j$) have been solved in average time of 0.1 seconds (sorting time included) on a CDC-6600.

The *Bounded Change-Making Problem* arises when it is imposed that

$$x_j \leq b_j \qquad (j=1,\ldots,n).$$

The only algorithm for the solution of this problem is the branch and bound procedure of Martello and Toth (1977c); this algorithm solves randomly generated problems up to 1000 variables (with w_j in the range (1, 2000), b_j in (1, 5) and $W = 0.5\sum_{j=1}^{n} b_j w_j$) in average time of 0.11 seconds (sorting time included) on a CDC-6600.

References

Ahrens, J. H., and G. Finke (1975). Merging and Sorting Applied to the Zero–One Knapsack Problem. *Operations Research,* **23,** 1099–1109.

Balas, E. and E. Zemel (1977). *Solving Large Knapsack Problems.* GSIA, Carnegie-Mellon University, September 1977.

Barr, R. S., and G. T. Ross (1975). A Linked List Data Structure for a Binary Knapsack Algorithm. *Research Report CCS 232, Centre for Cybernetic Studies, University of Texas.*

Bellman, R., and S. E. Dreyfus (1962), *Applied Dynamic Programming*, Princeton University Press, Princeton, N. J.

Brooks, L., and A. Geoffrion (1966). Finding Everett's Lagrange Multipliers by Linear Programming. *Operations Research*, **14**, 1149–1153.

Cabot, A. V. (1970). An Enumeration Algorithm for Knapsack Problems. *Operations Research*, **18**, (1970). 306–311.

Chang, S. K., and A. Gill (1970). Algorithmic Solution of the Change-Making Problem. *Journal of ACM*, **17**, 113–122.

Cord, J. (1964). A Method for Allocating Funds to Investment Projects when Returns are Subject to Uncertainty. *Management Science*, **10**, 335–341.

Dantzig, G. B. (1957). Discrete Variable Extremum Problems. *Operations Research*, **5**, 266–277.

Dembo, R. S., and P. L. Hammer (1975). A Reduction Algorithm for Knapsack Problems. *Research Report CORR 75-6, Department of Combinatorics and Optimization, University of Waterloo.*

Everett, H., III (1963). Generalized Lagrange Multiplier Method for Solving Problems of Optimum Allocation of Resources. *Operations Research*, **11**, 399–417.

Faaland, B. (1973). Solution of the Value-Independent Knapsack Problem by Partitioning. *Operations Research*, **21**, 332–337.

Fayard, D., and G. Plateau (1975). Resolution of the 0–1 knapsack problem: Comparison of methods. *Math. Programming*, **8**, 272–307.

Frieze, A. M. (1976). Shortest Path Algorithms for Knapsack Type Problems. *Mathematical Programming*, **11**, 150–157.

Gilmore, P. C., and R. E. Gomory (1961). A Linear Programming Approach to the Cutting Stock Problem I. *Operations Research*, **9**, 849–858.

Gilmore, P. C., and R. E. Gomory (1963). A Linear Programming Approach to the Cutting Stock Problem II. *Operations Research*, **11**, 863–888.

Gilmore, P. C., and R. E. Gomory (1965). Multi-Stage Cutting Stock Problems of Two and More Dimensions. *Operations Research*, **13**, 94–120.

Gilmore, P. C., and R. E. Gomory (1966). The Theory and Computation of Knapsack Functions. *Operations Research*, **14**, 1045–1074.

Greenberg, H., and R. L. Hegerich (1970). A Branch Search Algorithm for the Knapsack Problem. *Management Science*, **16**, 327–332.

Gulley, D. A., H. S. Swanson and R. E. D. Woolsey, (1972). On Not Searching for the Multipliers in Knapsack Problems. *Obtainable from R. E. D. Woolsey, Colorado School of Mines, Golden, Colorado.*

Horowitz, E., and S. Sahni (1974). Computing Partitions with Applications to the Knapsack Problem. *Journal of ACM*, **21**, 277–292.

Hu, T. C. (1969). *Integer Programming and Network Flows*, Addison-Wesley, New York.

Hudson, P. D. (1977). Improving the Branch and Bound Algorithms for the Knapsack Problem, *Queen's University Research Report, Belfast.*

Ingargiola, G. P., and J. F. Korsh (1973). A Reduction Algorithm for Zero–One Single Knapsack Problems. *Management Science*, **20**, 460–463.

Kaplan, S. (1966). Solution of the Lorie–Savage and Similar Integer Programming Problems by the Generalized Lagrange Multiplier Method. *Operations Research*, **14**, 1130–1136.

Kolesar, P. J. (1967). A Branch and Bound Algorithm for the Knapsack Problem. *Management Science*, **13**, 723–735.

Martello, S., and P. Toth (1977a). An Upper Bound for the Zero–One Knapsack Problem and a Branch and Bound Algorithm. *European Journal of Operational Research*, **1**, 169–175.

Martello, S., and P. Toth (1977b). Computational Experiences with Large-Size Unidimensional Knapsack Problems. *Presented at the TIMS/ORSA Joint National Meeting, San Francisco.*

Martello, S., and P. Toth (1977c). Solution of the Bounded and Unbounded Change-Making Problem. *Presented at the TIMS/ORSA JOINT National Meeting, San Francisco.*

Martello, S., and P. Toth (1977d). Branch and Bound Algorithms for the solution of the General Unidimensional Knapsack Problem. In M. Roubens (Ed.), *Advances in Operations Research*, North-Holland, Amsterdam.

Martello, S., and P. Toth (1977). Optimal and Canonical Solutions of the Change-Making Problem. *European Journal of Operational Research* (to appear).

Morin, T. L., and R. E. Marsten (1976). Branch and Bound Strategies for Dynamic Programming. *Operations Research*, **24**, 611–627.

Müller–Merbach, H. (1978). An Improved Upper Bound for the Zero–One Knapsack Problem; a Note on the Paper by Martello and Toth. *European Journal of Operational Research*, 2, 212–213.

Nauss, R. M. (1976). An Efficient Algorithm for the 0–1 Knapsack Problem. *Management Science*, **23**, 27–31.

Nemhauser, G. L., and Z. Ullmann (1968). A Note on the Generalized Lagrange Multiplier Solution to an Integer Programming Problem. *Operations Research*, **16**, 450–453.

Nemhauser, G. L., and Z. Ullmann (1969). Discrete Dynamic Programming and Capital Allocation. *Management Science*, **15**, 494–505.

Salkin, H. M. (1975). *Integer Programming*, Addison-Wesley, New York.

Salkin, H. M., and C. A. de Kluyver (1975). The Knapsack Problem: a Survey. *Naval Research Logistics Quarterly*, **22**, 127–144.

Shapiro, G. F. (1968). Dynamic Programming Algorithms for the Integer Programming Problem I; The Integer Programming Problem Viewed as a Knapsack Type Problem. *Operations Research*, **16**, 103–121.

Shapiro, G. F. (1961). Generalized Lagrange Multipliers in Integer Programming. *Operations Research*, **19**, 68–76.

Shapiro, G. F., and H. M. Wagner (1966). A Finite Renewal Algorithm for the Knapsack and Turnpike Models. *Operations Research*, **14**, 319–341.

Suhl, U. (1977). An Algorithm and Efficient Data Structures for the Binary Knapsack Problem. *Freie Universität Berlin Research Report.*

Toth, P. (1976). A New Reduction Algorithm for 0–1 Knapsack Problems. *Presented at the ORSA/TIMS Joint National Meeting, Miami.*

Toth, P. (1977). A Dynamic Programming Algorithm for the Zero–One Knapsack Problem. *Presented at the TIMS/ORSA Joint National Meeting, San Francisco.*

Weingartner, H. M. (1963). *Mathematical Programming and the Analysis of Capital Budgeting Problems*, Prentice Hall, Englewood Cliffs, N. J.

Weingartner, H. M. (1968). Capital Budgeting and Interrelated Projects: Survey and Synthesis. *Management Science*, **12**, 485–516.

Weingartner, H. M., and D. N. Ness (1967). Methods for the Solution of the Multi-Dimensional 0–1 Knapsack Problem. *Operations Research*, **15**, 83–103.

Wright, J. W. (1975). The Change-Making Problem. *Journal of ACM*, **22**, 125–128.

Zoltners, A. A. (1978). A Direct Descent Binary Knapsack Algorithm. *Journal of ACM*, **25**, 304–311.

CHAPTER 10

Complexity and Efficiency in Minimax Network Location

GABRIEL Y. HANDLER
Tel-Aviv University

10.1 Introduction

Let $G = (V, A)$ be an undirected graph with a finite vertex set $V = \{v_1, v_2, \ldots, v_n\}$ and a finite set A of arcs. To every arc $(i, j) \in A$, connecting v_i to v_j, associate a finite positive real number $b(i, j)$ representing its length. An arc may be considered as an infinite set of points. A point in arc (i, j) is specified by its distance from either of the end-points v_i, v_j. Let G denote also the set of all points in the graph including both the set of vertices and all points in all arcs. Define $d(x, y)$ as the length of a shortest path between $x, y \in G$. Let $X_m = \{x_1, x_2, \ldots, x_m\}$ denote a set of m points in G and let the quantity $d(X_m, y)$ be the generalization of $d(x, y)$ defined by

$$d(X_m, y) = \min_{x \in X_m} d(x, y). \tag{10.1}$$

Finally, let $l(X_m)$ be defined by

$$l(X_m) = \max_{y \in V} d(X_m, y). \tag{10.2}$$

We are now ready to formulate the 'absolute m-centre problem.'

Definition 10.1 For $m = 1, 2, \ldots$ a set of points $X_m^* \subseteq G$ is a set of *absolute m-centres* of G if for every $X_m \subseteq G$,

$$l(X_m^*) \leq l(X_m)$$

where $l(X_m)$ is given by (10.2). Furthermore, the quantity $r_m = l(X_m^*)$ is known as the *absolute m-radius* of G.

A common application of the absolute m-centre problem is to the location of emergency service facilities on a transportation network, where x represents the location of such a facility and y represents the location of a demand point. For example, we may wish to locate m fire stations in a rural community in a manner that minimizes the maximum response time from the closest station to any farm house.

The absolute m-centre problem defined above is but one example of a variety of minimax location problems on a network, some of which will be introduced below. The common features of these 'centre' problems include the network representation and the minimax criterion. The network assumption often allows for more precise modelling in comparison with the use of rectilinear or euclidean distance metrics. While the minimax criterion is particularly appropriate in emergency service scenarios, other objectives are frequently utilized, for example, the minisum criterion of the 'median' problems.

This paper is devoted to a methodological discussion of computational procedures for centre problems. As our point of departure we recall a statement by Hu (1971) to the effect that (by 1970) multi-facility location problems on a network was one of the major areas of discrete optimization about which very little was known. Hu refers specifically to the centre and median problems. It is noteworthy that of some 100 references in the literature directly related to these problems, about 84 appeared subsequent to Hu's writing (see Handler and Mirchandani, 1979). Furthermore, the earlier papers were mostly addressed to the special and much simpler case of single-facilities. It thus appears that much more is known today about network location, and our general aim here is to demonstrate that this is indeed the case with respect to centre problems.

Specifically, our objectives are two-fold. First, employing the concepts of computational complexity, we shall indicate what is known about the complexity of centre problems. In particular, we shall note that m-centre problems are generally NP-complete. Our second objective is to present an exact procedure for finding m-centres of a graph. This procedure is often effective for large-scale problems. Thus, we shall suggest that this class of problems is not particularly difficult or intractable in practice, despite its NP-complete property, which indicates that these are 'hard' problems in an asymptotic worst-case context.

As we indicated above, the absolute m-centre problem introduced by Definition 10.1 is one of a variety of minimax network location models. The following categories of model assumptions encompass much of this variety.

(i) *Facility location sets* Facilities may be situated on all points of G (vertices and interior points of the arcs) as in Definition 10.1, or they may be restricted to the set of vertices. In practice, facilities may well be restricted to a given finite set of possible locations.

(ii) *Demand location sets* In Definition 10.1 demand is assumed to occur only at the vertices. A natural extension provides for demand generation at any point on the network. This would be appropriate, for example, in determining locations for emergency service stations along a highway network, since breakdowns and accidents can occur anywhere on the network.

(iii) *Inverse centre problems* Instead of looking for a set of locations which minimizes the maximum distance for a given number of facilities, it may be appropriate to answer the 'inverse' question, namely: 'what is the least number of facilities and their locations such that the maximum distance from a random incident to the closest facility is less than or equal to a specified value?' Referring to such problems as *inverse centre* problems, we can formulate an inverse corresponding to each of the four multi-centre problems introduced above. For example, the inverse associated with the absolute m-centre problem of Definition 10.1 is to solve the optimization problem

$$h_\lambda = \min\{h : X_h \subseteq G; l(X_h) \leq \lambda; h \geq 0 \text{ and integer}\} \quad (10.3)$$

where h_λ is the minimum number of facilities needed for a maximum distance of λ and $l(\cdot)$ is given by (10.2).

(iv) *Tree networks* An interesting special case of the centre problems formulated above occurs when the network contains no cycles, that is, when the network is a tree, say T. This distinction presents a useful hierarchy in developing computational procedures for the general (cyclic) case. Furthermore, for those physical networks which do exhibit a tree-like structure, the special results for trees can be applied directly.

In order to facilitate identification of the center problems formulated above, we shall adopt the shorthand classification scheme shown in Table 10.1. The scheme is best explained by some examples. Consider the absolute m-centre problem associated with Definition 10.1. This would be identified as $A/V/m/G$ since facilities may be established anywhere on the arcs, demand occurs only at the vertices, m centres are to be located, and the network is a general (cyclic) graph. Now consider the inverse centre problem defined by (10.3). That problem is identified as $A/V/\lambda^{-1}/G$, where the '-1' is used in 'λ^{-1}' to distinguish λ from m when numbers replace symbols. As a

Table 10.1 Classification scheme for centre problems

Facility location Set $\{V \atop A\}$	Demand location Set $\{V \atop A\}$	Number of centres Maximum distance $\{m \atop \lambda^{-1}\}$	Network $\{G \atop T\}$
V = vertex set	A = arc set		
m = number of centres	λ = maximum distance		
G = general network	T = tree network		

final example, the 2-centre problem on a tree network where facilities and demand can exist anywhere on the network would be denoted $A/A/2/T$.

The variety of centre models may be further enriched by relaxing a host of implicit and explicit assumptions in the foregoing models. For example, facilities and demand points may be restricted to subsets of A or V; the network may be oriented (in which case we should differentiate between 'out-centres' and 'in-centres'); weights may be attached to demand points in multiplicative or additive fashion (so far we have implicitly assumed uniform weights); stochastic elements (representing variable travel time, for example) may be introduced; and so on. We shall restrict the discussion to the basic models introduced before, though much of it extends directly to the generalized models indicated above. For a fuller discussion of those and other extensions the interested reader is referred to Handler and Mirchandani (1979).

The literature on centre problems is quite extensive and for a comprehensive survey the reader is again referred to the previous reference. While the mathematical formulation and rudimentary analysis of location problems on a network can be traced back at least to the nineteenth century, the source of the recent research efforts in this area is generally accredited to the seminal papers by Hakimi (1964, 1965). With respect to those references directly pertaining to the subject of this paper, we note here that Christofides and Viola (1971) and Garfinkel, Neebe, and Rao (1977) also present algorithms for $A/V/m/G$. Other relevant references will be cited during the discussion.

In additon to the notation and definitions introduced previously we shall make use of the following terminology in our discussion. With regard to network constructs the terms graph/network, arc/link, will be used synonymously. A minimum path spanning tree of a graph G rooted at a point $x \in G$ will be denoted MPT(x) and the matrix of shortest distances between all pairs of vertices in V will be denoted D. To evaluate the last two constructs, standard algorithms as reported in Dijkstra (1959) and in Dreyfus (1969) can be used. We note here that for nonnegative link lengths, MPT(x) and D may be evaluated in $O(n^2)$ and $O(n^3)$ operations, respectively. The symbol VC will be used to identify an optimal location in the 'vertex centre problem' $V/V/1/G$. The cardinality of any set S will be denoted $|S|$ and a bold symbol will indicate a vector, for example $\mathbf{x}$.

We shall often use the following mathematical programming terminology. Consider the finite-dimensional optimization problem

$$z(C) = \inf_{\mathbf{x} \in C} f(\mathbf{x}) \qquad (10.4)$$

where $f: E^k \to E$, $\mathbf{x} \in E^k$, $C \subseteq E^k$ and E^k denotes k-dimensional euclidean space. A *relaxation* of (10.4) is any new problem with $C' \supset C$ replacing C in

Complexity and Efficiency in Minimax Network Location 285

(10.4) while a *restriction* of (10.4) is defined by substituting $C'' \subset C$ for C in (10.4). If $z(C'') = z(C)$ we shall refer to C'' as a *dominant set* for C in (10.4).

Finally, we wish to distinguish between our usage of the terms 'complexity' and 'efficiency' in assessing the computational effort in solving optimization problems. In specifying the *complexity* of an algorithm (or of a problem) we have in mind the maximum number of operations needed to process the problem. This is a worst-case measure of computational effort. On the other hand, the *efficiency* of an algorithm will have the, less precise, connotation of the computational effort involved in solving the problem in practice.

For expository purposes we shall initially confine our discussion to the class of absolute m-centre problems $A/V/m/G$. This class can be considered methodologically generic to the other centre models, which we consider subsequently. In Section 10.2 we summarize existing results concerning the complexity of absolute centre problems, and in Section 10.3 we describe an exact procedure for solving these problems which is often effective for large networks. A final section is devoted to the remaining classes of centre problems introduced above.

10.2 Complexity of Absolute Centre Problems

In this section we summarize existing results concerning the computational complexity of absolute centre problems, denoted as $A/V/m/G$ in the classification scheme of the previous section. Categorizing this class of problems according to the number of facilities and the type of network, we obtain a broad spectrum of complexities ranging from $O(n)$ for the simplest problems to *NP*-completeness for the most general ones.

The simplest problems occur when the network is a tree. In particular, the following algorithms, reproduced from Handler (1973, 1978) demonstrate that $O(n)$ algorithms are available for $m \leq 2$.

Algorithm 10.1 $(A/V/1/T)$

Step 1 Choose any point x on the tree and find a furthest point away, say e_1.

Step 2 Find a furthest point away from e_1, say e_2.

Step 3 The absolute centre is at the midpoint of the path connecting e_1 to e_2.

The major computational effort involves finding the longest distance from each of two points (x and e_1). This requires sequentially labelling nodes by their distance from the given point x or e_1. Thus, computational complexity of the algorithm is of the order $O(n)$.

Algorithm 10.2 (*A/V/2/T*)

Step 1 Using Algorithm 10.1, find the absolute single-centre of T.

Step 2 Delete from T any arc on the path connecting e_1 to e_2 containing the absolute centre, forming two subtrees.

Step 3 An optimal pair of locations is given by the absolute single-centres of these two subtrees.

Since the procedure is composed of three applications of Algorithm 10.1, computational complexity of the algorithm is $O(n)$.

Unfortunately, a direct extension of Algorithm 10.2 to the cases where $m > 2$ is not possible, and the latter problems appear to be significantly more complex than for $m \leq 2$. However, Hakimi, Schmeichel, and Pierce (1976) have devised a procedure for these problems which recursively makes use of the previous two algorithms. The resultant algorithm processes $A/V/m/T$ in $O(n \log^{m-2} n)$ time (incorporating an observation from Kariv and Hakimi (1976)).

Turning now to the general case of cyclic networks, Kariv and Hakimi (1976) have devised an $O(n^3)$ algorithm for $A/V/1/G$ and an $O(|A|^m n^{2m-1}/(m-1)!)$ algorithm for $A/V/m/G$, where $m > 1$. Thus, for a given value of m, a polynomial-bounded algorithm exists for the most general problem. However, when m too is taken as a varying input, such an algorithm is highly unlikely to exist. Kariv and Hakimi obtain this result by demonstrating that $A/V/m/G$ is *NP*-complete.

In conclusion, existing results concerning the complexity of absolute centre problems may be summarized as shown in Table 10.2. Of these results, Algorithms 10.1 and 10.2 also provide practically effective procedures for finding absolute single and two-centres of a tree. The remaining results provide the best known upper bounds on the computational effort for the corresponding absolute centre problems. However, the associated algorithms may be cumbersome from a practical standpoint, particularly for large problems. Furthermore, since the general problem is known to be *NP*-complete, it may be expected that these are inherently 'hard' problems.

Table 10.2 Complexity results for absolute centre problems.

Number of centres	Type of network	
	Tree	Cyclic graph
$m = 1$	$O(n)$	$O(n^3)$
$m = 2$	$O(n)$	$O(|A|^m n^{2m-1}/(m-1)!)$
$m \geq 3$	$O(n \log^{m-2} n)$	*NP*-complete

In the following section we present a relaxation strategy which appears promising for finding centres of large-scale networks in reasonable time. Indeed it would appear that, within the class of NP-complete problems, centre problems are quite tractable in practice.

10.3 A Relaxation Algorithm for Absolute Centre Problems

In this section we describe an efficient methodology for locating m-centres in $A/V/m/G$ and determining the m-radius, r_m, defined by

$$r_m = \min_{X_m \subseteq G} \max_{y \in V} d(X_m, y). \tag{10.5}$$

All of the existing algorithms operate on the same principle, generating and solving a series of set-covering problems (CP). The major computational difficulty with these algorithms has to do with the size of the generated CP. The approach we describe here is a problem-oriented relaxation technique which is particularly appropriate for the large-scale CP's encountered in multi-centre problems.

We describe first a rudimentary algorithm for $A/V/m/G$ which serves to introduce the CP framework and provides a point of departure for the subsequent developments. Consider first the following definition.

Definition 10.2 $\langle X, c, y \rangle$, denoting a *local centre* at a point $c \in G$ with respect to a pair of vertices $x, y \in V$, is said to exist if the minimum distance from c to x equals the minimum distance from c to y and there is no direction from c in which the minimum distances to both x and y are decreasing. Formally, the conditions are:

(i) $\qquad\qquad\qquad d(x, c) = d(y, c)$

(ii) $\qquad \left\{ a \in A_c : \lim_{\theta \to 0} [(d(z, x_\theta^a) - d(z, c))/\theta] < 0, \text{ for } z = x, y \right\} = \emptyset$

where A_c is the set of links or partial links incident at c and x_θ^a is a point distant θ from c on link $a \in A_c$. Note that $\langle x, c, x \rangle$ exists if and only if $c = x$; in this case the local centre $\langle x, x, x \rangle$ is referred to as the *null centre* at x.

Let C denote the set of all points in G corresponding to local centres, that is

$$C = \{C \in G : \langle x, c, y \rangle \text{ exists for some } x, y \in V\}, \tag{10.6}$$

Note that several local centres may coincide at a point $c \in G$, so that the cardinality of C is less than or equal to the number of locals centres.

We now present a fundamental 'mid-point property' for the case of single-centres. We shall see that this simple observation for the case $m = 1$ will be crucial in subsequent developments for $m > 1$.

Theorem 10.1 For the problem $A/V'/1/G$, $V' \subseteq V$, a global solution exists at a local centre $\langle x, c, y \rangle$, where $x, y \in V'$ and $d(c, x) = d(c, y) = max\{d(c, z): z \in V'\}$.

Proof Reasoning informally, this theorem follows because a location which does not fulfil the stated conditions can be shifted in the direction of a furthest node, thus reducing the maximum distance.

Minieka's Approach

Noting that the cardinality of C is finite, Minieka (1970) developed a finite procedure for determining m-centre locations. The following result enables the search for m-centres to be restricted from the infinite set of points in G to the finite set in C. We present a formal proof of this result, somewhat different from Minieka's analysis, to aid subsequent developments. Because the proof is based upon Theorem 10.1, we refer to the result as a corollary.

Corollary 10.1a (Minieka, 1970) The finite set C defines a dominant set for the location of centres in (10.5), which can now be reformulated as

$$r_m = \min_{X_m \subseteq C} \max_{y \in V} d(X_m, y). \tag{10.7}$$

Proof The reasoning is similar to that for Theorem 10.1. An m-centre $x \notin C$ can be perturbed in the direction of any furthest vertex being 'served' by it, without loss of optimality. The following formal proof will be useful further on. Given a set of m-centres of G, any node of G can be associated arbitrarily with an m-centre nearest to it, thus partitioning the vertex set into m subsets $\{V_i\}_{i=1}^m$. Let Π denote the set of all such partitions of V. Then (10.5) may be reformulated as

$$r_m = \min_{\{V_i\}_{i=1}^m \in \Pi} \max_{i \in \{1, 2, \ldots, m\}} \min_{x \in G} \max_{y \in V_i} d(x, y). \tag{10.8}$$

The final two operators in (10.8) define a 1-centre problem as in Theorem 10.1, from which it follows that the set C is indeed a dominant location set for m-centres.

A solution strategy for (10.5) can now be devised by reformulating (10.7). Form the matrix $F = \|f_{ij}\|$, with n rows and $|C|$ columns, where f_{ij} denotes the shortest distance from node v_i to the jth candidate centre. Then (10.7) can be reformulated as

$$r_m = \min_{H_m \subseteq S} \max_{v_i \in V} \min_{j \in H_m} f_{ij} \tag{10.9}$$

where H_m is the index set of a set of m columns in F and $S = \{1, 2, \ldots, |C|\}$.

The following algorithm solves (10.9), and hence (10.5), in a finite number of operations.

Algorithm 10.3 (A/V/m/G) (Minieka, 1970)

Step 0 Choose an arbitrary initial solution, H_m.

Step 1 (i) Let $d = \max_{v_i \in V} \min_{j \in H_m} f_{ij}$
(ii) Update the matrix $B = \|b_{ij}\|$, where

$$b_{ij} = \begin{cases} 0 & \text{if } f_{ij} \geq d \\ 1 & \text{otherwise.} \end{cases} \quad \begin{array}{l} i = 1, 2, \ldots, n \\ j = 1, 2, \ldots, |C| \end{array}$$

Step 2 Solve the CP

$$h = \min \mathbf{e}^t \mathbf{x} \tag{10.10}$$

$$B\mathbf{x} \geq \mathbf{e} \tag{10.11}$$

$$x_j \in \{0, 1\}, j = 1, 2, \ldots, |C| \tag{10.12}$$

where $\mathbf{e}^t = (1, 1, \ldots, 1)$.

Step 3 If $h > m$, stop; $r_m = d$ is the value of an optimal solution. Otherwise, an improved solution has been obtained. Update H_m and go to Step 1.

Column Elimination

While the preceding algorithm locates m-centres in a finite number of steps, in practice the procedure is prohibitively time-consuming except for the smallest graphs. As n increases, $|C|$ rapidly becomes extremely large. The result below has the effect of significantly reducing the number of columns in successive iterations of Step 2. Assume for the present that a candidate centre $c \in C$ is generated uniquely by just one pair of vertices $x, y \in V$. Relaxing this assumption involves a notational complication which we address later. We define first an important quantity associated with any local centre.

Definition 10.3 For any local centre $\langle x, c, y \rangle$, let d_c denote its *range*, defined as

$$d_c = d(x, c) = d(y, c). \tag{10.13}$$

Corollary 10.1b In Algorithm 10.3, if d is the value of the current solution for m-centres, then all candidate centres (columns) in the set $K = \{c \in C: d_c \geq d\}$ can be eliminated without loss of optimality.

Proof Let V_c be the set of vertices associated with the m-centre at $\langle x, c, y \rangle$ for which $d_c \geq d$. We assert first that $r_m < d$ implies $x, y \notin V_c$. This statement follows since otherwise, from (10.8), $r_m \geq d_c \geq d$. Furthermore, since

$$\min_{x \in G} \max_{y \in V_c} d(x, y) \leq \max_{y \in V_c} d(c, y) \tag{10.14}$$

the m-centre at c can be replaced by an optimal location in the problem defined in the left-hand side of (10.14) without loss of optimality. By Theorem 10.1, such an optimal location, c', exists at a local centre $\langle x', c', y' \rangle$, where $x', y' \in V_c$. From the first statement $c \neq c'$, so that the candidate centre at c may be discarded.

The following result, again following directly from Theorem 10.1, renders the CP matrix considerably more sparse. Furthermore, together with Corollary 10.1b, it eliminates the need for storing and working with the matrix F after the algorithm has been initiated.

Corollary 10.1c In Algorithm 10.3, for any column j and any row i, the initialization procedure $b_{ij} = 0$ if $f_{ij} > d_j$, where d_j is the range of the candidate centre associated with column j, can be made without loss of optimality.

Proof Consider the reformulation of (10.5) given by (10.8). From Theorem 10.1, there exists an optimal solution to any of the 1-centre problems in (10.8) at a local centre $c \in C$, where $d_c = \max \{d(c, y): y \in V_c\}$. For such an optimal solution, V_c does not contain nodes y such that $d(c, y) > d_c$.

Corollaries 10.1b and 10.1c lead directly to the following improved version of Algorithm 10.3.

Algorithm 10.4 $(A/V/m/G)$

Step 0 (i) Choose an arbitrary initial solution, H_m.
(ii) Let $d = \max_{v_i \in V} \min_{j \in H_m} f_{ij}$.
(iii) Set up matrix $B = \|b_{ij}\|$, where

$$b_{ij} = \begin{cases} 0 & \text{if } f_{ij} > d_j \\ 1 & \text{otherwise.} \end{cases} \quad \begin{array}{l} i = 1, 2, \ldots, n \\ j = 1, 2, \ldots, |C| \end{array}$$

Step 1 Eliminate from B all columns j such that $d_j \geq d$.

Step 2 Solve the CP ((10.10)–(10.12)).

Step 3 If $h > m$, stop; $r_m = d$ is the value of an optimal solution. Otherwise, an improved solution has been obtained. Update H_m; let $d = \max_{j \in H_m} d_j$ and return to Step 1.

In the revised algorithm columns of B are eliminated, whereas previously elements of B were zeroed, leaving B, in general, with its original dimensions. The effect of this reduction in size becomes increasingly significant as m increases. Notice, however, that for small m we still encounter a great number of columns which, coupled with large n, renders the algorithm impractical for large problems. Before we proceed to tackle this issue we need to relax the restrictive assumption made earlier that only one local centre may exist at a point in G.

Complexity and Efficiency in Minimax Network Location 291

Consider now the general case where k_c local centres $\{\langle x_i, c, y_i \rangle\}_{i=1}^{k_c}$ coincide at a point $c \in G$. Similar to C of (10.6), let C' denote the set of all local centres, that is

$$C' = \{\langle x_i, c, y_i \rangle: i = 1, 2, \ldots, k_c, \text{ for all } c \in C\}. \tag{10.15}$$

To generalize the column elimination procedure to this case, assume that the set C in Corollaries 10.1a and 10.1b is replaced by C', so that a single point in G may be associated with several candidate centres.

The theory and algorithm apply as before. Notice, however, that the number of columns in the CP matrices may exceed the number of columns in Minieka's procedure. In particular, the initial B matrix now contains $|C'| \geq |C|$ columns. However, it is simple to show that all local centres at a point on G may be represented by at most one column in the CP matrix, and hence the CP matrix need never contain more than $|C|$ columns.

Suppose that after Step 1 of the algorithm the matrix B contains a set of k columns corresponding to k local centres at a point in G. Suppose the columns are indexed $j = 1, 2, \ldots, k$ so that $d_1 \geq d_j$, $j = 2, 3, \ldots, k$. Then, by construction, $b_{i1} \geq b_{ij}$, $i = 1, 2, \ldots, n$, $j = 2, 3, \ldots, k$, and the structure of (10.11) renders columns $j = 2, 3, \ldots, k$ redundant in the CP. Notice, however, that these columns must be stored for future use in case column 1 is eliminated at some later stage.

A Relaxation Strategy

The size of the CP matrices arising in Algorithm 10.3 is n rows by $|C'|$ columns. For networks with hundreds of nodes, the corresponding matrices are very large indeed. While in Algorithm 10.4 the number of columns is generally reduced, the CP matrices may still be large, particularly for small values of m. In both algorithms, solving the CP's in Step 2 is a major computational burden severely limiting the size of problems which can be optimally solved.

Comprehensive reviews of CP's appear in Garfinkel and Nemhauser (1972) and Christofides and Korman (1975). While significant algorithmic advances have been achieved, the state of the art does not permit efficient solution of truly large-scale general CP's. The following strategy for multi-centre location constitutes a problem-oriented technique for solving the very large CP's arising in this context.

The 'relaxation strategy' for $A/V/m/G$ is based upon two fundamental observations.

(i) for any number of centres, m, there exists a set of critical vertices, a 'relaxed' set, $R \subseteq V$, which determine the optimal location of centres. Moreover, $|R|$ will be relatively unaffected by n, and instead will be fairly closely related to m.

In relation to other mathematical programming problems, this observation and implied strategy is akin to a relaxation approach for minimax formulations where, again, the observation is that most of the constraints in the derived constrained problem will be nonbinding at an optimal solution.

(ii) Unlike general mathematical programming relaxation strategies, in this case both rows and columns are eliminated as a result of the manner in which columns (candidate centres) are generated by rows (vertices). Thus, a special feature here is that advantages which generally result from restrictions are obtained as well, rendering the relaxation strategy particularly effective.

The following list of definitions together with previously defined quantities identifies notation used in the relaxation algorithm. We shall assume that a candidate centre $c \in G$ is generated uniquely by one pair of vertices. As before, this restriction may be relaxed by replacing C of (10.6) with C' of (10.15).

$$C(x, y) = \{c \in C : \langle x, c, y \rangle \text{ exists for a given pair } x, y, \varepsilon V\}.$$

$$C[(x, y) : d] = \{c \in C(x, y) : d_c < d\}.$$

$$C\{R : d\} = \bigcup_{x,y \in R} C[(x, y) : d].$$

$\{h, \hat{X}_h, \hat{d}\}$ = optimal solution quantities for the CP, where

h = value of objective function

$\hat{X}_h$ = identity of the h centres in the solution

$$\hat{d} = \max_{c \in \hat{X}_h} \{d_c\}.$$

Following is a statement of a procedure for determining m-centres, which combines Algorithms 10.3 and 10.4 in a relaxation framework. We shall state the algorithm in the context of solving $A/V/m/G$ for $m = 1, 2, \ldots, M$. Minor adjustments for the case $m = k, k+1, \ldots, M$, where $M \geq k > 1$, are made later. A flowchart of the algorithm appears in Figure 10.1

Algorithm 10.5 ($A/V/m/G$)

Step 0 Initialization Set $m = 1$ and select arbitrarily a vertex x. Let $R = \{x\}$, $d = \infty$, $C\{R : d\} = \langle x, x, x \rangle$, $\hat{d} = 0$ $\hat{X}_1 = x$. Proceed to Step 1.

Step 1 Improvement/Vertex-to-enter If every vertex in $V-R$ is within a range of $\hat{d}$ from some $x \in \hat{X}_m$, then $\hat{X}_m$ is an improved solution; let $X_m = \hat{X}_m$ be the new current solution with $d = \hat{d}$ an upper bound on r_m; update $C\{R : d\}$ by eliminating candidate centres whose ranges exceed or equal d and go to Step 3. Otherwise, designate as vertex-to-enter a vertex v, which is furthest away from $\hat{X}_m$. If $\hat{d} < l(\hat{X}_m) < d$, then $\hat{X}_m$ is still an improved solution; set $d = l(\hat{X}_m)$ and let $X_m = \hat{X}_m$. In both cases proceed to Step 2.

Complexity and Efficiency in Minimax Network Location

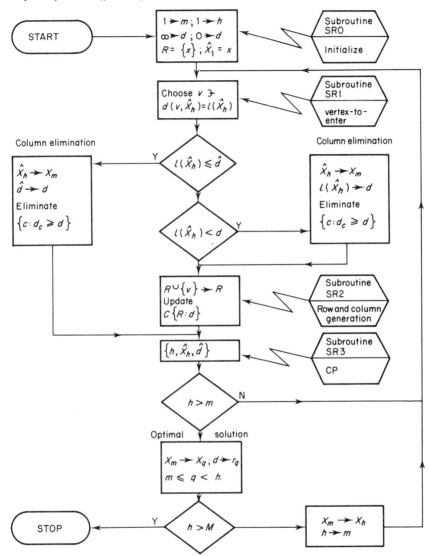

Figure 10.1 Flow-chart for relaxation algorithm

Step 2 Column generation Add v to R and update $C\{R:d\}$ by adding candidate centres $C[(v,r):d]$ for all $r \in R$. Proceed to Step 3.

Step 3 Covering problem Select from $C\{R:d\}$ an arbitrary set of m candidate centres, $\hat{X}_m$, 'covering' all vertices in R and go to Step 1. If this is impossible, proceed to Step 4.

Step 4 Optimality The current solution, X_m, constitutes a set of m-centres, with value $r_m = d$. If $m = M$, stop. Otherwise, let $m = m+1$ and return to Step 3.

The basic computational building blocks, depicted as subroutines in the flow-chart of Figure 10.1, need to be clarified.

A central issue in implementation of the algorithm has to do with the computation of minimum path distances. For large-scale problems it is expedient to distinguish two modes of operation. In the first, minimum path distances are computed as needed. In the second, the matrix D is initially computed and stored for subsequent use as needed. The first option is generally preferable where M is small in relation to n. Otherwise, the second option is more efficient.

SR0-initialization

The procedure in the flow-chart is the simplest, though not necessarily the most efficient, for initiating the algorithm. The set R comprises initially any vertex x with upper bound $d = \infty > r_1$. It is often possible to speed up the process by deriving better (lower) initial values for d. For example, when D is initially computed, an optimal location (vc) in $V/V/1/G$, found by inspection, provides an upper bound on r_1, namely $d = l(\text{vc}) \geq r_1$. Furthermore, it is simple to show that a furthest apart pair of nodes p, q satisfying $d(p, q) \geq d(x, y)$ for all $x, y \in V$, provides a useful lower bound $\frac{1}{2}d(p, q) \leq r_1$. With this information it also seems appropriate to choose p or q as the initial element of R.

Consider now the case where $m = k(>1), \ldots, M$ centres are to be located. One possibility is to build up the solutions by first generating solutions for $m = 1, 2, \ldots, k-1$. But a useful upper bound for r_k is easily obtained. Selecting arbitrarily a set of k vertices, X_k, an upper bound for r_k is given by $d = l(X_k) \geq r_k$, enabling the algorithm to commence directly with $m = k$.

SR1 Vertex-to-Enter

Assuming prior computation of D, it is straightforward to compute $l(\hat{X}_h)$. Alternatively, in the absence of this matrix, it is necessary to compute MPT(x), for all $x \in \hat{X}_h$. Note that since x is not necessarily a vertex, a slight modification is required in any standard MPT algorithm.

SR2 Row and Column Generation

The CP matrix needs to be updated by adding a row corresponding to the vertex chosen to enter and a set of columns corresponding to the new set of

Complexity and Efficiency in Minimax Network Location

candidate centres generated by this vertex. Consider first the issue of column generation. $C\{R:d\}$ requires updating according to the expression

$$C\{R\cup\{v\}:d\} = C\{R:d\} \bigcup_{r\in R} C[(v, r):d] \cup C(v, v). \tag{10.16}$$

A procedure for generating all candidate centres generated by a given pair of vertices is described below. This procedure is performed for pairs $\{v, r\}$ for all $r \in R$. Note also the inclusion of the null centre $c(v, v)$ in (10.16). For each candidate centre c generated, we need to record its range, d_c, and the subset $R_c \subseteq R$ covered by it, namely $R_c = \{r \in R : d(c, r) \leq d_c\}$. For every new candidate centre c, we add a new column to the CP matrix. An element of such a column equals 1 if the vertex corresponding to that row is a member of R_c; otherwise, the element equals 0.

To determine the row to be added to the CP matrix, we need to identify the subset $Q \subseteq C\{R:d\}$ covering vertex v, namely

$$Q = \{c \in C\{R:d\} : d(c, v) \leq d_c\}.$$

Then an element of this row equals 1 if the candidate centre associated with that column is a member of Q, as well as when the column is among the newly generated columns. Otherwise, the element equals 0.

We turn now to the issue of generating $C[(x, y):d]$ for a given pair of vertices $\{x, y\}$, $x \neq y$. For a generating pair of vertices $\{x, y\}$ and a generic link $(i, j) \in A$, define the following terms:

$$C^{ij}(x, y) = \{c \in C(x, y) : c \text{ is on } (i, j)\}.$$
$$C^{ij}[(x, y):d] = \{c \in C[(x, y):d] : c \text{ is on } (i, j)\}.$$
$$S^k(x, y) = \min\{d(k, x), d(k, y)\}, \ k \in \{i, j\}.$$

Definition 10.4 $\langle x, ij, y \rangle$, denoting a *flip-flop* condition on link (i, j) with respect to a pair of vertices $\{x, y\}$, is said to exist if and only if

$$(S^i(x, y) - d(i, x))(S^j(x, y) - d(j, x)) + (S^i(x, y) - d(i, y))(S^j(x, y) - d(j, y)) = 0.$$

The following theorem leads to an efficient algorithm for generating candidate centres.

Theorem 10.2 $C^{ij}(x, y) \neq \emptyset$ only if $\langle x, ij, y \rangle$ exists, in which case the following are mutually exclusive and exhaustive cases:

(i) $\langle x, c, y \rangle$ exists at an interior point of (i, j), $|C^{ij}(x, y)| = 1$ and the range and location of c are, respectively,

$$d_c = \tfrac{1}{2}(b(i, j) + S^i(x, y) + S^j(x, y))$$

and

$$b(i, c) = \tfrac{1}{2}(b(i, j) + S^j(x, y) - S^i(x, y))$$

(ii) $\langle x, c, y \rangle$ exists at v_k, $k = i$ or j, $|C^{ij}(x, y)| = 1$, and $d_c = S^k(x, y)$
(iii) $\langle x, c_i, y \rangle$ and $\langle x, c_j, y \rangle$ exist at vertices v_i and v_j respectively, $|C^{ij}(x, y)| = 2$, and $d_{c_k} = S^k(x, y)$ for $k = i, j$
(iv) $C^{ij}(x, y) = \emptyset$.

Furthermore, when all links $(i, j) \in A$ are investigated, no candidate centres in $C(x, y)$ are lost by ignoring case (iii) for all links.

Proof We shall refer to the diagrams in Figure 10.2. Consider first the general form of the function $f(\theta) = d(x, z_\theta)$, where z_θ is a point on link (i, j), θ units from v_i. It is readily established and illustrated in the example in Figure 10.2a that $f(\theta)$ is a concave, one or two-piece linear function with magnitude of slope equal to unity.

We consider the relationship between $d(x, z_\theta)$ and $d(y, z_\theta)$ and distinguish three mutually exclusive and exhaustive cases:

(i) $d(x, k) \neq d(y, k)$, $k = i, j$. A necessary and sufficient condition for $\langle x, z_\theta, y \rangle$ to exist in the range $\theta \in (0, b(i, j))$ is the existence of a flip-flop, $\langle x, ij, y \rangle$, as illustrated in Figure 10.2b. Furthermore, such a local centre, c, is unique. Finally, it is evident from the diagram that its range is given by $d_c = \frac{1}{2}(S^i(x, y) + S^j(x, y) + b(i, j))$ and that the location of c is given by $b(i, c) = \frac{1}{2}(b(i, j) + S^j(x, y) - S^i(x, y))$.

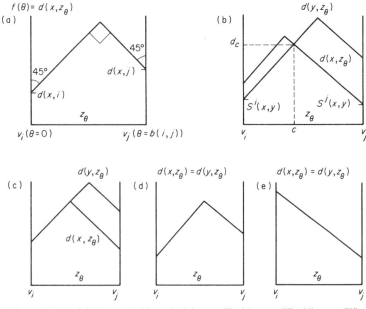

Figure 10.2 (a) Form of $d(x, z_\theta)$; (b) case (i); (c) case (ii); (d) case (iii); (e) case (iii)

(ii) $d(x, k) = d(y, k)$ for $k = i$ or $k = j$ but not both. Without loss of generality assume $k = i$ and consider the generic case depicted in Figure 10.2c. Only the point $c = v_i$ can possibly satisfy $\langle x, c, y \rangle$ though information from adjacent links is required to confirm this. Note that $d_c = S^i(x, y)$.

(iii) $d(x, k) = d(y, k)$, $k = i, j$. Consider the generic case depicted in Figure 10.2d. $d(x, z_\theta) = d(y, z_\theta)$ for all $\theta \in [0, b(i, j)]$ and only the points v_i and v_j are potential local centres, with ranges $S^i(x, y)$, $S^j(x, y)$ respectively.

Note that $\langle x, ij, y \rangle$ exists whenever $d(x, k) = d(y, k)$ for $k = i$ or $k = j$ or both, as well as in the situation illustrated in Figure 10.2b, thus establishing the four cases of the theorem.

We now wish to show that nothing is lost by ignoring local centres in case (iii) providing all links are investigated. Assuming case (iii) obtains, consider the two possibilities illustrated in Figures 10.2d and 10.2e. Without loss of generality consider vertex v_i and let $\Delta(i)$ denote the set of vertices adjacent to v_i. All links incident to v_i are in cases (ii) or (iii). In Figure 10.2d, existence of $\langle x, i, y \rangle$ implies there exists a vertex $v_k \in \Delta(i)$ such that link (i, k) is in case (ii). The contrary implies $x = y(=v_i)$ which, by assumption, is impossible. In Figure 10.2e, $\langle x, i, y \rangle$ does not exist. In conclusion, to generate $C(x, y)$ it suffices to inspect all links $(i, j) \in A$ in cases (i) and (ii) alone.

Theorem 10.2 leads directly to an efficient algorithm for generating $C[(x, y): d] = \cup_{(i,j) \in A} C^{ij}[(x, y): d]$ as described in the flow-chart in Figure 10.3.

SR3 Set-Covering Problem

In this subroutine, the CP defined in (10.10)–(10.12) is solved, with matrix B updated in SR2. We have already indicated that the relaxation strategy in its entirety can be viewed as a problem-oriented technique for solving the large-scale CP's arising in minimax facility location problems. Operationally, we have transformed a succession of large-scale CP's into a succession of small CP's, each of which requires relatively minor computational effort for its solution.

The size of a generic CP matrix in Algorithm 10.5 is $|R|$ rows by $|C\{R:d\}|$ columns, compared with n by $|C|$ and n by $|C\{V:d\}|$ in Algorithms 10.3 and 10.4, respectively. Since $|R|$ is generally related to m (independently of n), it is evident that for small m the CP is indeed manageable and often amenable to hand solution. This is particularly true as a consequence of the sequential nature of the algorithm, enabling the CP to be solved as the following feasibility problem: find a binary vector $\mathbf{x}$ satisfying $B\mathbf{x} \geq \mathbf{e}$ and $\mathbf{ex} = m$, or determine that no such vector exists. For further discussion of CP algorithms, the reader is referred to the survey by Christofides and Korman (1975) and to the text by Garfinkel and Nemhauser (1972).

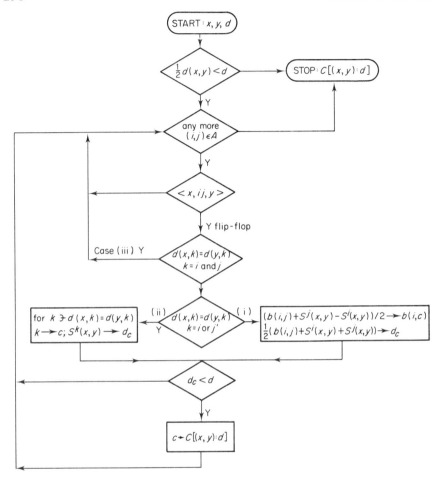

Figure 10.3 Flow-chart for column generation

The examples below illustrate the relaxation algorithm for location of m-centres. In the first example we consider a tree network and in the second a general cyclic network.

Example 10.1 $(A/V/m/T, m = 1, 2, \ldots, 7)$ The tree network with 60 vertices is shown in Figure 10.4. The example illustrates how a nontrivial tree-based problem can be solved manually. We shall let x_{pq} denote the midpoint of the path connecting p to q and, for compactness, we shall refer to x_{pq} as pq in labelling columns of CP matrices. The null centre at a vertex x and its associated column will be denoted as x.

Complexity and Efficiency in Minimax Network Location 299

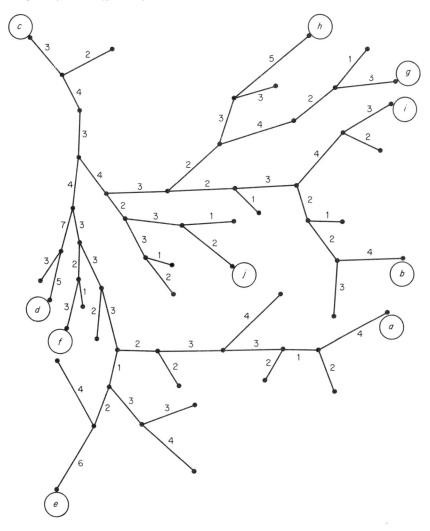

Figure 10.4 Network for Example 10.1. Numbers represent link lengths. Encircled vertex labels $a, b, \ldots, j$ represent successive additions to R

$m = 1$ Use Algorithm 10.1 for $A/V/1/T$ to obtain $X_1^* = x_{ab}$ and $r_1 = 23$.
$m = 2$ Use Algorithm 10.2 for $A/V/2/T$ to obtain $X_2^* = \{x_{bc}, x_{ad}\}$ and $r_2 = 17$.
$m = 3$ Initiate the relaxation algorithm arbitrarily, with the following upper bound solution derived by breaking up the maximal range in X_2^*:

$$X_3 = \{x_{bc}, x_{ae}, x_{df}\},\ d = 15,\ R = \{a, b, c, d, e, f\}.$$

Update $C(R:d)$ to obtain the following CP matrix:

	ae,	af,	cd,	cf,	df,	ef,	a,	b,	c,	d,	e,	f
a	1	1					1					
b								1				
c			1	1					1			
d			1		1					1		
e	1	1				1					1	
f			1	1	1	1						1
d_c [11	12	13	11	10	10	0	0	0	0	0	0]

Obtain, by inspection, $h = 3$, $\hat{X}_3 = \{x_{af}, x_{cd}, b\}$, and $\hat{d} = 13$. Now $l(\hat{X}_3) = d(g, \hat{X}_3) = d(g, x_{cd}) = 21 \geq 15 = d$. Adding g to R and updating $C\{R:d\}$, obtain the following enlarged CP matrix:

	ae,	af,	cd,	cf,	df,	ef,	gb,	gc,	a,	b,	c,	d,	e,	f,	g,
a	1	1							1						
b							1			1					
c			1	1				1			1				
d			1		1							1			
e	1	1				1							1		
f			1	1	1	1								1	
g							1	1							1
d_c [11	12	13	11	10	10	12	14	0	0	0	0	0	0	0]

Obtain, by inspection, $h = 3$, $\hat{X}_3 = \{x_{af}, x_{cd}, x_{gb}\}$ and $\hat{d} = 13$. Since $l(\hat{X}_3) = 13$, we have an improved global solution with $d = 13$. Attempting to improve upon this, eliminate columns with $d_c \geq 13$, thus obtaining the following CP matrix:

	ae,	af,	cf,	df,	ef,	gb,	a,	b,	c,	d,	e,	f,	g
a	1	1					1						
b						1		1					
c			1						1				
d				1						1			
e	1	1			1						1		
f			1	1	1							1	
g						1							1
d_c [11	12	11	10	10	12	0	0	0	0	0	0	0]

We find $h = 4$, $\hat{X}_4 = \{x_{af}, x_{cf}, x_{df}, x_{gb}\}$ and $\hat{d} = 12$. Hence, the current solution is optimal for $m = 3$, namely,

$$X_3^* = \{x_{af}, x_{cd}, x_{gb}\} \quad \text{and} \quad r_3 = 13.$$

Complexity and Efficiency in Minimax Network Location

$m = 4$ Since $l(\hat{X}_4) = 12$, we have an improved solution $X_4 = \hat{X}_4$ with $d = 12$. Eliminating columns with $d_c \geq 12$, the following CP matrix is obtained:

	ae,	cf,	df,	ef,	a,	b,	c,	d,	e,	f,	g
a	1				1						
b						1					
c		1					1				
d			1					1			
e	1				1				1		
f		1	1	1						1	
g											1
d_c [	11	11	10	10	0	0	0	0	0	0	0]

We find $h = 5$, $\hat{X}_5 = \{x_{ae}, x_{cf}, x_{df}, b, g\}$ and $\hat{d} = 11$. Hence, the current solution is optimal for $m = 4$, namely,

$$X_4^* = \{x_{af}, x_{cf}, x_{df}, x_{gb}\} \quad \text{and} \quad r_4 = 12.$$

$m = 5$ $l(\hat{X}_5) = d(h, \hat{X}_5) = 12 = d$. Adding h to R and updating $C\{R:d\}$, we obtain the following enlarged CP matrix:

	ae,	cf,	df,	ef,	hy,	hb,	a,	b,	c,	d,	e,	f,	g,	h
a	1						1							
b						1		1						
c		1							1					
d			1							1				
e	1			1							1			
f		1	1	1								1		
g					1								1	
h					1	1								1
d_c [	11	11	10	10	$8\frac{1}{2}$	$11\frac{1}{2}$	0	0	0	0	0	0	0	0]

We find $h = 5$, $\hat{X}_5 = \{x_{ae}, x_{cf}, x_{df}, x_{hg}, x_{hb}\}$ and $\hat{d} = 11\frac{1}{2}$. Since $l(\hat{X}_5) = 11\frac{1}{2}$, $X_5 = \hat{X}_5$ is an improved global solution for $m = 5$ with $d = 11\frac{1}{2}$. Eliminating columns with $d_c \geq 11\frac{1}{2}$ and solving the revised CP, we obtain $h = 5$, $\hat{d} = 11$ and $l(\hat{X}_5) = d(i, \hat{X}_5) \geq d$. Adding vertex i to R and updating $C\{R:d\}$, we obtain a new CP of size (9×7) (excluding null variables) yielding $h = 5$, $\hat{d} = 11$ and $l(\hat{X}_5) = d(j, \hat{X}_5) \geq d$. Adding vertex j to R and updating $C\{R:d\}$, we obtain an enlarged CP of size (10×10) yielding $h = 5$, $\hat{d} = 11$ and $l(\hat{X}_5) = 11$, so that $X_5 = \hat{X}_5$ is an improved solution with $d = 11$. Eliminating columns with $d_c \geq 11$ and resolving the CP of size (10×7) yields $h = 6$ and $d = 10\frac{1}{2}$ so that X_5 is an optimal solution with $r_5 = 11$.

$m = 6$ Since $l(\hat{X}_6) = 10\frac{1}{2}$, $X_6 = \hat{X}_6$ is an improved global solution for $m = 6$ with $d = 10\frac{1}{2}$. Eliminating columns with $d_c \geq 10\frac{1}{2}$ results in a CP of size (10×5), which yields $h = 7$ and $\hat{d} = 10$, so that X_6 is optimal and $r_6 = 10\frac{1}{2}$.

$m = 7$ $l(\hat{X}_7) = 20$ so that $X_7 = \hat{X}_7$ is an improved global solution for $m = 7$ with $d = 10$. Eliminating columns with $d_c \geq 10$, we obtain a CP of size (10×2) yielding $h = 8$, $\hat{d} = 8\frac{1}{2}$, so that X_7 is optimal and $r_7 = 10$.

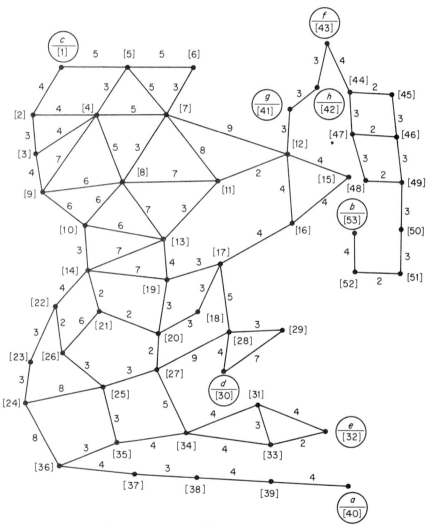

Figure 10.5 Network for Example 10.2. Numbers on links represent link lengths. Numbers in brackets identify vertices. Encircled vertex labels $a, b, \ldots, h$ represent successive additions to R

Complexity and Efficiency in Minimax Network Location 303

Example 10.2 $(A/V/m/G, m = 1, 2, 3, 4)$ The network, with 53 vertices and 81 links, is shown in Figure 10.5. For the general graph problem the task of updating $C\{R:d\}$ is tedious for a manual mode while the CP's are often amenable to manual solution by inspection. In this example an interactive approach was adopted with successive updates of $C\{R;d\}$ by computer and manual solutions of CP's by inspection. Details are reproduced only for $m = 1$ and $m = 2$. Computational data for this problem appear in Table 10.3.

$m = 1$ Computing first the matrix D, we locate a vertex centre, yielding an upper bound solution $l(16) = 37$. Hence, $r_1 \leq 37$. A most distant pair of vertices is $\{40, 53\}$. Initiate the algorithm by setting $d = 37$, $R = \{40\}$, $\hat{X}_1 = 40$ and $\hat{d} = 0$. Then $l(40) = d(40, 53) = 71$. Let $R = \{40, 53\}$ and generate $C[(40, 53):37]$ to obtain the following CP matrix:

$$\begin{array}{c c c c c} & x_{40,53}^{11,13} & x_{40,53}^{12,16} & 40 & 53 \\ 40 & \left[\begin{array}{cccc} 1 & 1 & 1 & \\ 1 & 1 & & 1 \end{array}\right] \\ 53 & & & & \\ d_c & [\ 35\tfrac{1}{2} & 36\tfrac{1}{2} & 0 & 0\ ] \end{array}$$

Here, $x_{p,q}^{i,j}$ denotes the candidate centre on link (i, j) generated by the pair of vertices p, q. The column associated with a null centre at p is labelled p. The set of columns associated with the null centres is termed the null set.

Choosing $\hat{X}_1 = x_{40,53}^{11,13}$, $\hat{d} = 35\tfrac{1}{2}$, we find $l(\hat{X}_1) = 35\tfrac{1}{2}$, so that $X_1 = \hat{X}_1$ and $d = 35\tfrac{1}{2}$. Eliminating columns with $d_c \geq 35\tfrac{1}{2}$ reduces the CP matrix to the null set, from which one column is insufficient to cover the elements in R. Hence, a 1-centre is given by $X_1^* = x_{40,53}^{11,13}$ and $r_1 = 35\tfrac{1}{2}$.

$m = 2$ The solution to the reduced CP yields $h = 2$, $\hat{X}_2 = \{40, 53\}$ and $\hat{d} = 0$. Since $l(\hat{X}_2) = d(1, \hat{X}_2) > 0$, vertex v_1 is added to R. Updating $C\{R:d\}$ yields the new CP matrix:

$$\begin{array}{c c c c c c c c c c} & x_{40,1}^{22,26} & x_{40,1}^{20,27} & x_{40,1}^{21,26} & x_{53,1}^{42,43} & x_{40,1}^{22,23} & x_{40,1}^{27,28} & 40 & 53 & 1 \\ 40 & \left[\begin{array}{ccccccccc} 1 & 1 & 1 & 0 & 1 & 1 & 1 & & \\ 0 & 0 & 0 & 1 & 0 & 0 & & 1 & \\ 1 & 1 & 1 & 1 & 1 & 1 & & & 1 \end{array}\right] \\ 53 & & & & & & & & & \\ 1 & & & & & & & & & \\ d_c & [\ 25 & 25 & 26 & 26 & 27 & 32\tfrac{1}{2} & 0 & 0 & 0\ ] \end{array}$$

Solving the CP yields $h = 2$, $\hat{X}_2 = \{x_{40,1}^{20,27}, 53\}$, $\hat{d} = 25$. Since $l(\hat{X}_2) = 25$, we have an improved global solution $X_2 = \hat{X}_2$, $d = 25$. Eliminating all columns with $d_c \geq 25$ reduces the CP matrix to the null set. Since three columns are now required to cover all elements of R, a set of 2-centres is given by $X_2^* = \{x_{40,1}^{20,27}, 53\}$ and $r_2 = 25$.

$m = 3$ Proceeding as before, an initial solution is given by the null set and a most distant vertex, v_{30}, is added to R. After five iterations an optimal solution for $m = 3$ is achieved with $R = \{40, 53, 1, 30, 32\}$ and $r_3 = 18$.

$m = 4$ Five iterations are needed to achieve optimality with $R = \{40, 53, 1, 30, 32, 43, 41, 42\}$ and $r_4 = 12\frac{1}{2}$.

To validate the relaxation algorithm we state and prove the following theorem.

Theorem 10.3 Algorithm 10.5 locates a set of m-centres in A/V/m/G in a finite number of steps.

Proof (i) *Optimality* A simple relaxation argument suffices. The algorithm finds an optimal solution, X_m^*, to $A/R/m/G$ for some $R \subseteq V$ according to Algorithm 10.4. Futhermore,

$$\min_{X_m \subseteq G} \max_{y \in R} d(X_m, y) = \max_{y \in V} d(X_m^*, y) \qquad (10.17)$$

according to the algorithm. Now consider $A/R/m/G$ reformulated as

$$\min \{d : X_m \subseteq G; d \geq d(X_m, y) \text{ for all } y \in R\} \qquad (10.18)$$

indicating that $A/R/m/G$ is a relaxation of $A/V/m/G$. Expressions (10.17) and (10.18) establish that X_m^* is also optimal for $A/V/m/G$.

(ii) *Finite Convergence* An iteration of the algorithm consists of a CP and subsequent row/column generation, clearly a finite operation. At each iteration one and only of the following occurs:

(a) One row is added with possible column additions and eliminations.
(b) At least one column is eliminated and none added.

Since n is finite, a necessary condition for an infinite number of iterations is that (b) is repeated infinitely for a given R. Since $|C\{R:d\}|$ is finite this is impossible and finite convergence is guaranteed.

Computational Efficiency and Experience

The critical factor determining computational efficiency is the cardinality of R at the optimal solution for a given m. In the worst case, Algorithm 10.5 will perform roughly as Algorithm 10.4. For example, consider $A/V/1/G$ where G is a single loop and vertices are distributed at equal intervals. Then $|R| \approx n$ at the optimal solution (see the example in Figure 10.6 and Table 10.6). However, apart from such pathological cases yielding rather unattractive computational upper bounds, the algorithm can be expected to perform efficiently. Preliminary computational experience suggests $|R| \leq 5m$ is a good approximation. This indicates a significant relative advantage over Algorithm 10.4 when m is small. Since the latter algorithm is particularly efficient for large m, it turns out that the proposed algorithm, by combining both features, is well-suited for all values of m.

We now wish to focus attention on the number of candidate centres $|C\{R:d\}|$ which, apart from the direct dependence on $|R|$, is based upon the generic quantities $|C[x, y):d]|$. From Theorem 10.2 we can assert that $|C[(x, y):d]| \leq |A|$. However, the following observations indicate why this bound is usually a gross over-estimate of the true number. Note first that in the previous inequality A may be replaced by the set of links derived by amalgamating adjacent links at vertices with degree 2. This observation is particularly important for $A/A/m/G$ (Section 10.4). Consider next the following theorem.

Theorem 10.4 *The existence of $\langle x, c, y \rangle$ implies that $\langle x, z, y \rangle$ does not exist at any point $z \neq c$ on a minimal distance path from c to x or to y.*

Proof Suppose, for example, that $z \neq c$ is a point on a minimum path from c to y, and assume $\langle x, c, y \rangle$ exists. Then by the latter assumption,

$$d(c, x) < d(c, z) + d(z, x) = d(c, y) - d(z, y) + d(z, x)$$

so that $d(z, y) < d(z, x)$, which implies that $\langle x, z, y \rangle$ cannot exist.

The preceding comments apply to $|C(x, y)|$. The effect of d in reducing $|C[(x, y):d]|$ is all-important. For small m, the induced 'spread' of nodes in R sharpens this effect. For example, note that $d(x, y) \geq 2d$ implies $|c[(x, y):d]| = 0$. For large m, small values of d again contribute to its effectiveness in reducing the cardinality of the set of candidate centres.

While $|C\{R:d\}|$ represents the number of distinct candidate centres at any iteration, the number of distinct columns in the corresponding CP matrix may generally be smaller. This follows since several candidate centres can give rise to an identical column in the CP matrix. In solving the CP, it is clearly desirable to ignore duplicate columns. To illustrate the point, consider Example 10.2. In the matrix for $m = 2$, five candidate centres produce an identical column so that the effective number of columns is two instead of six (exclusive of null columns).

Finally, note the special case of trees where $|C(x, y)| = 1$. Since now both CP and matrix generating efforts are minimal, it appears that tree-based problems can be solved manually for quite large networks.

We now present some computational results for the relaxation algorithm. Table 10.3 summarizes computational data for Example 10.2.

The advantages accruing from the relaxation scheme can be readily seen in this example. The number of vertices in R, identical to the number of rows in the CP's, is about twice the number of centres, m, in place of the full set of vertices, here 53. This represents a saving in orders of magnitude with respect to the resultant CP's and matrix generation. Indeed, it is doubtful whether Algorithm 10.3 can reasonably handle this problem, given the enormous number of generated columns. Employing Algorithm 10.5, the

Table 10.3 Computational data for Example 10.2 for network with 53 vertices and 81 links

	D matrix	cpu (seconds) 0.24
Number of centres, m	CP (rows × columns)	
1	(2×2)	0.03
2	(3×6)	0.04
3	$(4 \times 10); (4 \times 9); (4 \times 2); (5 \times 2); (5 \times 4)$	0.09
4	$(6 \times 16); (6 \times 7); (7 \times 18); (8 \times 26); (8 \times 16)$	0.22

(1) The program was written in Fortran IV, compiled under level G1, and run on an IBM-370/165.
(2) Times are incremental. (To solve for a given number of centres, M, would not necessitate solving for all previous $m < M$.)
(3) A large fixed cost was incurred in this example by first computing D. This is very inefficient for small values of m. Thus, for $m = 1$, total time would be approximately 0.05 seconds instead of 0.27 as indicated.
(4) CP's were solved manually by inspection.
(5) The number of columns in the CP's does not include the $|R|$ columns associated with the null centres at the vertices.

major computational effort is due to minimum path computations for matrix generation, while the CP's are small enough to allow solution by hand.

In the case of tree networks, since minimum path computations are particularly simple, the whole problem is often amenable to manual solution. Example 10.1, summarized in Table 10.4, illustrates this for a tree with 60 vertices. Notice that, in contrast, Algorithm 10.3 involves generating and solving a sequence of CP's of size (60×1770) or, in general for trees, $(n \times \frac{1}{2}n(n-1))$, excluding n null columns.

The effectiveness of the relaxation procedure is best demonstrated with large-scale networks, which are altogether beyond the computational

Table 10.4 Size of CP's for Example 10.1 for tree with 60 vertices

Number of centres, m	CP (rows × columns)
1, 2	Direct techniques (Section 10.2)
3	$(6 \times 6); (7 \times 8); (7 \times 6)$
4	(7×4)
5	$(8 \times 6); (8 \times 5); (9 \times 7); (10 \times 10); (10 \times 7)$
6	(10×5)
7	(10×2)

Complexity and Efficiency in Minimax Network Location 307

Table 10.5 Computational data for some random networks

Number of centres, m	number of vertices, n number of arcs, $|A|$	200 305	300 442	400 591	800 1216
$m = 1$	cpu seconds CP$_f$ rows × columns CP's	2.62 3×2 5	14.67 5×14 10	10.37 3×2 5	69.45 6×27 11
$m = 2$	cpu seconds CP$_f$ rows × columns CP's	4.31 5×3 5	16.14 7×17 8	50.53 8×31 12	211.07 12×82 18
$m = 3$	cpu seconds CP$_f$ rows × columns CP's	2.8 6×4 2	5.7 7×12 3	54.71 11×58 8	
$m = 4$	cpu seconds CP$_f$ rows × columns CP's	6.75 8×6 4		83.54 15×112 8	

(1) The program was written in Fortran IV, compiled under level 4.1 and run on a CDC-6600 in an interactive mode.
(2) cpu times are incremental.
(3) CP$_f$ indicates the size of the final CP matrix excluding null centres and duplicate columns.
(4) CP's indicates the number of CP's solved for the current value of m. These were solved manually by inspection.

bounds of rudimentary procedures such as Algorithm 10.3. Table 10.5 summarizes computational data for several problems with the number of vertices ranging from 200 to 800. The examples were constructed by generating random transit-like networks, with the probability of link length decreasing as a function of link length. To avoid the large fixed cost of computing D, minimum path distances were computed as needed.

10.4 Extensions to other Centre Problems

So far we have focused our discussion of complexity and efficiency in minimax network location on the class of absolute centre problems denoted as $A/V/m/G$ in the classification scheme of Section 10.1. We now briefly consider the remaining varieties of centre problems introduced in that section.

Recalling the results concerning the complexity of absolute centre problems from Section 10.2, we note first that less is known about the complexity of the other centre problems. While some of our comments below will be of a conjectural nature, requiring further verification, it appears that a similar spectrum of complexities exists for those other problems as was indicated for

$A/V/m/G$. For the simplest cases of single-centres of a tree $O(n)$ procedures are available. Specifically, Algorithm 10.1 solves $A/A/1/T$ and a variant thereof solves $V/V/1/T$ and $V/A/1/T$ (Handler, 1973). For $A/A/2/T$ a procedure similar to Algorithm 10.2 also involves $O(n)$ operations, while $V/V/2/T$ and $V/A/2/T$ can be processed in $O(n^2)$ time by an iterative version of Algorithm 10.2 (Handler, 1974, 1978; Handler and Mirchandani, 1979).

Turning now to cyclic networks, the single-centre problems $V/V/1/G$ and $V/A/1/G$ are essentially solved once the shortest-distance matrix is available, so that these can be processed in $O(n^3)$ time. Frank's (1967) procedure for $A/A/1/G$ appears also to be bounded by a low order polynomial in n. Finally, the general multi-centre problems $A/A/m/G$, $V/V/m/G$ $V/A/m/G$, and all the inverse centre problems appear to be at least as 'hard' as $A/V/m/G$, namely, at least as complex as NP-complete. As before, however, this observation does not imply that these problems are necessarily difficult to solve in practice. As we shall see next, the relaxation strategy of the previous section can be modified to provide effective algorithms for these problems, which we consider in turn.

$A/A/m/G$

This problem is perhaps the most challenging of the variety of minimax location problems we are considering. Physically, it represents what is often the most realistic situation while $A/V/m/G$ is some discretized approximation of it. One way of approaching this problem is indeed to discretize the set of demand generating points as finely as is desired and then solve as $A/V/m/G$. Without the relaxation technique, such a scheme becomes unwieldy as the approximation is tightened because of the critical effect of a large number of vertices in other techniques.

Consider now the relaxation approach described in the previous section. All along we have been solving relaxed problems, $A/R/m/G$, $R \subseteq V$, as surrogates for $A/V/m/G$. The only conceptual change here is that V is now the infinite set of points in G. The basic strategy remains unaltered. However, since demand can occur anywhere along the network, the changed definition of $l(X_m)$, namely,

$$l(X_m) = \max_{y \in G} d(X_m, y) \tag{10.19}$$

necessitates an extension in the MPT computations in subroutine SR1.

Although $A/A/m/G$ utilizes to the fullest extent the capabilities of the relaxation method and provides the most striking context for its application, the algorithm nevertheless remains elusive in its convergence properties when applied to $A/A/m/G$. We have shown that for $A/V/m/G$ the algorithm converges finitely (Theorem 10.3). The same is not true for $A/A/m/G$, and

Complexity and Efficiency in Minimax Network Location

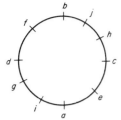

Figure 10.6 A single cycle network to illustrate $A/A/1/G$

the following, possibly pathological, example illustrates this. Consider the network in Figure 10.6 composed of a single cycle, and solve the problem $A/A/1/G$. Then application of the algorithm beginning at some arbitrary point, a, will lead to an infinitely convergent series of intermediate solutions. The labels $a, b, c, \ldots$ in Figure 10.6 represent succesive additions to R as determined by (10.19). Successive iterations of the algorithm yield intermediate results as shown in Table 10.6.

When the algorithm fails to terminate finitely it is also possible that the limiting solution will not be optimal. This difficulty may be overcome by a modified version of the relaxation algorithm.

Table 10.6 Intermediate results for $A/A/1/G$ in the single cycle network example

R	$\hat{X}_1$	$2\hat{d}$	$2l(\hat{X}_1)$
a	a	0	1
a, b	d	1/2	1
a, b, c	c	1/2	1
a, b, c, d	f	3/4	1
a, b, c, d, e	e	3/4	1
a, b, c, d, e, f	h	3/4	1
a, b, c, d, e, f, g	g	3/4	1
a, b, c, d, e, f, g, h	j	7/8	1
etc.			

$V/V/m/G$

A method for determining optimal locations in $V/V/m/G$ due to Toregas et al. (1971) corresponds to Minieka's scheme for $A/V/m/G$ (Algorithm 10.3) with the dominant set of candidate centres C replaced by the set V. The midpoint property embodied in Corollary 10.1a no longer holds, though centres are now, by definition, restricted to a finite set of points. Indeed, with respect to the rudimentary framework of Algorithm 10.3 for solving $A/V/m/G$, $V/V/m/G$ is by far the easier of the two problems since, in general, $|C| > \binom{n}{2} + n \gg n$. In fact, we shall see that this observation, though intuitively appealing, is quite misleading.

The relaxation approach can be applied here with some modification. Note first that the column elimination scheme utilized in Algorithm 10.4 is no longer applicable since that scheme is inextricably tied up with the midpoint property. Consequently, columns cannot be eliminated, and the full set of n candidate centres must be carried explicitly throughout. Otherwise, the relaxation scheme applies as before and, clearly, finite convergence is guaranteed in this case.

Though the improvement over the rudimentary procedure is not as dramatic as for $A/V/m/G$ and $A/A/m/G$, substantial savings can be expected because of the importance of the number of rows to the solution of

Table 10.7 The minimum-distance matrix for 30 locations in New York State.

	1	2	3	4	5	6	7	8	9	10	11	12	13	14	15
1	0	244	140	128	281	196	181	51	248	167	338	54	203	146	295
2	244	0	158	359	37	111	66	268	60	112	101	278	272	328	51
3	140	158	0	202	194	56	92	170	117	46	215	137	256	170	209
4	128	359	202	0	395	258	294	179	319	248	416	90	331	61	410
5	281	37	194	395	0	145	102	305	92	148	69	317	309	366	19
6	196	111	56	258	145	0	60	229	61	34	159	189	269	226	162
7	181	66	92	294	102	60	0	208	67	47	157	220	225	262	117
8	51	268	170	179	305	229	208	0	275	195	366	105	152	197	319
9	248	60	117	319	92	61	67	275	0	87	111	254	292	287	111
10	167	112	46	248	148	34	47	195	87	0	185	179	235	216	163
11	338	101	215	416	69	159	157	366	111	185	0	348	373	381	88
12	54	278	137	90	317	189	220	105	254	179	348	0	257	95	329
13	203	272	256	331	309	269	225	152	292	235	373	257	0	349	323
14	146	328	170	61	366	226	262	197	287	216	381	95	349	0	379
15	295	51	209	410	19	162	117	319	111	163	88	329	323	379	0
16	211	222	206	339	259	219	175	180	242	185	323	250	66	343	273
17	295	77	160	361	104	114	322	56	130	55	293	339	326	93	
18	78	200	62	176	236	118	134	108	175	93	273	86	236	179	251
19	169	106	114	290	143	112	59	186	126	79	207	205	168	284	157
20	38	281	177	100	318	233	218	81	285	204	375	65	233	144	332
21	167	332	279	295	369	315	279	116	346	281	433	221	60	313	383
22	112	263	105	106	299	161	197	163	222	151	316	58	315	69	314
23	71	294	136	70	330	192	228	124	253	182	351	20	274	75	345
24	220	33	136	337	70	100	46	242	60	90	134	254	239	306	84
25	157	284	239	285	321	274	237	106	304	240	385	211	46	303	335
26	16	233	129	143	272	185	170	41	237	156	327	69	193	161	284
27	135	109	78	254	146	91	49	159	116	57	206	169	178	248	160
28	7	248	144	133	285	200	185	48	252	171	342	61	200	151	299
29	90	161	92	211	198	128	101	107	168	94	258	126	171	219	212
30	165	164	148	293	201	161	117	175	184	127	265	204	108	297	215

CP's. Furthermore, even in the single-centre case where no CP need be solved, substantial saving are obtained. To illustrate this point consider the example given by Toregas *et al.* (1971). Table 10.7 reproduces the D matrix for 30 locations in New York State. Consider now the problem of locating a solution vc to $V/V/1/G$. The classical approach is initially to compute D and inspect the matrix for an optimal solution. Employing instead the relaxation technique and beginning arbitrarily with $R=\{1\}$, results in the series of solutions to $V/R/1/G$ shown in Table 10.8, derived by inspection of the relevant columns and rows in Table 10.7. Thus, vc = 27 and $l(vc) = 254$ is a solution to $V/V/1/G$. Notice that to arrive at the solution we only require computation of MPT(x), for $x \in \{1, 11, 7, 4, 3, 21, 27\}$, in place of the full D matrix.

Although, as we have seen, $V/V/m/G$ can be profitably solved with the relaxation algorithm, the advantages due to this approach are not quite as

(Distances are in miles)

16	17	18	19	20	21	22	23	24	25	26	27	28	29	30
211	295	78	169	38	167	112	71	220	157	16	135	7	90	165
222	77	200	106	281	332	263	294	33	284	233	109	248	161	164
206	160	62	114	177	279	105	136	136	239	129	78	144	92	148
339	361	176	290	200	295	106	70	337	285	143	254	133	211	293
259	74	236	143	318	369	299	330	70	321	272	146	285	198	201
219	104	118	112	233	315	161	192	100	274	185	91	200	128	161
175	114	134	59	218	279	197	228	46	237	170	49	185	101	117
180	322	108	186	81	116	163	124	242	106	41	159	48	107	175
242	56	175	126	285	346	222	253	60	304	237	116	252	168	184
185	130	93	79	204	281	151	182	90	240	156	57	171	94	127
323	55	273	207	375	433	316	351	134	385	327	206	342	258	265
250	293	86	205	65	221	58	20	254	211	69	169	61	126	204
66	339	236	168	233	60	315	274	239	46	193	178	200	171	108
343	326	179	284	144	313	69	75	306	303	161	248	151	219	297
273	93	251	157	332	383	314	345	84	335	284	160	299	212	215
0	289	192	118	248	126	293	270	189	92	198	128	213	129	58
289	0	218	173	332	393	261	296	106	351	284	163	299	215	231
192	218	0	130	118	221	124	106	178	194	67	94	82	64	146
118	173	130	0	206	228	219	225	73	180	156	36	171	79	60
248	332	118	206	0	191	123	76	257	187	52	172	37	127	202
126	393	221	228	191	0	279	238	299	52	157	230	162	187	168
293	261	124	219	123	279	0	60	241	269	127	183	119	164	247
270	296	106	225	76	238	60	0	272	228	86	189	76	146	224
189	106	178	73	257	299	241	272	0	251	209	85	224	135	131
92	351	194	180	187	52	269	228	251	0	147	188	154	147	120
198	284	67	156	52	157	127	86	209	147	0	124	15	77	152
128	163	94	36	172	230	183	189	85	188	124	0	139	52	70
213	299	82	171	37	162	119	76	224	154	15	139	0	92	167
129	215	64	79	127	187	164	146	135	147	77	52	92	0	83
58	231	146	60	202	168	247	224	131	120	152	70	167	83	0

Table 10.8 Intermediate solutions for $V/V/1/G$ using relaxation technique in the New York State example

R	$\widehat{vc}$	$\hat{d}$	$l(\widehat{vc}) = d(\widehat{vc}, v)$
1	1	0	$338 = d(1, 11)$
1, 11	7	181	$294 = d(7, 4)$
1, 11, 4	3	215	$279 = d(3, 21)$
1, 11, 4, 21	27	254	254

dramatic as for $A/V/m/G$ and $A/A/m/G$. This serves to emphasize the special features of the relaxation scheme when applied to these problems. With respect to $V/Vm/G$, the relaxation technique can be viewed as a classical application of relaxation concepts to a minimax problem with corresponding elimination of constraints. But for $A/V/m/G$ and $A/A/m/G$ the technique is ideally suited because the linkage between rows and columns results in both suited because the linkage between rows and columns results in both row and column eliminations. Thus, problem relaxation here carries the advantages of both relaxation and restriction.

Finally, note the relative computational difficulty for these problems. $V/V/m/G$ can often be the more difficult problem to solve, contrary, perhaps, to intuition.

$V/A/m/G$

Once again, the relaxation approach is well-suited to this problem. The procedures outlined for $V/V/m/G$ and $A/A/m/G$ can be readily combined to form a solution strategy for $V/A/m/G$. An important distinction from $A/A/m/G$ is contained in the following theorem.

Theorem 5 *The relaxation algorithm, appropriately modified to solve $V/A/m/G$, converges to an optimal solution in a finite number of steps.*

Proof An iteration of the algorithm results in one of the following two cases:

(i) A new artificial node is added to R.
(ii) At least one nonzero element of the CP matrix is changed to zero.

To prove finite convergence it suffices to show that R has finite maximal cardinality. According to the algorithm, $R \subseteq O$ where

$$P = \{p \in G : d(X_m, p) = l(X_m) \text{ for some } X_m \subseteq V\}.$$

Since n is finite so is $|P|$ and convergence is guaranteed.

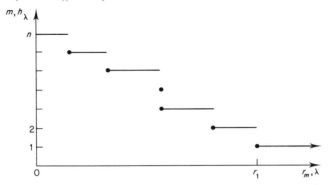

Figure 10.7 Relationship between $A/V/m/G$ and $A/V/\lambda^{-1}/G$.
• represents solution to $A/V/m/G$; — represents solution to $A/V/\lambda^{-1}/G$

Inverse Centre Problems

Consider first $A/V/\lambda^{-1}/G$ and let h_λ denote the minimal number of centres needed to insure a maximum distance not exceeding λ from any vertex to a nearest centre. The relationship between h_λ and the solution, r_m, to $A/V/m/G$ is illustrated in Figure 10.7.

Solving $A/V/\lambda^{-1}/G$ requires minor and generally simplifying modifications to any procedure that solves $A/V/m/G$. Thus, Algorithms 10.3 or 10.4 can be used on a 'one-shot' basis, solving just one CP. As before, however, the relaxation technique is more efficient. Setting $d = \lambda$ from the outset and ignoring intermediate updates of X_m where $h > m$, Algorithm 10.5 will yield $h_\lambda = h$ when $l(\hat{X}_h) = \lambda$ for the first time.

Extensions to the cases $A/A/\lambda^{-1}/G$, $V/V/\lambda^{-1}/G$, and $V/A/\lambda^{-1}/G$ are straightforward.

References

Christofides, N., and S. Korman (1975). A computational Survey of Methods for the Set Covering Problem. *Man. Sci.*, **21**, 591–599.

Christofides, N., and R. Viola (1971). The Optimum Location of Multi-Centers on a Graph. *Opns Res. Quart.*, **22**, 145–154.

Dijkstra, E. W. (1959). A Note on Two Problems in Connection with Graphs. *Numer. Mathematik*, **1**, 269–271.

Dreyfus, S. E. (1969). An Appraisal of Some Shortest-Path Algorithms. *Opns. Res.*, **17**, 395–412.

Frank, H. (1967). A Note on a Graph Theoretic Game of Hakimi's. *Opns. Res.*, **15**, 567–570.

Garfinkel, R. S., A. W. Neebe, and M. R. Rao, (1977). The m-Center Problem: Minimax Facility Location. *Man. Sci.*, **23**, 1133–1142.

Garfinkel, R. S., and G. L. Nemhauser (1972). *Integer Programming*, Wiley, New York.

Hakimi, S. L. (1964). Optimum Locations of Switching Centers and the Absolute Centers and Medians of a Graph. *Opns. Res.*, **12,** 450–459.

Hakimi, S. L. (1965). Optimum Distribution of Switching Centers in a Communication Network and Some Related Graph Theoretic Problems. *Opns. Res.*, **13,** 462–475.

Hakimi, S. L., E. F. Schmeichel, and J. G. Pierce (1976). On p-Centers in Networks. *Northwestern University, September,* 1975, *revised April,* 1976.

Handler, G. Y. (1973). Minimax Location of a Facility in an Undirected Tree Graph. *Trans. Sci.*, **7,** 287–293.

Handler, G. Y. (1974). Minimax Network Location: Theory and Algorithms. *Technical Report No.* 107, *Operations Research Center and FTL-R*74-4, *Flight Transportation Laboratory, M.I.T.*

Handler, G. Y. (1978). Finding Two-Centers of a Tree: The Continuous Case *Trans. Sci.*, **12,** 93–106.

Handler, G. Y., and P. B. Mirchandani (1979). *Location in Networks: Theory and Algorithms,* M.I.T. Press, Cambridge, Mass.

Hu, T. C. (1971). Some Problems in Discrete Optimization. *Mathematical Programming,* **1,** 102–112.

Kariv, O., and S. L. Hakimi (1976). An Algorithmic Approach to Network Location Problems—Part 1: the p-Centers. *Northwestern University, December,* 1976.

Minieka, E. (1970). The m-Center Problem. *SIAM Review,* **12,** 138–139.

Toregas, C., R. Swan, C. ReVelle, and L. Bergman (1971). The Location of Emergency Service Facilities. *Opns. Res.*, **19,** 1366–1373.

CHAPTER 11

The Vehicle Routing Problem

NICOS CHRISTOFIDES
Imperial College, London

ARISTIDE MINGOZZI
SOGESTA, Urbino, Italy

PAOLO TOTH
University of Bologna

11.1 Introduction

We consider a problem in which a set of geographically dispersed 'customers' with known requirements must be served with a fleet of 'vehicles' stationed at a central facility or depot in such a way as to minimize some distribution objective. It is assumed that all vehicle routes must start and finish at the depot.

The *vehicle routing problem* (VRP) is a generic name given to a whole class of problems involving the visiting of 'customers' by 'vehicles'. The VRP (also known in the literature as the 'vehicle scheduling' (Clarke and Wright, 1964; Eilon, Watson-Gandy, and Christofides, 1971; Gaskell, 1967), 'vehicle dispatching' (Christofides and Eilon, 1969; Dantzig and Ramser, 1959; Gillett and Miller, 1974; Pierce, 1970), or simply as the 'delivery' problem (Balinski and Quandt, 1964; Hays, 1967; Tillman and Cochran, 1969) appears very frequently in practical situations not directly related to the physical delivery of goods. For example, the collection of mail from mailboxes, the pickup of children by school buses, house-call tours by a doctor, preventive maintenance inspection tours, the delivery of laundry, etc. are all VRP's in which the 'delivery' operation may be a collection, collection and/or delivery, or neither; and in which the 'goods' and 'vehicles' can take a variety of forms, some of which may not even be of a physical nature. In view of the enormous number of practical situations which give rise to VRP's, it is not surprising to find that an equally large number of constraints and/or objectives appear in such problems. However, the basic VRP can be characterized as follows.

11.1.1 The Basic VRP

We consider the VRP for a given time period T. Let $X = \{x_i \mid i = 1, \ldots, N\}$ be a set of N customers and x_0 be the depot.

A set $V = \{v_k \mid k = 1, \ldots, M\}$ of vehicles stationed at the depot is given. A customer x_i has the following requirements:

C1. A quantity q_i of some product to be delivered by a vehicle.
C2. A time u_i required by a vehicle to visit the customer and to unload the quantity q_i.
C3. A priority δ_i.

A vehicle v_k has the following characteristics:

V1. A total capacity of Q_k for carrying products.
V2. A working period from time T_k^s to time T_k^f.
V3. A fixed cost of C_k.

We will assume that the cost of the least cost path from customer x_i to customer x_j is given as c_{ij} and that t_{ij} is the corresponding travel time. The costs c_{ij} will be referred to as the variable cost.

Objectives

The VRP is to design routes for vehicles to supply customers so as to minimize some objective. There are many possible objectives to the VRP. For example:

O_1. Maximize the sum of the priorities of the customers that can be supplied by the set V of vehicles.
O_2. Minimize the fixed cost of vehicles used in the routing.
O_3. Minimize the total variable cost of the routes.

Quite clearly objectives O_2 and O_3 above cannot be stated in isolation (since the value of the optimum solutions is then zero), but must be stated in conjunction with objective O_1. We will use the terminology $O_x(O_y(O_z))$ to mean: apply objective O_z; if more than one optimum solution to O_z exists, apply to them objective O_y; if more than one optimum solution still exists, apply to them objective O_x.

All objectives normally used in practice and found in the literature can be represented in this way. Some of these are:

(i) Minimize total cost of vehicles used (assuming that all customers must be routed). Set $\delta_i = \bar{M}$ (large positive constant) for all i, $C_k = $ cost of vehicle k. Objective is: $O_2(O_1)$.

(ii) Minimize total route distance (or time) travelled (assuming that all customers must be routed). Set $\delta_i = \bar{M}$ for all i, $c_{ij} = $ distance (or time) from x_i to x_j. Objective is: $O_3(O_1)$.

The Vehicle Routing Problem

(iii) Minimize total number of vehicles used (assuming that all customers must be routed) and using this number of vehicles minimize the distance travelled. Set $\delta_i = \overline{MM}$ (large positive constant) for all i, $C_k = \bar{M}$ (large positive constant) $\ll \overline{MM}$ for all k, set c_{ij} ($\ll \bar{M}$) the distance from x_i to x_j. Objective is: $O_3(O_2(O_1))$.

11.2 Formulations and Exact Methods for the Basic VRP

The basic VRP defined above allows a vehicle to be used first for one route and then for another. In formulations 11.2.1 and 11.2.2 that follow each vehicle is assumed to be used for only one route.

11.2.1 Formulation Related to the TSP

The relationship between the VRP and the TSP has been pointed out in Hays (1967), Christofides (1976), and Eilon, Watson-Gandy, and Christofides (1971). A formulation of the VRP with objective (ii) was first given (as an integer program closely related to the travelling salesman problem—TSP) by Golden (1975). This formulation with a more general objective is given below.

Let $\xi_{ijk} = 1$ if vehicle k visits customer j immediately after visiting customer i, $\xi_{ijk} = 0$ otherwise.

The VRP is:

$$\min z = \alpha \sum_{i=0}^{N} \sum_{j=0}^{N} \left(c_{ij} \sum_{k=1}^{M} \xi_{ijk} \right) + \beta \sum_{k=1}^{M} \left(C_k \sum_{j=1}^{N} \xi_{0jk} \right) + \gamma \sum_{j=1}^{N} \left(\delta_j \sum_{i=0}^{N} \sum_{k=1}^{M} \xi_{ijk} \right) \tag{11.1}$$

subject to

$$\sum_{i=0}^{N} \sum_{k=1}^{M} \xi_{ijk} \leq 1 \qquad j = 1, \ldots, N \tag{11.2}$$

$$\sum_{i=0}^{N} \xi_{ipk} - \sum_{j=0}^{N} \xi_{pjk} = 0 \qquad k = 1, \ldots, M; \quad p = 0, \ldots, N \tag{11.3}$$

$$\sum_{i=1}^{N} \left(q_i \sum_{j=0}^{N} \xi_{ijk} \right) \leq C_k \qquad k = 1, \ldots, M \tag{11.4}$$

$$\sum_{i=0}^{N} \sum_{j=0}^{N} t_{ij} \xi_{ijk} + \sum_{i=1}^{N} \left(u_i \sum_{j=0}^{N} \xi_{ijk} \right) \leq T_k^f - T_k^s \qquad k = 1, \ldots, M \tag{11.5}$$

$$\sum_{j=1}^{N} \xi_{0jk} \leq 1 \qquad k = 1, \ldots, M \tag{11.6}$$

$$y_i - y_j + N \sum_{k=1}^{M} \xi_{ijk} \leq N - 1 \qquad i \neq j = 1, \ldots, N \tag{11.7}$$

$$\xi_{ijk} \in \{0, 1\} \text{ for all } i, j, k$$
$$y_i \text{ arbitrary.} \tag{11.8}$$

In expression (11.1), α, β, and γ are constants for weighting the terms corresponding to objectives O_3, O_2, and O_1 respectively. Expression (11.2) states that a customer can be visited at most once. Expression (11.3) states that if a vehicle visits a customer, it must also depart from it. Expressions (11.4) and (11.5) are the capacity and working time limitations on the vehicle. Expression (11.6) states that a vehicle can be used at most once. Expression (11.7) is the subtour-elimination condition derived for the travelling salesman problem by Miller, Tucker, and Zemlin (1960), and which also forces each route to pass through the depot; (11.8) are the integrality conditions.

The main shortcoming of the above formulation—other than the obvious one of size—is the fact that it is virtually impossible to introduce any of the additional practical constraints (mentioned later) which are not part of the basic VRP defined above. This second shortcoming is avoided by another exact formulation based on the set partitioning model.

11.2.2 Set Partitioning Formulation

The method starts by assuming that the totality of routes which a single vehicle can operate feasibly can be generated. Thus, if $S \subseteq X$ is a subset of the customers which can be supplied feasibly on a single route by a vehicle v_k, then it is assumed that the total variable cost associated with the optimal way of routing the customers in S can be calculated. Since the problem of routing optimally the customers in S is a TSP, this is not a trivial task if $|S|$ happens to be large.

For each vehicle v_k a family S_k of all feasible single routes for this vehicle is generated. A matrix $G = [g_{ij}]$ is then produced with row i corresponding to customer x_i and with M blocks of columns. The kth block of columns corresponds to vehicle v_k and column j_k of this block corresponds to a feasible single route S_{j_k} of this vehicle. Let $g_{ij_k} = 1$ or 0 depending on whether customer x_i is an element of S_{j_k} or not, respectively, and let $C(S_{j_k})$ be the 'cost' associated with the operation of this route by vehicle v_k. This 'cost' can be suitably chosen depending on whether objective (i), (ii), or (iii) above is applicable. (If the most general version of O_1 is required—i.e. not every customer needs to be routed—a slightly more general model than set partitioning can be used.)

The VRP now becomes the problem of choosing at most one column from each block of G so that every row of G has an entry of 1 under exactly one of the chosen columns, and the total cost of columns chosen is minimized. The problem can be easily modified to become a set partitioning problem and the set of columns in the solution contains the optimal routes in the VRP (Christofides and Korman, 1975).

The Vehicle Routing Problem

This model can accommodate any number of additional constraints (since these only serve to reduce the number of columns of G) but suffers from the obvious shortcoming that the number of columns of G can be enormous (Golden, 1976).

11.2.3 A Direct Tree Search Algorithm

In addition to the method described in 11.2.2 above, where all feasible single routes must be generated *a priori*, a depth-first tree search could be employed in which feasible single routes are generated as and when required. Thus, let us define a node of the search tree to correspond to a single route S_j, which can be feasibly operated by at least one vehicle v_k. The state of the search at stage h can then be represented by an ordered list:

$$L = \{S_{j_1}(x_{i_1}), S_{j_2}(x_{i_2}), \ldots, S_{j_h}(x_{i_h})\}$$

where $S_{j_r}(x_{i_r})$ is a feasible single route (the j_rth) which includes a specified customer x_{i_r} and other customers which are not already included in the previous routes $S_{j_1}(x_{i_1}), \ldots, S_{j_{r-1}}(x_{i_{r-1}})$. The state represented by list L is shown diagrammatically in Figure 11.1, where

$$F_h = X - \bigcup_{r=1}^{h} S_{j_r}(x_{i_r})$$

is used for the set of 'free' (i.e. as yet unrouted) customers following stage h.

In order for the state represented by L to be feasible, there must exist a feasible assignment of vehicles to the list $S_{j_1}(x_{i_1}), \ldots, S_{j_h}(x_{i_h})$ of routes. This can be checked quite easily by the solution of a linear rectangular assignment problem (Christofides, 1976) at the node representing L.

Once the bottom of the tree is reached, say at stage m when $F_m = \emptyset$, or when no more customers can be routed, the list L contains a solution to the VRP consisting of m routes. Backtracking must then occur in order to consider the possibilities not yet considered. A backtracking step at some general stage h involves the removal of the last set from L [i.e. the removal of $S_{j_h}(x_{i_h})$] and its replacement by another (as yet unconsidered) feasible single route (say the j'_hth) passing through customer x_{i_h} to form the new state $L = \{S_{j_1}(x_{i_1}), \ldots, S_{j'_h}(x_{i_h})\}$. If no feasible single route through x_{i_h} remains unconsidered, the last-but-one set in L is replaced to form $L = \{S_{j_1}(x_{i_1}), \ldots, S_{j'_{h-1}}(x_{i_{h-1}})\}$, etc. Forward branching is then continued from the new state.

A forward branching from some stage h involves the choice of a customer $x_{i_{h+1}} \in F_h$ and the generation of a list $P(x_{i_{h+1}})$ of all single routes (feasible for some vehicle) passing through this customer. It is quite apparent that the smaller the number of branching possibilities at any stage h the more efficient the tree search, and it is therefore obvious that $x_{i_{h+1}}$ should be

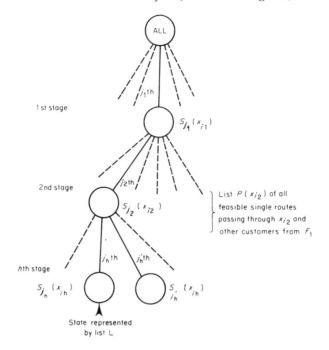

Figure 11.1 Direct tree-search algorithm

chosen from F_h so as to make the list $P(x_{i_{h+1}})$ as short as possible. This would tend to be the case if $x_{i_{h+1}}$ is chosen to be an 'isolated' customer far from the depot. Moreover, not all nodes produced by $P(x_{i_{h+1}})$ need be kept, and, in general, it may be possible to show that a specific node represented by $S_{j_{h+1}}^\alpha(x_{i_{h+1}}) \in P(x_{i_{h+1}})$ can be rejected. This could be done in the following circumstances.

(i) If the set of routes already in L together with route $S_{j_{h+1}}^\alpha(x_{i_{h+1}})$ fails the feasible-vehicle-assignment test mentioned above.

(ii) Let F_{h+1}^α be the set of free customers implied by the routes in list L and route $S_{j_{h+1}}^\alpha(x_{i_{h+1}})$ and let LB(F_{h+1}^α) be a lower bound on the total cost needed to supply F_{h+1}^α. Then, if the total cost of the routes in L plus the cost of $S_{j_{h+1}}^\alpha(x_{i_{h+1}})$ plus LB (F_{h+1}^α) is greater than z^*.

(iii) If in the case of objectives (i), (ii), and (iii) it could be shown that the remaining free vehicles i.e. those not used for the routes in L or for route $S_{j_{h+1}}^\alpha(x_{i_{h+1}})$, are not capable of supplying the customers in F_{h+1}^α (e.g. because of insufficient capacity).

The Vehicle Routing Problem

(iv) Let $\text{UB}(F^{\alpha}_{h+1})$ be an upper bound on the total cost needed to supply the customers in F^{α}_{h+1}. Let $S_{j_{h+1}^{\beta}}(x_{i_{h+1}}) \in P(x_{i_{h+1}})$ be another node of the tree produced by $P(x_{i_{h+1}})$. If

$$C[S_{j_{h+1}^{\beta}}(x_{i_{h+1}})] + \text{UB}(F^{\beta}_{h+1}) \leq C[S_{j_{h+1}^{\alpha}}(x_{i_{h+1}})] + \text{LB}(F^{\alpha}_{h+1}) \qquad (11.9)$$

where $C(S)$ is the cost of the single feasible route S, then node $S_{j_{h+1}^{\beta}}(x_{i_{h+1}})$ dominates node $S_{j_{h+1}^{\alpha}}(x_{i_{h+1}})$ and the latter can be removed from the list $P(x_{i_{h+1}})$.

In the above tests it was assumed that upper and lower bounds on the remaining subproblem defined by the free customers could be calculated at any stage. Upper bounds could be calculated by 'solving' this subproblem by one of the heuristic methods described in the following sections. This provides, in addition to the bound, a possibility of improving the best solution obtained so far, and also other information which could aid the forward branching. Lower bounds could be calculated by relaxing some or all of the constraints and either solving the problem optimally or calculating a lower bound to the solution of the relaxed problem. Any of the available lower bounds for the TSP could be used in this respect (Christofides, 1975; Eilon, Watson-Gandy, and Christofides, 1971; Held and Karp, 1970, 1971).

In the tree search algorithm described above, $C(S)$ has been used to denote the 'cost' of route S. Once more, a suitable choice of $C(S)$ enables the algorithm to solve the basic VRP with any of the objectives mentioned in the Introduction.

11.3 General Principles of Heuristic Procedures and Practical VRP's

Almost all the heuristic methods suggested for solving the vehicle routing problem are constructive in nature in the sense that at any given stage one (or more) incomplete route exists which is extended in the following stages until it becomes the final route. This approach is dictated to a large extent by the very large number of constraints that apply to the vehicle routing problem and is in complete contrast to the majority of procedures—both exact and many of the approximate ones—that are used in the case of the unconstrained travelling salesman problem. However, once a constructive method is used to produce a solution to the vehicle routing problem, this solution can be improved by applying local optimization techniques which maintain the feasibility of the solution.

Before proceeding with a description and computational comparison of the various heuristic procedures available for solving VRP's, we will extend the basic VRP described in Section 11.1.1 by introducing a few of the most usual additional constraints that are found in real-world problems. For any heuristic procedure to be practically useful, it must be able to deal with most (if not all) of the following list of constraints. It perhaps should be pointed out here that this list is by no means exhaustive. In order to distinguish this problem from the basic VRP, we will refer to it as the *extended,* VRP.

11.3.1 The Extended VRP

Again we consider a set X of customers, a depot x_0, and a set V of vehicles.

A customer x_i has the following requirements:

C1. A set of L_i different types of products to be delivered by a vehicle. With each product type (i, l), $l = 1, \ldots, L_i$ is associated a quantity q_{il} and a label π_i (See later for the meaning of this label).
C2. A time u_i required to visit the customer and to unload the total quantity

$$\bar{q}_i = \sum_{l=1}^{L_i} q_{il}.$$

C3. A set of P_i time periods during which delivery can be made; the pth time period is defined by the initial and final times $\tau_{i,p}$ and $\tau'_{i,p}$, $p = 1, \ldots, p_i$. Outside these periods the customer cannot accept delivery.
C4. A subset $V_i \subseteq V$ of vehicles that can be used to deliver to the customer.
C5. A priority δ_i.

A vehicle v_k has the following characteristics:

V1. A set of H_k compartments for carrying products. With each compartment (k, h), $h = 1, \ldots, H_k$ is associated a capacity Q_{kh} and a label Π_{kh}.
V2. A set of G_k working periods; the gth working period being defined by the initial and final times T_{kg} and T'_{kg}, $g = 1, \ldots, G_k$. Outside these periods the vehicle is idle.
V3. A time θ_k required to load the vehicle at the depot.
V4. A rule for determining whether r quantities $q_{i_1 l_1}, \ldots, q_{i_r l_r}$ with labels $\pi_{i_1 l_1}, \ldots, \pi_{i_r l_r}$ can be loaded or not into a compartment of capacity Q_{kh} having label Π_{kh}.

Note This rule may require not only that $\sum_{j=1}^{r} q_{i_j l_j} \leq Q_{kh}$ but also that some function $f(\pi_{i_1 l_1}, \ldots, \pi_{i_r l_r}, \Pi_{kh})$ has the value 0 (say).

For example, in the distribution of frozen ($l = 1$, say) and nonfrozen ($l = 2$, say) goods, with vehicles having refrigerated ($h = 1$, say) and nonrefrigerated ($h = 2$, say) compartments, we can set $\pi_{i1} = 1$, $\pi_{i2} = 2$, $\forall i$ and $\Pi_{k1} = 1$, $\Pi_{k2} = 2$, $\forall k$ with the function $f(\pi_{i_1 l_1}, \ldots, \pi_{i_r l_r}, \Pi_{kh}) = 0$ if and only if $\pi_{i_1 l_1} = \pi_{i_2 l_2} = \cdots = \Pi_{kh}$.

The Vehicle Routing Problem

As another example, in the distribution of gasoline of different types, the types obviously cannot be mixed but otherwise it may be immaterial into which compartment each type is loaded.

(i) If the requirements of customers for the same gasoline type can be mixed in the same compartment, we set each π_{i,l_i} equal to the type of gasoline (i_j, l_j) with the function $f(\pi_{i_1,l_1}, \ldots, \pi_{i,l_i}, \Pi_{kh}) = 0$ if and only if $\pi_{i_1,l_1} = \ldots = \pi_{i,l_i}$.

(ii) If the requirements of each customer for each gasoline type must be kept in separate compartments, we set each π_{i,l_i} equal to a different integer number with $f(\pi_{i_1,l_1}, \ldots, \pi_{i,l_i}, \Pi_{kh}) = 0$ if and only if $\pi_{i_1,l_1} = \ldots = \pi_{i,l_i}$.

11.3.2 Constructive Methods

Constructive methods can be classified according to: (a) the criterion used for expanding the routes and (b) whether the routes are constructed simultaneously or in sequence.

(a) Criteria for Route Expansion

A criterion (or score) is a function defined over the customers and which is used to determine which customer should enter the route(s) being constructed and in which position. That customer is chosen (for expanding the route) which optimizes the criterion function. Some of the more often used criteria are as follows.

Consider an ordered triplet (x_i, x_j, x_l) of customers (possibly including the depot) where x_i and x_j are in a route being constructed and x_l is an as yet unrouted customer.

(i) *Savings* The 'saving' of a customer x_l with respect to x_0 and x_j is given by

$$s_0(l, j) = c_{l0} - c_{lj} + c_{0j}. \tag{11.10}$$

$s_0(l, j)$ is the saving in mileage of supplying x_l and x_j on one route as opposed to supplying them individually directly from the depot.

(ii) *Extra-mileage* The 'extra-mileage' of a customer x_l with respect to two consecutive customers x_i and x_j is given by

$$m(i, l, j) = c_{il} + c_{lj} - c_{ij}. \tag{11.11}$$

(iii) *Radial position* The angle $\theta_0(l, j)$ that the ray (x_0, x_l) forms with the ray (x_0, x_j) can be used as a criterion function defined over the customers x_l. (Note that this criterion requires customer coordinates to be specified.)

(iv) *Composite criteria* These are composite functions of savings, extra-mileage and radial criteria and in addition of the quantity $\bar{q}_l$ to be delivered to a customer x_l, the number of other unrouted customers (n_l say) remaining in the 'neighbourhood' of x_l, etc. The functions are such that the larger the value of s, $1/m$, $\bar{q}$, $1/n$, etc. are, the larger the criterion value of the customer. The above measures are in most cases specialized to ease computations.

(b) *Sequential and Parallel Methods*

In a sequential procedure one route is constructed at a time until all the customers are routed. At no time is the question raised whether a customer x_l should be placed on route R or route S. This consideration is made implicitly by deciding whether to include x_l on route R or not. However, when such a decision has to be made an alternative route for x_l is not available.

Parallel procedures fall into two classes:

(i) Procedures in which the number of routes being formed in parallel is fixed *a priori* to some number, say K. In this case the K routes are grown simultaneously by entering a customer x_l into one of the K routes at each stage. At the end of the procedure K routes exist.

(ii) Procedures in which smaller routes (initially routes consisting of one customer only) are coalesced into larger ones until the routes can grow no further. The number of routes remaining at the end of such an algorithm is not predictable.

In a sequential algorithm a route R is grown (according to an 'extra-mileage' criterion, for example) by choosing a pair of customers (x_i, x_j) in R (not necessarily the same pair for every x_l) and choosing that x_l which minimizes $m(i, l, j)$ for insertion into the route between x_i and x_j. Similarly in the cases where $s_0(l, j)$, $\theta_0(l, j)$, etc. are used as criteria.

In a parallel algorithm of type (i); for each of the K routes ($p = 1, \ldots, K$), that pair of customers (x_{i_p}, x_{j_p}) is chosen in route R_p which optimizes the criterion function for a given customer x_l. For each customer x_l the criterion is then a K-dimensional vector. In the case of 'extra-mileage' for example, this vector would be $m(i_1, l, j_1), \ldots, m(i_K, l, j_K)$, and a decision as to what customer to insert in what route is made according to a scalar defined over this vector. Such a scalar could, for example, be

$$\min_{p=1,\ldots,K} [m(i_p, l, j_p)].$$

In a parallel algorithm of type (ii), if x_i is chosen as the depot, x_j as a customer in one route, and x_l as a customer in another route, the criteria could be used to determine whether the route containing x_j should be joined to the route containing x_l to form a larger route.

One should note that any constructive heuristic requires initialization.

The Vehicle Routing Problem

11.3.3 The Effectiveness of Simple Criteria

It is quite easy to show that even for the basic VRP none of the criteria listed above is uniformly better than the others. Consider for example a parallel approach of type (ii) initialized with four routes each starting from the depot to a customer and back (it is assumed that a vehicle can take at most 2 customers). In the example in Figure 11.2, we see the results of the savings and extra-mileage, indicating that the extra-mileage measure is better for this example, whereas for the example in Figure 11.3 the comparison is in favour of the savings. Moreover, even for the same example all of the above criteria may produce bad solutions. Consider, for example, the problem in Figure 11.4 where $\bigcirc$ is the depot and customer c is on the line $(a, \bigcirc)$. We will use a sequential procedure, and assume that a vehicle can take at most 5 customers. If the procedure is initialized with route $(\bigcirc, b, \bigcirc)$ and a savings criterion is used, the solution in Figure 11.4b is obtained. If the procedure is initialized with route $(\bigcirc, a, \bigcirc)$ and the extra-mileage criterion is used, the same solution is obtained. However, a better solution is shown in Figure 11.4a which can be obtained if the procedure is initialized with $(\bigcirc, b, \bigcirc)$ and an extra-mileage criterion is used, or if it is initialized with $(\bigcirc, a, \bigcirc)$ and a savings criterions is employed. This example illustrates the importance of initializing the routes.

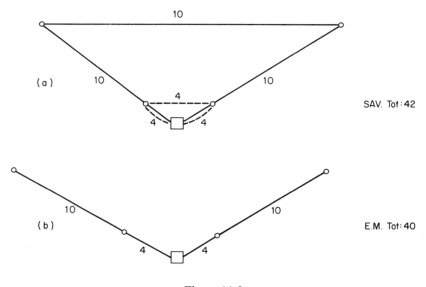

Figure 11.2

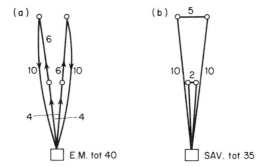

Figure 11.3

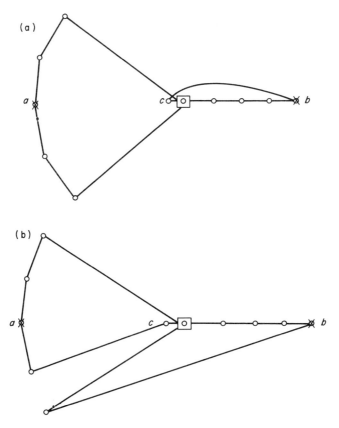

Figure 11.4

The Vehicle Routing Problem

11.4 Heuristic Algorithms

In this section we describe some published and some new heuristics for solving VRP's. The section is not meant to be a survey of all known procedures for VRP's but includes an analysis for many (though not all) of the published algorithms that have been demonstrated by the original authors to hold some promise. In particular, the number of heuristics investigated was limited by the fact that we have undertaken independent coding and extensive testing of each one of the heuristics described in this section. The results of these tests, described and analysed in the next section, form the first reasonable basis of conclusions and comparisons of these heuristics. It is very important to note here that although the descriptions of algorithms in this section are in a skeleton form (in order to avoid repeating the original descriptions), the algorithms as coded and tested in the next section include all the sophistications described in the originals. It should also be noted here that heuristic algorithms in general 'aim' for objective (ii) of Section (11.1.1). This is the most often required objective in practice, which is rather fortunate since the heuristics cannot be modified to distinguish in any effective way between objectives: for example between (ii) and (iii).

11.4.1 The Algorithm of Clarke and Wright (1964)

This algorithm is mainly included for comparison purposes. It is without doubt the most widely known heuristic and has formed the basis of many programs including the IBM-VSPX (IBM, 1970) package.

The algorithm proceeds as follows:

Step 1 Calculate the 'savings' $s_0(i, j)$—as defined by (11.10)—for all pairs of customers x_i, x_j.

Step 2 Arrange the savings in a list in descending order of magnitude.

Step 3 Starting from the top of the list do either of the following:

Parallel version:
(i) If making a given link results in a feasible route according to the constraints of the VRP, then add this link to the solution; if not, reject the link.
(ii) Try the next link in the list and repeat (i) until no more links can be chosen.

Sequential version:
(i) Find the first feasible link in the list which can be used to extend one of the two ends of the currently constructed route.

(ii) If the route cannot be expanded, or no route exists, choose the first feasible link in the list to start a new route.
(iii) Repeat (i) and (ii) until no more links can be chosen.

Step 4 The links picked form a solution to the VRP.

A number of comments about the above procedure must be made.

(a) In the parallel version, the final number of routes produced—and their feasibility in terms of the vehicles available—cannot be guaranteed. Instead, one could impose the restriction that no more than M routes are being formed in parallel at any one time and test for a feasible vehicle assignment to these M (or less) routes at each stage. If the addition of a link would cause the number of routes to exceed M, the link is rejected.

(b) In the sequential procedure the partially completed list of routes must also be tested at each step for a feasible vehicle assignment.

If at the end of either sequential or parallel procedures some customers remain unrouted, new starting times can be computed for the vehicles and the procedure repeated. Alternatively, in the case of a sequential procedure a new starting time can be computed for a vehicle whenever its corresponding route is completed.

11.4.2 The Algorithm of Mole and Jameson (1976)

The algorithm of Mole and Jameson is a sequential procedure in which, for a given value of the two parameters λ and μ, the following two criteria are used to expand a route under construction:

$$e(i, l, j) = c_{il} + c_{lj} - \mu c_{ij}$$
$$\sigma(i, l, j) = \lambda c_{0l} - e(i, l, j).$$

The algorithm then proceeds as follows:

Step 1 For each unrouted customer x_l compute the feasible insertion in the emerging route R as:

$$e(i_l, l, j_l) = \min \left[e(r, l, s) \right]$$

for all adjacent customers $x_r, x_s \in R$, where x_{i_l} and x_{j_l} are customers between which x_l has the best insertion.

Step 2 The best customer x_{l*} to be inserted in the route is computed as the one for which the following expression is maximized:

$$\sigma(i_{l*}, l^*, j_{l*}) = \max \left[\sigma(i_l, l, j_l) \right]$$

for x_l unrouted and feasible.

Step 3 Insert x_{l*} in route R between $x_{i_{l*}}$ and $x_{j_{l*}}$.

Step 4 Optimize route R using r-optimal methods (Lin and Kernighan, 1973).

Step 5 Return to Step 1 to start a new route R (see Note (a)), either until all customers are routed or no more customers can be routed.

It is easy to see in the above definition of $\sigma(i, l, j)$ and $e(i, l, j)$ that by changing the values of λ and μ it is possible to obtain different criteria to choose the best customer for insertion. For example:

$\lambda = 0, \mu = 1$ insertion of the customer with minimum extra-mileage;
$\lambda = 0, \mu = 0$ insertion of the customer with the minimum sum of distances to the two nearest neighbours;
$1 \leq \lambda \leq 2, \mu = \lambda - 1$ generalized Gaskell's criterion (Gaskell, 1967);
$\lambda \to \infty, 0 < \mu < \infty$ insertion of the furthest customer from the depot.

Generally, as λ grows, the shape of the emerging route tends to be circumferential, and as μ grows the presence of long links is discouraged.

Notes

(a) The above description explains how a route R is expanded by the addition of customers. Initially (and each time a new route is to be started) some cutomer x_s must be chosen to initialize the route R as (x_0, x_s, x_0). Customer x_s may be chosen in a variety of ways, e.g. the furthest unrouted customer, the customer with the largest demand $\bar{q}_s$, the customer with the most stringent delivery time restrictions $(\tau_{s,p}, \tau'_{s,p}, p = 1, \ldots, P_s)$, etc.

(b) Mole and Jameson describe the procedure for a fleet of identical vehicles. In this case the assignment of a vehicle to an emerging route is trivial except where vehicles are used for second, third, etc., trips—in which case the departure times of vehicles from the depot (for these additional trips) will be different for each vehicle and a choice exists as to what vehicle to assign to the current route. More generally, at some stage when a route R is being constructed, different size vehicles with different starting and ending times and different allowable working periods will be available and an assignment of vehicles to routes must be made.

11.4.3 The Sweep Algorithm of Gillet and Miller (Gillett and Miller, 1974; Gillett, 1976)

In contrast to the previous two algorithms, which require only cost and time matrices $[c_{ij}]$ and $[t_{ij}]$, the present algorithm requires geographical coordinates for each customer. The procedure can again be classified as a sequential method.

We will assume that the customers are ordered according to increasing values of their polar coordinate angles $\theta_i = \theta_0(i, 1)$ (taking customer 1 as 'reference'). The construction of an emerging route R is as follows:

Step 1 Starting from the unrouted customer x_{i_1} with the smallest polar coordinate angle, include in the route consecutive customers $x_{i_2}, x_{i_3}, \ldots$, as long as the set $\{x_{i_1}, x_{i_2}, x_{i_3}, \ldots\}$ of customers included can be routed into a feasible route R.

The emerging route R is periodically 'optimized' by applying 2-opt or 3-opt local optimization procedures as customers enter R. Let x_l be the last customer that has entered R.

Step 2 Find the unrouted customer x_j nearest to customer x_l of the route. If the insertion of x_j in the route is feasible and 'worthwhile', insert x_j in the route, update x_l and repeat Step 2.

Step 3 Find the 'best' customer x_r to be removed from the route. If the replacement of x_r by x_j is feasible and worthwhile, remove x_r from the route, insert x_j in it, update x_l, and go to Step 2.

Step 4 Find the unrouted customer x_s nearest to x_j. If the replacement of x_r by x_j and x_s is feasible and 'worthwhile', remove x_r from the route, insert x_j and x_s in it, update x_l, and go to Step 2. Otherwise the route is complete.

Step 5 Return to Step 1 repeat until either all customers are routed or no more customers can be routed.

Notes

(a) In order to evaluate in Steps 2, 3, and 4 whether a change in the current route R is 'worthwhile' the following technique has been used:
(i) build the set U formed by the next 5 (say) unrouted customers in the ordered list;
(ii) compute the minimum distances $D1$ and $D2$ respectively of the current route and of the route formed by the customers of U;
(iii) compute the minimum distances $D1'$ and $D2'$ respectively of the changed route R' and of the corresponding projected route $R + U - R'$ through the unrouted customers;
(iv) if $D1 + D2 > D1' + D2'$ the change is 'worthwhile' and the new current route is R'. Otherwise the change is not 'worthwhile'.

(b) In Step 3 the choice of the 'best' customer x_r to be removed from the current route R is done so as to minimize the function $c_{0j} - \theta_j$ (average cost of all customers from the depot) over all the customers x_j of R.

(c) When the VRP is finished the algorithm restarts after a rotation of all the customers (i.e. changing the 'reference' customer). The best solution is obtained over all the possible rotations.

The Vehicle Routing Problem 331

11.4.4 The Tree Search of Section 11.2.3 as a Heuristic

The exact tree search algorithm described in Section 11.2.3 for the basic VRP can be modified to produce a heuristic procedure which is equally applicable to the extended VRP.

The main limitations of the tree search procedure are:

(i) The set $P(x_{i_h})$ contains many routes.

(ii) Generating a feasible route $S_{j_h}(x_{i_h})$ is computationally expensive.

(iii) Good upper and lower bounds ($\text{UB}(F_h)$ and $\text{LB}(F_h)$ respectively) on the remaining subproblem defined by the free customers F_h are difficult to obtain.

(iv) A good initial value z^* is not available.

The heuristic to be described below is an attempt at overcoming these difficulties. In particular:

(a) $P(x_{i_h})$ is not generated completely but a subset $\bar{P}(x_{i_h}) \subset P(x_{i_h})$ is generated which contains only a few (hopefully good) routes generated by applying different criteria (mentioned later) to an initial route (x_0, x_{i_h}, x_0).

(b) $C(S)$ is not the cost of an optimal route S but is the cost of a route obtained by a heuristic.

(c) Condition (ii) of Section 11.2.3 is not used at all and condition (iv) of that section is relaxed by replacing $\text{UB}(F)$ and $\text{LB}(F)$ by an estimate $E(F)$. This relaxation implies that at any stage h only one of the branchings will remain, i.e. that node of the tree corresponding to $S^* \in \bar{P}(x_{i_h})$ for which $C(S^*) + E(F^*)$ is minimum.

The Heuristic

If the criteria of savings and extra-mileage are particularized so that $s_0(l, j) = s_0(l, h)$ and $m(i, l, j) = m(0, l, h)$, then it is quite clear that (in the case of a euclidean cost metric $[c_{ij}]$), with respect to x_0 and x_h the equisavings lines form a family of hyperbolae with foci at x_0 and x_h, and the equi-extra-mileage lines form a family of ellipses also with foci at x_0 and x_h. From Figure 11.5, which shows an equisaving and equi-extra-mileage line it is clear that the saving is a criterion favouring circumferential, whilst extra-mileage is a criterion favouring radial, routes. Thus the two criteria are essentially orthogonal and varying shape routes (passing through x_h) can be obtained by using a criterion which is a varying linear combination of saving and extra-mileage.

The procedure is then essentially as follows:

Step 0 Set $h = 1$.

Step 1 Choose a free customer $x_{i_h} \in F_{h-1}$ to generate the hth set of routes $\bar{P}(x_{i_h})$. If none can be found, exit.

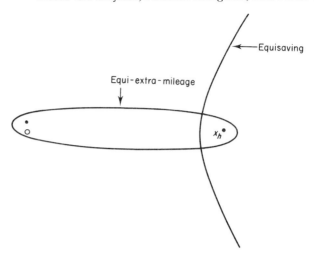

Figure 11.5

Step 2 Generate the set $\bar{P}(x_{i_h})$ by using a varying linear combination of the criteria described earlier.

Step 3 Evaluate a route $S^\alpha \in \bar{P}(x_{i_h})$ by:

$$V(S^\alpha) = C(S^\alpha) + E(F^\alpha)$$

where $E(F^\alpha)$ is computed as the shortest spanning tree of the customers contained in $F^\alpha = F_{h-1} - S^\alpha$ (i.e. all as yet unrouted customers excluding those in S^α).

Step 4 Choose the best route S^* according to evaluation $V(S^\alpha)$ above. If all the customers are routed, exit; if not, update $F_h = F_{h-1} - S^*$, set $h = h+1$ and return to Step 1.

11.4.5 A Two-Phase Algorithm

This algorithm consists of two phases. In the first phase a solution is produced in a sequential way according to the following steps:

Phase 1

Step 1 Set $h = 1$.

Step 2 Choose an unrouted customer x_{i_h} to initialize an emerging route R_h. Compute for all the unrouted customers x_l the following score

$$\delta_l = c_{0l} + \lambda c_{l i_h} \quad (\lambda \geqslant 1).$$

The Vehicle Routing Problem

Step 3 Insert into R_h the feasible customer x_{l*} such that

$$\delta_{l*} = \min [\delta_l], \qquad x_l \text{ unrouted and feasible.}$$

Optimize route R_h using r-optimal methods and repeat Step 3 until no more customers can enter R_h.

Step 4 Set $h = h+1$ and return to Step 2 to start a new route R_h until all customers are routed or no more customers can be routed.

In the second phase, h emerging routes are considered where the rth route is $\bar{P}_r = (x_0, x_{i_r}, x_0)$ and x_{i_r} is the same customer chosen at Step 2 in the first phase to initialize route R_r. Let S be the set of routes $\bar{R}_r = (x_0, x_{i_r}, x_0)$, $r = 1, \ldots, h$.

Phase 2

Step 1 Compute for each unrouted customer x_l and for each route $\bar{R}_r \in S$

$$\varepsilon_{rl} = c_{0l} + \mu c_{li_r} - c_{0i_r} \qquad (\mu \geq 1).$$

Associate x_l with that route $\bar{R}_{r*}$ such that

$$\varepsilon_{r*l} = \min_{R_r \in S} [\varepsilon_{rl}].$$

Step 2. Choose any route $\bar{R}_{\bar{r}} \in S$, set $S = S - \bar{R}_{\bar{r}}$ and compute for each customer x_l associated with this route the following score

$$\bar{\delta}_l = \varepsilon_{r'l} - \varepsilon_{\bar{r}l}$$

where r' is given by

$$\varepsilon_{r'l} = \min_{R_r \in S} [\varepsilon_{rl}].$$

Step 3 Insert into $\bar{R}_{\bar{r}}$ the feasible customer x_{l*} associated with $\bar{R}_{\bar{r}}$ such that

$$\delta_{l*} = \max [\delta_l], \qquad x_l \text{ feasible and associated with } \bar{R}_{\bar{r}}.$$

Optimize the route $\bar{R}_{\bar{r}}$ using r-optimal methods and repeat Step 3 until there are no more feasible customers associated with $\bar{R}_{\bar{r}}$.

Step 4 If $S \neq \emptyset$, go to Step 1 of Phase 2. Otherwise, if all the customers are routed, stop; or, if unrouted customers remain, go to Step 1 of Phase 1.

11.5 Computational Results

The five heuristic algorithms described in the last section i.e.:

A. The Clarke and Wright algorithm;
B. The Mole and Jameson algorithm;
C The sweep algorithm;
D. The tree algorithm of Section 11.4.4;
E. The two-phase algorithm of Section 11.4.5;

have been coded in Fortran to deal with the extended VRP described in Section 11.3, and in addition could deal with several other constraints found in practical problems and not described here.

Computational tests were undertaken on a number of randomly generated problems from the literature and on some structured problems. The results are shown in Table 11.1. The computational times shown in this table are in seconds on the CDC-6600 using the FTN compiler of that machine. These times exclude the computation of intercustomer distances and times.

Problems 1, 2, and 3 are the 50, 75, and 100 customer problems, respectively, whose details are given in Christofides and Eilon, 1969. Problem 4 is a 150 customer problem produced by adding the customers of problems 1 and 3 to the depot and vehicle list of problem 3. Problem 5 is a 199 customer problem produced by adding the customers of problem 4 to the first 49 customers of problem 2, with the depot and vehicle list as for problem 4. Problems 6- to 10 are the same as problems 1 to 5 with additional restrictions on the maximum allowable route time for each problem assumed as follows:

>Problem 6: Max. time 200 units per route
>Problem 7: Max. time 160 units per route
>Problem 8: Max. time 230 units per route
>Problem 9: Max. time 200 units per route
>Problem 10: Max. time 200 units per route.

In all cases the travel times are assumed to be equal to the distances, and an unloading time of 10 units is considered for each call on the route.

Problems 11 and 12 are structured in the sense that customers appear in clusters and may represent practical cases more accurately than problems 1 to 10, which have customers generated by a random uniform distribution. The data for these problems are given in Tables 11.2 and 11.3.

Problems 13 and 14 are the same as problems 11 and 12, respectively, but with a limitation on the maximum allowable time per route of 720 units and unloading time of 50 units for problem 13, and maximum allowable time of 1040 units and unloading time of 90 units for problem 14.

From Table 11.1 it can be seen that, in terms of the quality of the solution, for the random uniformly distributed problems algorithm C performs much better than algorithms A and B and slightly better than algorithms D and E, whereas for the structured problems the reverse is true, i.e. algorithm C is worse than the others and in particular is much worse than algorithms D and E. From what has been said above it is clear that algorithms D and E are the most stable—in the sense that their performance is not very data dependent—and may perform better than the other algorithms on practical problems.

The Vehicle Routing Problem

Table 11.1

Problem	A		B		C		D		E	
1	585	6 / 8	(575)	5 / ~5.0	532 (524)	5 / 12.2	534	5 / 7.1	547	5 / 2.5
2	900	11 / 1.7	(910)	10 / ~11.0	874 (865)	11 / 24.3	871	11 / 15.6	883	11 / 4.2
3	886	8 / 2.4	(882)	8 / ~36.0	851 (851)	8 / 65.1	851	8 / 38.2	851	8 / 9.7
4	1204	12 / 6.6	1259	12 / 71.7	1079	12 / 142.0	1064	12 / 81.1	1093	12 / 11.8
5	1540	17 / 11.0	1545	17 / 119.6	1389	17 / 252.2	1386	17 / 138.4	1418	17 / 16.7
6	619	6 / 1.3	599	6 / 5.1	560	6 / 11.4	560	6 / 5.3	565	6 / 2.6
7	976	12 / 2.6	969	12 / 10.1	933	12 / 23.8	924	12 / 13.6	969	12 / 4.4
8	973	9 / 3.5	999	9 / 28.6	888	9 / 58.5	885	9 / 33.4	915	9 / 7.0
9	1426	16 / 11.2	1289	15 / 63.6	1230	15 / 134.7	1217	15 / 74.0	1245	15 / 10.1
10	1800	21 / 14.0	1770	20 / 110.0	1518	19 / 238.5	1509	19 / 135.6	1508	19 / 15.8
11	1079	7 / 5.6	1100	7 / 68.9	1266	7 / 104.3	1092	7 / 51.1	1066	7 / 11.3
12	831	10 / 2.4	879	10 / 37.2	937	10 / 50.8	816	10 / 39.3	827	10 / 6.4
13	1634	11 / 10.4	1590	11 / 54.3	1776	12 / 85.5	1608	11 / 61.8	1612	11 / 8.7
14	877	11 / 2.9	883	11 / 35.3	949	11 / 53.6	878	11 / 45.2	876	11 / 6.3

1. The numbers to the left of each box give the total mileage. In the case of problems 1, 2, and 3, the mileages shown under B are those obtained by Mole and Jameson (1976); the mileages shown in parentheses under C are those obtained by Gillett (1976) using 3-optimal refining procedures. All other mileages shown use 2-optimal refining methods to improve the routes.
2. The top right-hand number in each box is the number of routes, and the bottom right-hand number is the computing time in seconds on the CDC-6600.

Table 11.2 Details of problem 11. Depot coordinates: (10, 45). Vehicle capacity 200 units.

No.	x	y	q	No.	x	y	q	No.	x	y	q	No.	x	y	q
1	25	1	25	31	84	5	10	61	93	82	17	91	20	40	4
2	25	3	7	32	84	9	3	62	93	84	7	92	18	41	16
3	31	5	13	33	85	1	7	63	93	89	16	93	20	44	7
4	32	5	6	34	87	5	2	64	94	86	14	94	22	44	10
5	31	7	14	35	85	8	4	65	95	80	17	95	16	45	9
6	32	9	5	36	87	7	4	66	99	89	13	96	20	45	11
7	34	9	11	37	86	41	18	67	37	83	17	97	25	45	17
8	46	9	19	38	86	44	14	68	50	80	13	98	30	55	12
9	35	7	5	39	86	46	12	69	35	85	14	99	20	50	11
10	34	6	15	40	85	55	17	70	35	87	16	100	22	51	7
11	35	5	15	41	89	43	20	71	44	86	7	101	18	49	9
12	47	6	17	42	89	46	14	72	46	89	13	102	16	48	11
13	40	5	13	43	89	52	16	73	46	83	9	103	20	55	12
14	39	3	12	44	92	42	10	74	46	87	11	104	18	53	7
15	36	3	18	45	92	52	9	75	46	89	35	105	14	50	8
16	73	6	13	46	94	42	11	76	48	83	5	106	15	51	6
17	73	8	18	47	94	44	7	77	50	85	28	107	16	54	5
18	24	36	12	48	94	48	13	78	50	88	7	108	28	33	12
19	76	6	17	49	96	42	5	79	54	86	3	109	33	38	13
20	76	10	4	50	99	46	4	80	54	90	10	110	30	50	7
21	76	13	7	51	99	50	21	81	10	35	7	111	13	40	7
22	78	3	12	52	83	80	13	82	10	40	12	112	15	36	8
23	78	9	13	53	83	83	11	83	18	30	11	113	18	31	11
24	79	3	8	54	85	81	12	84	17	35	10	114	25	37	13
25	79	5	16	55	85	85	14	85	16	38	8	115	30	46	11
26	79	11	15	56	85	89	10	86	14	40	11	116	25	52	10
27	82	3	6	57	87	80	8	87	15	42	21	117	16	33	7
28	82	7	5	58	87	86	16	88	11	42	4	118	25	35	4
29	90	15	9	59	90	77	19	89	18	40	15	119	5	40	20
30	84	3	11	60	90	88	5	90	21	39	16	120	5	50	13

In terms of computation times, algorithms E and D are the fastest, and algorithm C the slowest. Approximately the same effort was spent coding each of the above five algorithms, and although it may be possible to appreciably improve on the computation times given in Table 11.1 by employing different implementations, it is probably true to say that the conclusions drawn above will still be valid. In particular, it could not be over-emphasized that the codes from which the above timings have been derived are general codes that can deal with much more complex problems than the ones used in the tests and hence these codes may be appreciably

The Vehicle Routing Problem

Table 11.3. Details of problem 12. Depot coordinates: (40, 50). Vehicle capacity 200 units.

No.	x	y	q	No.	x	y	q	No.	x	y	q
1	45	68	10	33	8	40	40	67	47	40	10
2	45	70	30	34	8	45	20	68	45	30	10
3	42	66	10	35	5	35	10	69	45	35	10
4	42	68	10	36	5	45	10	70	95	30	30
5	42	65	10	37	2	40	20	71	95	35	20
6	40	69	20	38	0	40	30	72	53	30	10
7	40	66	20	39	0	45	20	73	92	30	10
8	38	68	20	40	35	30	10	74	53	35	50
9	38	70	10	41	35	32	10	75	45	65	20
10	35	66	10	42	33	32	20	76	90	35	10
11	35	69	10	43	33	35	10	77	88	30	10
12	25	85	20	44	32	30	10	78	88	35	20
13	22	75	30	45	30	30	10	79	87	30	10
14	22	85	10	46	30	32	30	80	85	25	10
15	20	80	40	47	30	35	10	81	85	35	30
16	20	85	40	48	28	30	10	82	75	55	20
17	18	75	20	49	28	35	10	83	72	55	10
18	15	75	20	50	26	32	10	84	70	58	20
19	15	80	10	51	25	30	10	85	68	60	30
20	30	50	10	52	25	35	10	86	66	55	10
21	30	52	20	53	44	5	20	87	65	55	20
22	28	52	20	54	42	10	40	88	65	60	30
23	28	55	10	55	42	15	10	89	63	58	10
24	25	50	10	56	40	5	30	90	60	55	10
25	25	52	40	57	40	15	40	91	60	60	10
26	25	55	10	58	38	5	30	92	67	85	20
27	23	52	10	59	38	15	10	93	65	85	40
28	23	55	20	60	35	5	20	94	65	82	10
29	20	50	10	61	50	30	10	95	62	80	30
30	20	55	10	62	50	35	20	96	60	80	10
31	10	35	20	63	50	40	50	97	60	85	30
32	10	40	30	64	48	30	10	98	58	75	20
				65	48	40	10	99	55	80	10
				66	47	35	10	100	55	85	20

slower (because of large fixed costs) than specialized versions dealing only with simple problems.

It should also perhaps be noted that algorithm C requires coordinates for each customer location and hence cannot be applied to problems when only distance/time matrices are available. Moreover, the performance is likely to suffer appreciably if the metric is noneuclidean (e.g. if geographical features, such as rivers with bridges, etc., are introduced).

References

Balinski, M. L., and R. E. Quandt (1964). On an Integer Program for a Delivery Problem. *Opns. Res.*, **12**, 300.
Christofides, N., and S. Eilon (1969). An Algorithm for the Vehicle Dispatching Problem. *Opl. Res. Q.*, **20**, 309.
Christofides, N., and S. Korman (1975). A Computational Survey of Methods for the Set Covering Problem. *Management Science (Theory)*, **21**, 591.
Christofides, N. (1975). *Graph Theory, An algorithmic approach*, Academic Press, London.
Christofides N. (1976). The Vehicle Routing Problem. *Rev. Frans. Res. Oper.*, **10**, 55.
Clarke, G., and J. W. Wright (1964). Scheduling of Vehicles from a Central Depot to a Number of Delivery Points. *Opns. Res.*, **12**, 568.
Dantzig, G. B. and K. H. Ramser (1959). The Truck Dispatching Problem. *Opns. Res.*, **12**, 80.
Eilon, S., C. Watson-Gandy, and N. Christofides (1971). *Distribution Management, Mathematical modelling and practical analysis*, Griffin, London.
Gaskell, T. J. (1967). Bases for Vehicle Fleet Scheduling. *Opl. Res. Q.*, **18**, 281.
Gillett, B. E., and L. R. Miller (1974). A Heuristic Algorithm for the Vehicle Dispatch Problem. *Opns. Res.*, **22**, 340.
Gillet, B. E. (1976). Vehicle Dispatching: Sweep Algorithm and Extensions. *ORSA-TIMS National Meeting, Miami, 1976*.
Golden, B. L. (1975). Vehicle routing problems: Formulations and Heuristic Solution Techniques. *ORC Tech. Rep. 113, August, 1975, MIT*.
Golden, B. L. (1976). Large-Scale Vehicle Routing and Related Combinatorial Problems, Ph.D. Dissertation, Operations Research Centre, M.I.T., Cambridge, Mass.
Hays, R. (1967). The Delivery Problem. *Carnegie Inst. of Tech., Man. Science Res., Report No. 106*.
Held, M., and R. Karp (1970). The TSP and Minimum Spanning Trees, I. *Opns. Res.*, **18**, 1138.
Held, M., and R. Karp (1971). The TSP and Minimum Spanning Trees, II. *Math. Prog.*, **1**, 6.
IBM Corp. (1970). System 360/Vehicle Scheduling Program Application Description-VSPX. *Report GH19-2000/0, White Plains, N.Y.*.
Lin, S., and B. W. Kernighan (1973). An Effective Heuristic Algorithm for the TSP. *Opns. Res.*, **21**, 498.
Miller, C., A. W. Tucker, and R. A. Zemlin (1970). Integer Programming Formulation of the Travelling Salesman Problem. *J. ACM*, **7**, 326.
Mole, R. H. and S. R. Jameson (1976). A Sequential Route-Building Algorithm Employing a Generalised Savings Criterion. *Opl. Res. Q.*, **27**, p. 503.
Pierce, J. F. (1970). A two Stage Approach to the Solution of Vehicle Dispatching Problems. Presented at *17th TIMS Int. Conf., London, 1970*.
Tillman, F. A. and H. Cochran (1969). A Heuristic Approach for Solving the Delivery Problem. *J. Indust. Eng*, **19**, 354.

CHAPTER 12

Loading Problems

NICOS CHRISTOFIDES
Imperial College, London

ARISTIDE MINGOZZI
Sogesta, Urbino, Italy

PAOLO TOTH
University of Bologna

12.1 Introduction

In this chapter we introduce the class of 'loading problems', so-called because they arise in situations where 'items' must be loaded into 'boxes'. The well-known knapsack problem—considered as a problem in its own right in Chapter 9 of this book—is a member of this class. We will, in particular, be considering loading problems in which liquids of different types are loaded into tanks of different capacities. We will be considering both static problems (i.e. ones in which only loading operations of the liquids into the tanks exist—in which case the order in which the liquids are loaded is immaterial, hence the name static) and dynamic problems (i.e. ones in which both loading of the liquids into tanks and unloading operations of liquids from tanks exist, in which case the order in which operations take place is important). In all cases it is required that different types of liquids must not be mixed in the tanks and some objective must be optimized.

Consider, for example, the loading-only problem with only one liquid (of quantity q) to be loaded into M tanks of capacity Q_j and value v_j, $j = 1, \ldots, M$. It may be required to load the liquid in such a manner so as to minimize the value of boxes used, i.e. we want to:

$$\min z = \sum_{j=1}^{M} v_j x_j$$

$$\sum_{j=1}^{M} Q_j x_j > q$$

where $x_j = 1$ if tank j is used and 0 otherwise.

The above problem can be clearly recognized as the knapsack problem. Consider now the dynamic case (also involving only one type of liquid in which an amount $q^{(1)}$ must be loaded into the tanks starting at time $t^{(1)}$ and requiring time $\Delta t^{(1)}$, followed by (say) another loading of amount $q^{(2)}$ starting at time $t^{(2)}$ and requiring time $\Delta t^{(2)}$, followed by (say) a third operation of unloading an amount $q^{(3)}$ starting at time $t^{(3)}$ and requiring time $\Delta t^{(3)}$, etc. It may be required to perform these operations using as few tanks as possible (or to minimize the cost of tanks used). These types of problem appear quite often in practical situations. For example, at an oil terminal or port, crude oil arrives on ships and is unloaded into storage tanks. In a separate operation batches of crude oil are unloaded from the storage tanks and transmitted via a pipeline network to refineries for conversion into finished products. In the long term a problem exists as to the numbers and sizes of storage tanks that make up the 'tank farm'. However, in the short term the tanks are given and what is required is the 'best' method of conducting the loading and unloading operations.

In most such practical problems, however, there are many types of liquids to be loaded/unloaded, each liquid being of a certain type, with the requirement that liquids of different types must not be mixed. The following example illustrates such a problem.

Let there be four tanks available, with capacities 20, 40, 60, and 70 units, two types of liquids, A and B, and three operations to be performed, i.e.

(1) Load 90 units of A
(2) Unload 50 units of A
(3) Load 150 units of B.

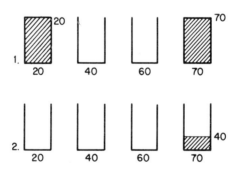

Figure 12.1 Example of dynamic loading and unloading problem. ▨ Liquid A; ▩ Liquid B

Loading Problems

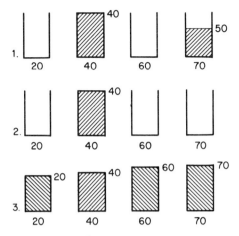

Figure 12.2 Example of dynamic loading and unloading problem. ▨ Liquid A; ▧ Liquid B

We will assume that none of these operations overlap in real time and that the tanks are initially empty.

Figure 12.1(1) shows a possible way to load 90 (which would result from an attempt to utilize tanks as much as possible). Figure 12.1(2) shows the state of the tanks after 50 units of A are unloaded (20 from tank 1 and 30 from tank 4). We have now arrived at a state from which it is impossible to continue. In contrast, Figure 12.2 shows a feasible way of performing the operations.

In Part 1 of this chapter we deal with the static problem of loading many types of liquid into tanks of different capacities and with a separate problem of unloading liquids from partially filled tanks. In both cases the objective is to minimize the value of tanks containing liquids at the end of the operations.

In Part 2 of the chapter we deal with the dynamic problem. The objective considered is a general integral criterion and includes a criterion based only on the terminal state of the tanks as a special case. For the dynamic problem the following assumptions are made:

(i) In the case of an oil tank farm, transfer of oil from one half-filled tank to another is considered as generally undesirable and to be avoided, firstly because the transfer operation itself is costly in terms of the energy required, and secondly because after every loading or unloading operation from a tank, some time (of the order of 2 hours) must elapse before another operation is performed on the same tank. Hence, any transfer operation leads to an under-utilization of available resources. Thus, in Part 2 of this chapter—dealing with this type of problem—we have excluded the possibility of transfers between tanks.

(ii) Depending on the physical layout of the pipeline system used for the loading and unloading of oil, it may or may not be possible for operations to overlap in time. In Part 2 we consider the case where overlaps of different types of crude are possible (in both the loading and unloading modes), but overlaps of operations involving the same type of crude are not. This situation corresponds to a physical system with 'by-passes'.

(iii) If l is a loading operation, we will assume that the tanks to be used in this operation are decided on at the beginning of the operation and are from that time assumed to be unavailable until they are fully unloaded in some future operation.

(iv) Similarly, if l is an unloading operation the tanks to be unloaded from are decided on at the beginning of the operation but are unavailable until the end of the operation. This assumption corresponds to the practical requirements of the real oil tank farm system.

This assumption can be removed at the expense of increasing the number of operations in the problem by dividing any unloading operation during which one or more other loading operations start (as shown in Figure 12.3a) into a number of unloading operations in series so that loading operations start only when unloading operations finish (as shown in Figure 12.3b).

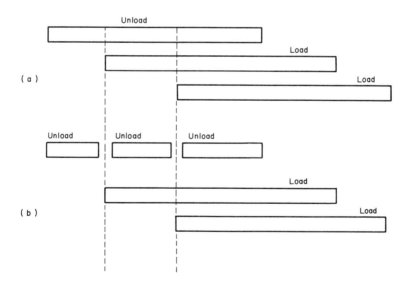

Figure 12.3

PART 1

STATIC LOADING AND STATIC UNLOADING PROBLEMS

12.1.1 General: Static Loading

We consider the problem of loading quantities $q_1, q_2, \ldots, q_N$ of N different types of liquid (which cannot be mixed) into M tanks of capacities Q_1, $Q_2, \ldots, Q_M$.

The objectives considered are:

(i) Maximize the number of storage tanks left totally empty after the loading operations; or
(ii) Maximize the net capacity of storage tanks left totally empty.

The tanks that are filled or only partly filled are not considered in the evaluation of the objective functions.

Let E_0 be the set of tanks that are initially empty, V_i be the set of tanks already containing liquid i, and α_j be the amount of liquid contained in tank j ($\alpha_j = 0$ if $j \in E_0$).

Let v_j be the 'value' of tank j. Depending on whether we take $v_j = 1$ or $v_j = Q_j$, maximizing the sum of the 'values' of the tanks remaining empty corresponds to objectives (i) and (ii) respectively.

The loading problem stated above is related to the multiple-knapsack or 'loading' problem (Eilon and Christofides, 1971; Johnson, 1974; Lev, 1972). The following simple observation reduces and simplifies the problem to be considered:

Lemma 12.1 An optimum loading of liquid into tanks exists in which liquid of type i is first loaded into the tanks of V_i and only when all tanks in V_i are full is any liquid loaded into tanks of E_0.

Proof In a solution let an amount x of liquid i be loaded into tank $a \in E_0$ and let some tank $b \in V_i$ have an amount of empty space y. If $x > y$ an equally good solution could be obtained by filling tank b first and only loading $(x - y)$ into a. If $x \leq y$ a better solution (by v_a) can be obtained by loading the amount x into tank b rather than a. Hence a solution which is at least as good can be obtained by first loading into a tank $b \in V_i$ until either b is full or all liquid of type i has been loaded.

In view of Lemma 12.1 the loading-only problem reduces to one of loading quantities $\bar{q}_i = q_i - \sum_{j \in V_i}(Q_j - \alpha_j)$, $i = 1, \ldots, N$ into the set of empty tanks E_0. We will assume $\bar{q}_i > 0$, $i = 1, \ldots, N$.

The problem can easily be formulated as an integer program as follows.

Let $x_{ij} = 1$ if crude i is loaded into tank j and $x_{ij} = 0$ otherwise. The problem is then:

$$(\text{P}) \begin{cases} \min z = \sum_i \sum_j v_j x_{ij} & (12.1) \\ \text{s.t.} \sum_i Q_j x_{ij} \geq \bar{q}_i & i = 1, \ldots, N \quad (12.2) \\ \sum_i x_{ij} \leq 1 & j = 1, \ldots, \bar{M} \quad (12.3) \\ x_{ij} \in \{0, 1\} & \end{cases}$$

where the tanks in E_0 are numbered $1, \ldots, \bar{M}$.

Problem P above does not have a sufficiently pronounced structure that can be exploited and hence P can only be solved as a general integer program for problems of small size.

12.1.2 Problem Formulation

Consider two sets of tanks X and Y. We say that X dominates Y if it is possible to map X one-to-one (but not necessarily into) Y by a function π so that:

$$Q_j \leq Q_{\pi(j)} \quad \forall j \in X$$

and (12.4)

$$v_j \leq v_{\pi(j)} \quad \forall j \in X.$$

Obviously, for such a mapping to be possible we must have $|X| \leq |Y|$. With arbitrary v_j and Q_j an assignment problem must be solved to decide whether X dominates Y or not. However, with the special case of $v_j = a + bQ_j$ ($a, b \geq 0$), corresponding to loading objectives (i) and (ii) mentioned earlier, the dominance test becomes very simple. One only has to map $j \in X$ onto that $j' \in Y$ for which the difference $Q_{j'} - Q_j \geq 0$ is smallest. If X can be mapped one-to-one into Y in this way then X dominates Y; otherwise it does not.

A subset $F_i \subseteq E_0$ of tanks into which $\bar{q}_i$ can be loaded, i.e. for which

$$\sum_{j \in F_i} Q_j \geq \bar{q}_i \quad (12.5)$$

is called minimal if there exists no $F_i' \subset F_i$ which also satisfies (12.5).

Let $\mathcal{F}_i$ be the family of such minimal sets.

Let S_i be a minimal set of tanks into which all liquids $1, 2, \ldots, i$ can be loaded and which is not dominated (in the above sense) by any other

Loading Problems

minimal set of tanks that can hold these liquids. Let $\mathscr{S}_i$ be the family of all such minimal undominated sets, and take $\mathscr{S}_0 = \emptyset$.

$\mathscr{S}_i$ can be calculated recursively as follows:

Step 1 Initialize $\mathscr{S}_i = \emptyset$

Step 2 For any $S_{i-1} \in \mathscr{S}_{i-1}$ and $F_i \in \mathscr{F}_i$ satisfying

$$S_{i-1} \cap F_i = \emptyset \tag{12.6}$$

form

$$X = S_{i-1} \cup F_i$$

Step 3
(i) If X is dominated by an element of $\mathscr{S}_i$ reject X.
(ii) If X dominates elements of $\mathscr{S}_i$, remove the dominated sets. Set $\mathscr{S}_i = \mathscr{S}_i \cup \{X\}$.
(iii) If X neither dominates nor is dominated by elements of $\mathscr{S}_i$ set $\mathscr{S}_i = \mathscr{S}_i \cup \{X\}$.

Step 4 Repeat Steps 2 and 3 for different $S_{i-1} \in \mathscr{S}_{i-1}$ and $F_i \in \mathscr{F}_i$ until all have been considered.

The final family $\mathscr{S}_i$ is the required family which can now be used to derive $\mathscr{S}_{i+1}$, etc., until $\mathscr{S}_N$ is computed.

The minimum value of tanks necessary for loading all N liquids is then:

$$z_{\min} = \min_{S \in \mathscr{S}_N} \left[\sum_{j \in S} v_j \right] \tag{12.7}$$

and hence the maximum value of tanks remaining empty at the end of loading is

$$\left[z_{\max} = \sum_{j=1}^{N} v_j - z_{\min}. \right]$$

The optimum loading itself is given by the set S^* which produces the minimum in expression (12.7). This set is composed of the union of N disjoint subsets $F_1^*, F_2^*, \ldots, F_N^*$, with liquid i loaded into the set of tanks F_i^* for $i = 1, \ldots, N$. The above description corresponds to a breadth-first tree search approach using only dominance conditions.

In the actual computer implementation a depth-first tree search is used, where a node of the tree at level i corresponds to an element of $\mathscr{S}_i$. At a general stage of a depth-first tree search the families $\mathscr{S}_i$, $i = 1, \ldots, N$, are only partially constructed. This implies that complete dominance tests cannot be applied before branching and it is possible that branching has been performed from a node at level i which may later be found to have been dominated by another node at this level.

Lower Bounds

Consider a node of the tree corresponding to $\tilde{S}_i \in \mathscr{S}_i$; liquids $i+1, \ldots, N$ must now be loaded into the set $E_0 - \tilde{S}_i$ of available tanks, and the subproblem remaining is of exactly the same form as the initial loading problem. Thus, in order to simplify the exposition we will describe the calculation of the lower bounds for the initial problem.

The total amount of liquid to be loaded into the tanks is $q = \sum_{i=1}^{N} q_i$. Thus, if the requirement that liquids cannot be mixed is removed we can formulate a knapsack problem:

(KP)
$$\begin{cases} \min \quad z = \sum_j v_j x_j \\ \text{s.t.} \sum_j Q_j x_j \geq q \\ x_i \in \{0, 1\}, \quad j = 1, \ldots, \bar{M}. \end{cases}$$

The solution of problem KP is therefore a lower bound (b_1, say) to the loading problem.

An alternative lower bound (b_2, say) can be derived by solving separately N knapsack problems of the above form, where for the ith problem q is replaced by q_i, $i = 1, \ldots, N$, and adding the values of the N separate solutions. A related bound can be obtained by solving N separate problems as above, but with $v_j = 1$ for all $j = 1, \ldots, \bar{M}$. If z_i^* is the solution to the ith problem then $\bar{z} = \sum_i z_i^*$ tanks must be used to load all the liquids and the sum of the $\bar{z}$ smallest values v_j is a lower bound (b_3).

If the well-known continuous solution is used instead of the exact one for the above knapsack problems, corresponding bounds b_1', b_2' and b_3' are derived. We should note here that b_1' is also the optimum solution to the continuous problem corresponding to problem P, and that $b_3 = b_3'$.

In the proposed tree search procedure, the computation of the families $\mathscr{F}_i$ of minimal sets for each liquid i is required at the beginning of the search. Numbering sequentially all the minimal sets of $\mathscr{F}_1$ from 1 to p_1, those of $\mathscr{F}_2$ from $p_1 + 1$ to p_2 etc., the loading problem can be reformulated as:

(P')
$$\begin{cases} \min z = \sum_{r=1}^{p_N} \lambda_r x_r \\ \text{s.t.} \sum_{r=1}^{p_N} t_r^j x_r \leq 1 \quad j = 1, \ldots, \bar{M} \quad (12.8) \\ \sum_{r=p_k+1}^{p_{k+1}} x_r = 1 \quad k = 0, \ldots, N-1 \quad (12.9) \\ x_r \in \{0, 1\} \quad r = 1, \ldots, p_N \end{cases}$$

Loading Problems

where $t_r^j = 1$ if tank j is an element of the rth minimal set and 0 otherwise, $\lambda_r = \sum_i t_r^i v_i$ and $p_0 = 0$.

The linear programming solution of the continuous problem corresponding to P' is a valid lower bound. We should note here that although the number of variables in P' can be very large, the problem has a very pronounced structure. In particular, if (as in real problems) the tanks are not all different but are divided into C classes with all tanks within a class being identical, problem P' contains C rows of type (12.8) (coupling constraints) and N totally decoupled rows of type (12.9), all with coefficients 1. Obviously LP decomposition is a very suitable means of solving P'. Because of this structure the solution of the linear program corresponding to P' is quite often integer, in contrast to the continuous version of problem P.

None of the lower bounds b_1, b_2, b_3, and b_4 described above dominate one another. However, b_4 dominates b_1' and b_2'.

In the depth-first search the nodes emanating from a given node are ordered in ascending order of the total value of the tanks used. The nodes are then considered for forward branching in the above order.

12.1.3 Example

Consider seven types of liquid in quantities of 40, 60, 70, 80, 80, 90, and 100 respectively. There are 14 tanks available of the following sizes: 4 tanks of size 20, 4 tanks of 50, and 6 tanks of 60. We want to solve the loading problem with objective (i), i.e. all $v_j = 1$.

We will use the nomenclature (abc) to mean the set containing a tanks of capacity 20, b tanks of capacity 50, and c tanks of capacity 60. The families $\mathscr{F}_i$, $i = 1, \ldots, 7$, can be immediately enumerated as:

$$\mathscr{F}_1 = \{(200), (010), (001)\}$$

$$\mathscr{F}_2 = \{(300), (110), (020), (001)\}$$

$$\mathscr{F}_3 = \{(400), (110), (020), (101), (011), (002)\}$$

$$\mathscr{F}_4 = \mathscr{F}_5 = \{(400), (210), (020), (011), (101), (002)\}$$

$$\mathscr{F}_6 = \{(210), (011), (020), (201), (002)\}$$

$$\mathscr{F}_7 = \{(020), (310), (201), (011), (002)\}.$$

From these families problem P' can be formulated with 35 variables, 3 constraints of type (12.8) and 7 constraints of type (12.9). The solution of P' as a linear program is integer, and is therefore the solution to the whole problem. Therefore, in order to illustrate the tree search algorithm we will not use bound b_4 and only use bound b_1. At the root of the tree (shown in Figure 12.4), $b_1 = 10$.

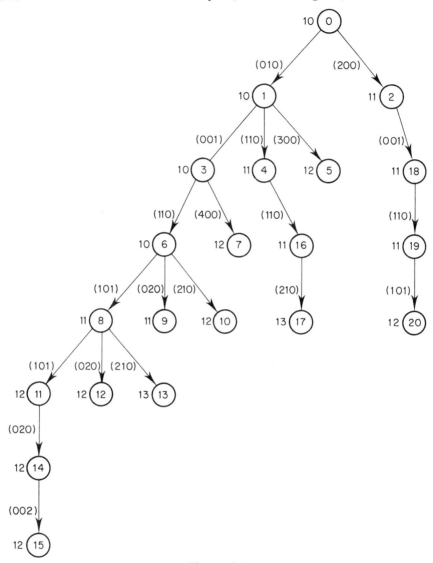

Figure 12.4

In Figure 12.4 the minimal set corresponding to the branching is shown next to the branching. The lower bounds are shown next to the nodes, and the order in which the nodes are generated is shown by the numbering of the nodes. Dominated nodes are not shown.

At level 1 three nodes are generated corresponding to the elements of $\mathscr{F}_1$. One of these nodes (corresponding to set (001)) is dominated and is not shown. Using the branching rule mentioned earlier, branching continues

Loading Problems

from node 1, etc. Of the 20 nodes shown in Figure 12.4, 7 have been rejected because of the bound, and 30 others not shown in Figure 12.4 have been rejected by the dominance tests. In the case of node 9, although the node itself is not dominated by another node at level 4, all nodes emanating from it are dominated at level 5 and hence 9 is rejected by the dominance tests. In the example, the dominance tests are applied prior to the calculation of the bound.

12.1.4 Computational Results

A computer code was developed to test the efficiency of the proposed algorithm and evaluate the effectiveness of the various bounds described in

Table 12.1 Computational performance of the tree search with various bounds

N	M	b_1	b'_1	b_2	b'_2	b_3	b_4	$P'+b_2$	P'
5	15	0.021 0.044	0.045 0.069	0.043 0.065	0.059 0.080	0.059 0.080	0.238 0.427	0.075 0.114	117(3) 0.042
10	20	0.119 0.161	0.123 0.191	0.052 0.135	0.151 0.238	0.146 0.234	0.232 0.939	0.059 0.100	114(7) 0.049
15	30	5.876 47.847	7.903 65.026	0.202 0.615	10.166 78.185	5.198 30.064	1.812 5.089	0.185 0.382	188(4) 0.091
20	30	0.901 1.667*	0.691 1.306*	0.025 0.049	0.761 1.421*	0.614 2.263	0.116 0.129	0.116 0.129	184(10) 0.116
20	40	—	—	0.112 0.376	—	—	1.968 8.151	0.275 0.453	325(5) 0.180
25	40	—	—	0.038 0.093	—	—	0.751 1.688	0.380 0.613	315(5) 0.361
30	40	—	—	0.025 0.029	—	—	0.380 0.942	0.294 0.379	294(7) 0.281
30	50	—	—	0.071 0.204	—	—	0.898 2.649	0.409 0.513	380(5) 0.367
30	60	—	—	3.495 20.452	—	—	4.471 4.471†	—	—
35	50	—	—	0.040 0.058	—	—	1.072 1.919	0.410 0.576	391(1) 0.386
35	70	—	—	9.714 40.216	—	—	core limit	core limit	—

Columns b_1, b'_1, b_2, b'_2, b_3, and b_4 contain the results of the tree search algorithm with the corresponding bound of Section 12.1.2. Column $P'+b_2$ contains the results of an algorithm in which the LP corresponding to problem P' is solved first, followed by the use of bound b_2 whenever the LP solution is not integer. Column P' gives information on the size of the LP tableau, the computation time, and the probability that the LP solution is integer.
* 3 problems solved.
† 1 problem solved.

Section 12.1.2. The experiments with the code were performed on randomly generated problems varying in size from 5 to 35 types of liquids (N) and from 15 to 70 tanks (M) of 5 different classes. For each value of N and M, 10 problems of that size were solved. All problems generated had a feasible loading of the liquids, and the total amount of liquid to be loaded was on average 70% of the total tank capacity.

The computational results are shown in Table 12.1. In this table, two numbers are shown for a given N and M for each column except the last. The first entry is the average solution time (CDC-7600 sec) for the 10 test problems and the second entry is the maximum solution time for these problems. From Table 12.1 it can be seen that bounds b_2 and b_4 are quite clearly better than bounds b_1, b'_1, b'_2, and b_3, and are also the ones which lead to the most stable tree search. The knapsack algorithm described in (Martello and Toth., 1977) was used for the computation of bounds b_1, b_2, and b_3, and a standard LP package (without decomposition) was used to compute bound b_4. Thus, the computational times for the tree search with bound b_4 can possibly be reduced quite appreciably by using a more specialized LP package. An indication of the reduction that can be achieved can be obtained from the last column of Table 12.1. For a given M and N, the first entry in this column gives the total number of variables in the LP problem corresponding to problem P', the second entry (in parentheses) gives the number of times (out of the 10 problems of size $M \times N$) that the LP solution was naturally integer, and the third entry gives the average solution time for one LP. It may be worth noting here that the tree search with bound b_4 almost never contained more than 10 nodes.

12.1.5 Sensitivity Results

Often it is required to know what is the maximum allowable increase of the quantity q_i of liquid of type i which would leave the solution value unchanged, or to know which tanks from the set E_0 could be removed and not affect this value.

Exact answers to these questions cannot be obtained from the results of the algorithm described in Section 12.1.2. In order to answer these questions exactly, it is necessary to (a) replace the definition of dominance as given by (12.4) with a much weaker condition, namely, X dominates Y if $X \subset Y$; and (b) reject a node only if its bound is strictly greater than the value of the currently best answer.

With these changes all solutions of the same value will be generated as elements of the final family $\mathscr{F}_N$ and sensitivity questions can be answered exactly quite simply. These changes, especially (a), are totally impractical from the computational viewpoint and approximate sensitivity results must, therefore, be derived from the (single) solution obtained by the use of the stronger conditions.

Loading Problems

Increase in q_i

The solution S^* as mentioned earlier is composed of N disjoint subsets F_i^*, $i = 1, \ldots, N$, with liquid i loaded into the set of tanks F_i^*. Hence q_i can be increased by

$$\Delta_i' = \sum_{j \in F_i^*} Q_j - \bar{q}_i \tag{12.10}$$

before all tanks in F_i^* are filled.

Consider now another liquid $k \neq i$ and a tank $r \in F_k^*$. The amount of liquid in tank r could be as low as

$$W_r = \bar{q}_k - \sum_{\substack{j \neq r \\ j \in F_k^*}} Q_j \tag{12.11}$$

by filling tank r last with liquid of type k.

If for any pair of tanks $t \in F_i^*$ and $r \in F_k^*$ we have

$$W_r \leq Q_t \leq Q_r$$

then tanks r and t can be exchanged and an additional amount $(Q_r - Q_t)$ of liquid i could be accommodated.

Now form a graph G as follows:

1. For every tank $t \in F_i^*$, introduce a vertex x_t, $t = 1, \ldots, |F_i^*|$.

2. For every set of tanks F_k^*, introduce a vertex x_k, $k = 1, \ldots, N-1$.

3. For every tank $h \in E_0 - S^*$, introduce a vertex x_h.

Let X_t, X_k, and X_h be the set of vertices introduced by Steps 1, 2, and 3 above, respectively. Introduce a source vertex x_0' and a sink vertex x_0'':

1. Add an arc from any $x_t \in X_t$ to any $x_k \in X_k$ whenever

$$c_{tk} = \max_{r \in F_k^*}[Q_r - Q_t | W_r \leq Q_t]$$

exists and is positive and let r_k be the corresponding tank. Let c_{tk} be the cost of arc (x_t, x_k).

2. Add an arc from any $x_t \in X_t$ to any $x_h \in X_h$ whenever $Q_h > Q_t$ and $v_h = v_t$ and let $c_{th} = Q_h - Q_t$ be the cost of link (x_t, x_h).

3. Add an arc from any $x_h \in X_h$ to any $x_k \in X_k$ whenever

$$\hat{Q}_k = \max_{r \in F_k}[Q_r | Q_r > Q_h, W_r \leq Q_h]$$

exists. Let $c_{hk} = \hat{Q}_k - Q_h$ be the cost of link (x_h, x_k).

4. Add arcs of cost 0 from x'_0 to every $x_t \in X_t$ and from every $x \in X_t \cup X_h \cup X_k$ to x''_0.

Set the capacity of every arc and vertex of G equal to 1, and solve the maximum flow maximum cost problem for graph G (Christofides, 1975). The maximum amount by which liquid of type i could be further increased is at least as large as the cost Δ''_i of this flow and the total maximum increase without changing the value of the solution is therefore at least as large as:

$$\Delta_i = \Delta'_i + \Delta''_i.$$

This follows from the fact that every feasible flow pattern in G corresponds to a feasible exchange (one-for-one) of tanks containing liquid i with tanks containing other liquids. The cost of this flow corresponds to the feasible increment in q_i produced by the tank exchange. Thus, a flow of one unit from x'_0 to x_t to x_h to x_k to x''_0 involves the exchange of tank t with tank h and the subsequent exchange of tank h with tank k. (Note that it is unnecessary to add arcs from vertices in X_k to those in X_h.)

It is possible to account for many-for-one and one-for-many tank exchanges in very much the same way but at the cost of a vastly increased size of network flow problem and computation times which may be unjustified for sensitivity purposes.

Removal of Tanks

Consider tank $r \in S^*$ and assume $r \in F_k^*$. The least amount of liquid of type k in tank r is W_r given by equation (12.11). If tank r is removed from the set E_0 of available tanks, then the solution S^* will no longer be valid. Once more we have to generate approximate sensitivity results from the single solution S^*. Form a graph G as follows:

1. Add a vertex for tank r (the tank which is to be removed from the solution) and a vertex for every other tank j with $W_j < W_r$.

2. Add an arc (x_{j_1}, x_{j_2}) between any two vertices j_1 and j_2 if

$$Q_{j_2} \geq W_{j_1} > W_{j_2} \tag{12.12}$$

and if j_1 and j_2 are not in the same set F_i^*. Set the 'length' $c_{j_1 j_2}$ of this arc to $W_{j_1} - W_{j_2}$.

The above graph G is acyclic because of conditions (12.12). A path $(x_r, x_{j_1}, x_{j_2}, \ldots, x_{j_t})$ in G represents the exchange of tank r with j_1 followed by the exchange of tank r with j_2, etc., until finally tank r is exchanged with tank j_t. Thus, finally tank j_1 contains an amount W_r of liquid, j_2 contains an amount W_{j_1}, etc., and tank r contains an amount W_{j_t}. The net amount by which the contents of tank r is reduced is therefore $W_r - W_{j_t}$ which is the sum of the 'lengths' of the arcs of this path as defined earlier. Thus, using the set S^* of tanks only, the smallest amount or liquid that tank r must hold is at least as low as $\bar{W}_r = W_r - L_r$, where L_r is the length of the longest path in G from vertex x_r. This longest path problem is very simple and can for example be solved as a CPM problem.

If tank r is removed, it must be replaced by one or more tanks from $\bar{S} = E_0 - S^*$. Let $x_j = 1$ if $j \in \bar{S}$ is one of the tanks used to replace tank r and $x_j = 0$ otherwise.

A knapsack problem can now be defined as:

$$\min z = \sum_{j \in \bar{S}} v_j x_j$$

s.t.
$$\sum_{j \in \bar{S}} Q_j x_j \geq \bar{W}_r \qquad (12.13)$$
$$x_j \in \{0, 1\}.$$

The reduction in the value of the optimal solution caused by the removal of tank r is therefore at most $(z^* - v_r)$ where z^* is the solution to the above knapsack problem.

12.1.6 Unloading Only

Consider the problem of unloading q_i', $i = 1, \ldots, N$ from M tanks of capacity $Q_1, Q_2, \ldots, Q_M$. Once more we will denote by E_0 the set of empty tanks and by V_i the set of tanks containing crude i, with α_j the amount of crude contained in tank j.

There is a fundamental difference between the loading and unloading problems because—contrary to the loading case—the unloading problem is a totally uncoupled N-stage problem which can, therefore, be decomposed into N separate one-stage problems. This is so because quantity q_i' must be unloaded from the tanks of V_i and therefore can only change the state of these tanks which are not involved in the unloading of any other crude $k \neq i$. Thus, we can consider N separate problems, the ith one being: unload q_i' from the set of tanks V_i to maximize the value of tanks left totally empty.

Let $f_j(x)$ be the value of tanks emptied by unloading amount x from tanks $1, \ldots, j$. We can then write:

$$f_j(x) = \max_{\lambda \in \{0, 1\}} [f_{j-1}(x - \lambda \alpha_j) + \lambda v_j] \qquad (12.14)$$

and $f_{|V_i|}(q_i')$ is the required maximum value of emptied tanks.

Equation (12.14) is the usual dynamic programming formulation of the one-dimensional knapsack problem (Gilmore and Gomory, 1966), and needs no further explanation except to note that for objective (i) the recursion simplifies to the rule: 'unload tanks in increasing order of the amount they hold'.

PART 2

DYNAMIC LOADING AND UNLOADING

12.2.1 General

A typical problem involving loading and unloading is shown diagrammatically in Figure 12.5. This figure will be used to explain the problem and introduce the terminology.

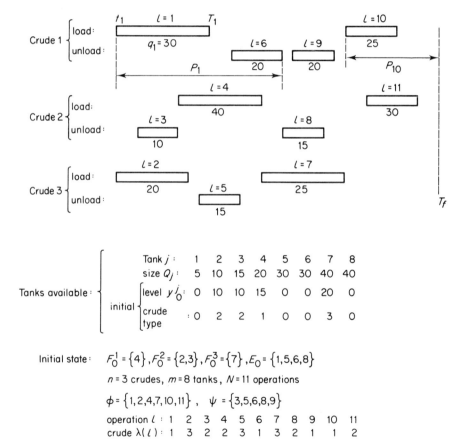

Figure 12.5 An example of a loading/unloading problem. Amount to be loaded/unloaded at each operation is shown under the corresponding bar

Loading Problems

We assume that there are N loading and unloading operations that must take place and that these are known. At the lth operation, $(l = 1, \ldots, N)$, either a loading or unloading of an amount q_l takes place starting at time t_l and finishing at time T_l (t_l and T_l are related by the amount to be loaded/unloaded and the rate of loading/unloading). In total there are n different types of liquid i $(i = 1, \ldots, n)$ each of which can appear in the sequence of loading/unloading operations more than once. Liquids of different types must be kept in different tanks and cannot be mixed.

Let ϕ and ψ be the index sets of the loading and unloading operations respectively, and let $\lambda(l)$ be the type of liquid involved in the lth operation. There are m tanks available of capacities $Q_1, \ldots, Q_m$ and corresponding costs $v_1, \ldots, v_m$.

Let E_l be the set of tanks that are empty after the lth operation takes place, F_l^i be the set of tanks holding liquid i and y_l^j be the amount of liquid in tank j at the end of the lth operation. When $l = 0$, E_0, F_0^i $(i = 1, \ldots, n)$, y_0^j $(j = 1, \ldots, m)$ define the initial condition of the tanks.

We first recall the two assumptions ((i) and (ii) in the Introduction) relating to the dynamic loading and unloading problem considered in this part. The objective considered is to minimize

$$z = \sum_{l=1}^{N} p_l \sum_{j \in F_l^{\lambda(l)}} v_j. \tag{12.15}$$

where p_l is a 'weight' given to the lth operation. If we indicate by k the first operation following l for which $\lambda(k) = \lambda(l)$, then one way of defining p_l is:

$$\begin{aligned}
p_l &= t_k - t_l & \text{if } l, k \in \phi \\
&= T_k - t_l & \text{if } l \in \phi, k \in \psi \\
&= T_k - T_l & \text{if } l, k \in \psi \\
&= t_k - T_l & \text{if } l \in \psi, k \in \phi
\end{aligned} \tag{12.15a}$$

In this case expression (12.15) becomes an integral criterion which gives the average cost of tanks holding crude during the period in which the loading and unloading operations take place.

If l is the last operation involving crude $\lambda(l)$ then p_l can be set to $T_f - t_l$ if $l \in \phi$ or $T_f - T_l$ if $l \in \psi$ where T_f is some final time at which the sequence of loading and unloading operations is considered finished. For the problem shown in Figure 12.5, p_1 and p_{10} (for example) are shown in that figure.

We should also, perhaps, note here that if p_l is set to 0 for all operations l except for the last operation of each crude, and set to 1 for all last operations, then expression (12.15) becomes an objective depending only on the terminal state.

It may be worthwhile to note here that with a given set of tanks, and initial conditions of the liquid in these tanks, it may not be feasible to execute a given sequence of loading and unloading operations.

12.2.2 The Problem

Let $\xi_l^j = 1$ if tank j contains liquid $\lambda(l)$ at the end of the lth operation and 0 otherwise, and let y_l^j be the amount of liquid in tank j. (We have $\xi_l^j = 1$ iff $j \in F_l^{\lambda(l)}$.)

A simple mixed integer programming formulation of the problem in terms of the above two vectors can be written down immediately. However, the structure of this formulation is not pronounced enough to lead to any special algorithm other than the general integer programming procedures. An alternative formulation—which is the one on which the tree search procedure of the following sections is based—is described below.

Definitions

A set $S \subseteq \{1, \ldots, m\}$ of tanks is called *weak minimal* with respect to quantities q and q' of liquid if:

$$q' \leq \sum_{j \in S} Q_j < q + \max_{j \in S} [Q_j]. \tag{12.16}$$

It is called *weak maximal* with respect to q, q', and a set $S' \supseteq S$ if either

$$\sum_{j \in S'} Q_j > q' \geq \sum_{j \in S} Q_j > q - \max_{j \in S'-S} [Q_j] \tag{12.17}$$

or

$$S = S' \text{ and } q' = \sum_{j \in S'} Q_j$$

Let the operations be numbered in ascending order of time τ_l where for $l \in \phi$, $\tau_l = t_l$ and for $l \in \psi$, $\tau_l = T_l$. Ties for cases where $\tau_l = \tau_k$ for two operations l and k are broken arbitrarily except when $l \in \phi$ and $k \in \psi$ when k is ordered before l.

Let α_l be the total quantity of liquid $\lambda(l)$ that exists in the system after operation l, that is:

$$\alpha_l = \sum_{j \in F_0^{\lambda(l)}} y_0^j + \sum_{\substack{k \in \phi \\ k \leq l \\ \lambda(k) = \lambda(l)}} q_k - \sum_{\substack{k \in \psi \\ k \leq l \\ \lambda(k) = \lambda(l)}} q_k. \tag{12.18}$$

Loading Problems

Let $\pi(l, i)$ be the maximum value of $k < l$ for which $\lambda(k) = i$. If no such k exists we will assume $\pi(l, i) = -i$. We will also write $\pi(l, \lambda(l))$ as $\pi(l)$ for short.

Consider the situation just before operation l. The set $S' = F_{\pi(l)}^{\lambda(l)}$ of tanks contains liquid $\lambda(l)$. If $l \in \phi$ only an amount $\bar{q}_l$, $q_l \geq \bar{q}_l \geq q'_l = q_l - [\sum_{j \in S'} Q_j - \alpha_{\pi(l)}]$ needs to be loaded into new tanks. It is quite obvious that for any sequence of operations to be optimal (for any objective function z for which: $S_1 \supseteq S_2 \rightarrow z(S_1) \geq z(S_2)$), the set of tanks into which $\bar{q}_l$ is loaded must form a weak minimal set as defined by (12.16) with $q' = q'_l$ and $q = q_l$.

Similarly, if $l \in \psi$ the set of tanks totally emptied at operation l must be a weak maximal set as defined by (12.17) with

$$q' = q_l + \left[\sum_{j \in S'} Q_j - \alpha_{\pi(l)}\right]$$

and $q = q_l$.

Based on the above, we will now use conditions (12.16) and (12.17) to construct for each level (operation) l a family $\mathcal{F}_l$ of sets that are possible candidates for containing the amount α_l of liquid $\lambda(l)$ after operation l. In other words, conditions (12.16) and (12.17) are used to reduce the number of possible states of the system at each stage. The families $\mathcal{F}_l$ are constructed as follows.

Assuming $\mathcal{F}_{\pi(l)}$ is known, and starting with $\mathcal{F}_l = \emptyset$ then for every $S' \in \mathcal{F}_{\pi(l)}$ generate the family $\mathcal{S}$ of all the sets S satisfying condition (12.16) if $l \in \phi$ (or satisfying condition (12.17) if $l \in \psi$). Form the family $\mathcal{U} = \{S \cup S' \mid S \in \mathcal{S}\}$ if $l \in \phi$, or the family $\mathcal{U} = \{S' - S \mid S \in \mathcal{S}\}$ if $l \in \psi$. Update $\mathcal{F}_l = \mathcal{F}_l \cup \mathcal{U}$ and continue until all $S' \in \mathcal{F}_{\pi(l)}$ have been considered.

The above procedure is initialized by setting $\mathcal{F}_i = F_0^i$.

Formulation

Let $x_l^r = 1$ if the rth weak minimal set $S^r \in \mathcal{F}_l$ is used to hold the quantity α_l of liquid $\lambda(l)$ at stage l, and 0 otherwise, and let y_l^j be the quantity of liquid $\lambda(l)$ held in tank j after the lth operation. Also let $t_{jr}(l) = 1$ if $j \in S^r$ and 0 otherwise, and $\beta_l = |\mathcal{F}_l|$.

The problem can now be stated as:

$$\min z = \sum_{l=1}^{N} p_l \sum_{r=1}^{\beta_l} V_l^r x_l^r \quad (12.19)$$

$$\text{s.t.} \sum_{r=1}^{\beta_l} x_l^r = 1 \quad l = 1, \ldots, N \quad (12.20)$$

$$\sum_{j=1}^{m} (y_l^j - y_{\pi(l)}^j) = \delta q_l \quad l = 1, \ldots, N \quad (12.21)$$

$$\delta\left(\sum_{r=1}^{B_0} t_{jr}(l)x_l^r - \sum_{r=1}^{B_{\pi(l)}} t_{jr}(\pi(l))x_{\pi(l)}^r\right) \geq 0 \qquad j=1,\ldots m; l=1,\ldots,n \tag{12.22}$$

$$\delta(y_l^j - y_{\pi(l)}^j) \geq 0 \qquad j=1,\ldots,m; l=1,\ldots,n \tag{12.23}$$

$$\sum_{\substack{i=1 \\ i \neq \lambda(l)}}^{n} \sum_{r=1}^{B_{\pi(l,i)}} t_{jr}(\pi(l,i))x_{\pi(l)}^r + \sum_{r=1}^{B_0} t_{jr}(l)x_l^r \leq 1 \qquad j=1,\ldots,m; l=1,\ldots,n \tag{12.24}$$

$$y_l^j \leq Q_j \sum_{r=1}^{B_0} t_{jr}(l)x_l^r \qquad j=1,\ldots,m; l=1,\ldots,n \tag{12.25}$$

where $\delta = +1$ if $l \in \phi$ and -1 if $l \in \psi$, and where

$$V_l^r = \sum_{j=1}^{m} v_j t_{jr}(l).$$

We must also take for initialization $\beta_{-i} = 1$, $y_{-i}^j = y_0^i$, $t_{jr}(-i) = 1$ if $j \in F_0^i$ and 0 otherwise. (Note that since $\beta_{-i} = 1$, r can only take the value 1, and $x_{-i}^r = 1$.)

Equation (12.20) says that only one S^r must be chosen from each family $\mathcal{F}_l$ at stage l. Equation 12.21 says that the total increase (decrease) of the levels of tanks holding liquid $\lambda(l)$ is equal to the amount loaded (unloaded) at stage l. Conditions (12.22) and (12.23) are the continuity conditions: (12.22) for the tanks used and (12.23) for the amounts of liquid in these tanks. Condition (12.24) says that a tank can be used at most once for any liquid at any one time (i.e. it is the no-mixing requirement). Condition (12.25) is the capacity constraint on the tanks used at stage l. (Note also that this condition sets to zero the level of liquid in any tank not used at stage l.)

12.2.3 The Algorithm

In this section we describe a depth-first tree search algorithm for the solution of the problem. A branching of this tree at level k represents a set of tanks $S^r \in \mathcal{F}_k$ chosen for holding the quantity α_k of liquid $\lambda(k)$ after operation k.

A node of the tree at level k is represented by an ordered list $L = (S^{r_{-n}}, \ldots, S^{r_{-1}}, S^{r_1}, \ldots, S^{r_k})$ where $S^{r_{-i}} = F_0^i$ (the initial set of tanks containing liquid i) and $S^{r_i} \in \mathcal{F}_i, \ldots, S^{r_k} \in \mathcal{F}_k$. Thus, the initial conditions are now introduced by assuming one loading operation (numbered $-i$) for each liquid i and this loading is into the set of tanks F_0^i which can also be considered as the only element of a family $\mathcal{F}_{-i}$. With this numbering all the tanks $j = 1, \ldots, m$ can be considered to be initially empty.

Loading Problems

A set S^{r_h} in L is then taken to mean that the liquid involved in the hth operation is held in the set S^{r_h} of tanks after the operation is complete. In the algorithm we will occasionally treat L as an (unordered) set, the meaning being obvious from the context.

The algorithm is initially described in a skeleton form, and parts (bounds, feasibility tests, etc.) of the algorithm are described in greater detail in the following sections.

The method proceeds as follows:

Step 1 (*Initialization*) Compute $\mathcal{F}_k$ for each $k = 1, \ldots, N$. Set $L = (S^{r_{-n}}, \ldots, S^{r_{-1}})$. Set z^* (the cost of the best answer so far) $= \infty$ and $k = 1$.

Step 2 (*Dominance tests*) Define the modified family $\bar{\mathcal{F}}_k$ as:

$$\bar{\mathcal{F}}_k = \{S \in \mathcal{F}_k \mid S \text{ satisfies conditions (12.26) and (12.27) below}\}$$

$$S \cap \bigcup_{i \neq \lambda(k)} S^{r_{\pi(k,i)}} = \emptyset \tag{12.26}$$

(constraint (12.24) satisfied) and:

| if $k \in \phi$ | $S \supseteq S^{r_{\pi(k)}}$ | (12.27a) |
| if $k \in \psi$ | $S \subseteq S^{r_{\pi(k)}}$ | (12.27b) |

(constraint (12.22) satisfied).

Remove from $\bar{\mathcal{F}}_k$ every element dominated by another element of $\bar{\mathcal{F}}_k$ (see Section 12.2.3.3). Let $\bar{\beta}_k = |\bar{\mathcal{F}}_k|$ and set $r_k = 0$.

Step 3 (*Forward branching*) Set $r_k = r_k + 1$. Choose the set $S^{r_k} \in \bar{\mathcal{F}}_k$ and form $L = L \cup S^{r_k}$.

Step 4 (*Feasibility test*) If $k \in \psi$, test that it is feasible to reach the state represented by L from the initial state, i.e. that variables y_l^j ($l = 1, \ldots, k$; $j = 1, \ldots, m$) satisfying constraints (12.21), (12.23), and (12.25) can be found (see section 12.2.3.1).

If state L is infeasible, backtrack. If not, go to Step 5. Note that if $k \in \phi$, any choice of S^k leads to a state for which constraints (12.21), (12.23), and (12.25) are always satisfied.

Step 5 (*Lower bound*) Calculate a lower bound LB(L) as mentioned in Section 12.2.3.2.

If LB(L) $\geq z^*$ backtrack, otherwise:
If $k < N$, set $k = k + 1$ and go to Step 2.
If $k = N$, then if $z(L) < z^*$ set $z^* = z(L)$.

Step 6 (*Backtrack*) Remove the last element S^{r_k} from L. If $r_k < \bar{\beta}_k$ go to Step 3. If $r_k = \bar{\beta}_k$ then: if $k = 1$ stop (the cost of the optimal solution is z^*) or else set $k = k - 1$ and go to Step 6.

12.2.3.1 Feasibility Test

Let us consider the situation for one type of liquid (say i). Assume there are 4 tanks of capacities 30, 30, 40, and 100, numbered 1, 2, 3, and 4 respectively. Let the initial levels y_0^i in the tanks be given by the vector $(20, 20, 0, 0)$. Let us assume three operations involving this liquid: a loading of 40, another loading of 40, and an unloading of 40. When 40 is loaded, the total amount of liquid i in the system is $\alpha_1 = 80$. The family $\bar{\mathcal{F}}_1$ is $\{\{1,2,3\},\{1,2,4\}\}$. Let us say that $S^{r_1} = \{1,2,3\}$ is chosen for branching.

When the second loading of 40 takes place, the total amount of liquid i in the system is $\alpha_2 = 120$. The family $\bar{\mathcal{F}}_2$ is $\{\{1,2,3,4\}\}$. $S^{r_2} = \{1,2,3,4\}$ is picked for branching at level 2.

When the unloading of 40 takes place, $\alpha_3 = 80$ and $\bar{\mathcal{F}}_3$ is $\{\{1,4\},\{2,4\},\{3,4\},\{4\},\{1,2,3\},\{1,2,4\},\{1,3,4\},\{2,3,4\},\{1,2,3,4\}\}$ if $S^r = \{1,4\} \in \bar{\mathcal{F}}_3$ is picked for branching, the node L of the tree produced by the three branchings S^{r_1}, S^{r_2}, and S^{r_3} is infeasible, in the sense that no variables y_k^i exist which satisfy constraints (12.21), (12.23), and (12.25). This can be seen by the fact that when S^{r_1} is chosen for the first branching the minimum amount of liquid that must be held in tanks 2 and 3 together is 50 (regardless of how the liquid is distributed). Thus, irrespective of how the liquid is distributed in the second loading operation, the third operation (of unloading 40) cannot empty both tanks 2 and 3 and hence the choice of $\{1, 4\}$ at level 3 is infeasible.

The test for feasibility of a node L of the branch and bound tree is as follows. Let us define a variable:

$$\xi_k^j = y_k^j - y_{\pi(k)}^j \quad \text{for} \quad k \in \phi, \quad j = 1, \ldots, m \tag{12.28}$$

and

$$\xi_k^j = y_{\pi(k)}^j - y_k^j \quad \text{for} \quad k \in \psi, \quad j = 1, \ldots, m.$$

Equation (12.21) then becomes:

$$\sum_{j=1}^{m} \xi_k^j = q_k \quad k = 1, \ldots, l. \tag{12.29}$$

Constraints (12.23) become simply the nonnegativity constraints on ξ_k^j.

Constraints (12.25) are:

$$\begin{aligned} y_k^j &\leq Q_j \quad \text{for} \quad j \in S^{r_k} \quad k = 1, \ldots, l \\ &= 0 \quad \text{for} \quad j \notin S^{r_k} \quad k = 1, \ldots, l \end{aligned} \tag{12.30}$$

Equations (12.28) define the network flow structure as shown in Figure 12.6 where the figure is drawn for the last tree stages assuming the

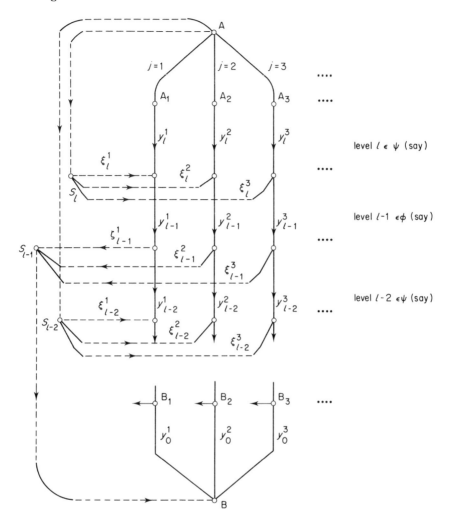

Figure 12.6 Graph for the feasibility test

operations l, $l-1$, and $l-2$ to be unloading, loading, and unloading operations, respectively.

Equations (12.24) specify the demand/supply at the terminals s_l, s_{l-1}, $s_{l-2}, \ldots$ to be q_l, $q_{l-1}, \ldots$.

Conditions (12.30) specify that the flow y_k^j for $j \notin S^{r_k}$ must be zero (i.e. the corresponding arcs in Figure 12.6 can be removed) and that arcs corresponding to y_k^j for $j \in S^{r_k}$ have a capacity Q_j placed on them. All other arcs corresponding to flows ξ_k^j in Figure 12.6 are not capacitated.

In Figure 12.6 we write y_0^j for the initial level of liquid in tank j. Let us now introduce a 'super-source' A and connect A to vertices $A_1, \ldots, A_m$ with arcs of capacity $Q_1, \ldots, Q_m$ respectively, and also connect A to the vertices s_k, $k \in \psi$ with arcs (A, s_k) having capacity q_k. In addition we introduce a 'supersink' B and connect vertices $B_1, \ldots, B_m$ to B with arcs of capacity $y_0^1, \ldots, y_0^m$ and also connect vertices s_k, $k \in \phi$ to B with arcs (s_k, B) having capacity q_k.

It is quite apparent that the current node $L_l = \{F_0^i, S^{r_1}, \ldots, S^{r_l}\}$ of the branch and bound tree is feasible if the maximum flow from A to B saturates all arcs terminating at B.

It is important for computational purposes to note the following: Consider a branching from node L_l (which has already been tested for feasibility) to a new node $L_{l+1} = \{F_0^i, S^{r_1}, \ldots, S^{r_l}, S^{r_{l+1}}\}$.

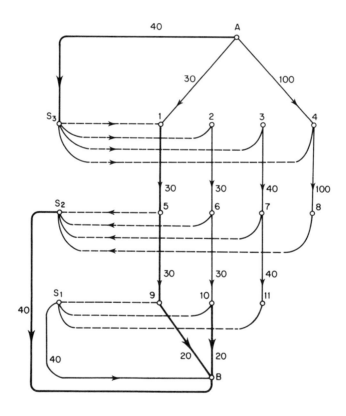

Figure 12.7   Graph for the example

Loading Problems

(i) If $(l+1) \in \phi$, then it follows directly from Figure 12.6 that the feasibility condition for the new graph will be automatically satisfied and no test on L_{l+1} is necessary.

(ii) If $(l+1) \in \psi$, then L_{l+1} must be checked for feasibility as mentioned above. However, it should be noted here that—by construction—the graph of Figure 12.6 remains a sequential acyclic graph at all times and the maximum flow calculation from A to B is extremely simple (Christofides, 1975), i.e.

Find any path from A to B along which flow could be sent, saturate this path with flow and reduce the capacity of the arcs along this path. Repeat until no more paths that can carry flow from A to B can be found. The net flow is then the maximum flow.

In order to illustrate the procedure let us again consider the example at the beginning of this section. Figure 12.7 shows the graph at the end of the unloading operation. Arcs with specified zero flow are not shown. Initially we find path $(A, 4, 8, s_2, B)$ and send an amount of 40, saturating arc (s_2, B). We now find path $(A, s_3, 1, 5, 9, B)$ and send a flow of 20, saturating arc $(9, B)$. We now find path $(A, s_3, 2, 6, 10, B)$ and send a flow of 20, saturating arcs $(10, B)$ and (A, s_3). We now find path $(A, 1, 5, 9, s_1, B)$ and send a flow of 10 to saturate arcs $(1, 5)$ and $(5, 9)$. No more paths—along which flow could be sent—can now be found and arc (s_1, B) remains unsaturated; hence the node of the search tree represented by $L = \{\{1, 2\}, \{1, 2, 3\}, \{1, 2, 3, 4\}, \{1, 4\}\}$ is infeasible as noted earlier.

12.2.3.2 Lower Bounds

A lower bound that can be used to limit the size of the tree search can be derived by making use of a relaxation of the integer programming formulation given in Section 12.2.2. For a node $L_k = \{F_0^n, \ldots, F_0^1, S^{r_1}, \ldots, S^{r_k}\}$ of the tree this relaxation involves ignoring constraints (12.21), (12.23), and (12.25), thus converting the problem from a mixed to a pure integer problem.

Consider a given crude i and let l^* be the last operation on crude i before level $k+1$, and $\bar{l}$ be the first operation after level k involving this crude. The subset of the set $S^{r_{l^*}}$ of tanks which held crude $i = \lambda(\bar{l})$ at level k and which will certainly continue to hold this type of crude up to level l is denoted by A_l. An upper bound on the total empty space in the set $S^{r_{l^*}}$ of tanks at level l is:

$$G = \sum_{h \in S^{r_l}} Q_h - \alpha_l + \sum_{\substack{h = k+1 \\ \lambda(h) = i \\ h \in \psi}}^{l} q_h.$$

Hence

$$A_l = \{j \in A_{\pi(l)} | G < Q_j\} \qquad (12.31)$$

where $A_{l^*} = S^{r_{l^*}}$. The sets A_l can be computed (for a given liquid i) recursively using equation (12.31).

Let E_l^* be an upper bound on the set of possibly empty tanks at level l. E_l^* can also be computed recursively as:

$$E_l^* = E_{l-1}^* \cup (A_{\pi(l)} - A_l) \qquad (12.32)$$

where $E_k^* = E_k$, the actual set of empty tanks corresponding to node L_k.

Using (12.31) and (12.32), it is now possible to define families ($\mathscr{F}_l'$, say) of weak minimal sets smaller than those defined in Section 12.2.2, and still ensure that at level l of the tree search only elements of $\mathscr{F}_k'$ need be considered. Thus

$$\mathscr{F}_l' = \{S \in \mathscr{F}_l | S \subseteq E_l \cup A_l\}. \qquad (12.33)$$

Consider a given liquid and form a graph G_i as follows. For each $l > k$ such that $\lambda(l) = i$ and for each set $S_l^r \in \mathscr{F}_k'$, add a vertex θ_l^r. In addition, add a 'source' vertex θ^+ and a 'sink' vertex θ^-. Add an arc $(\theta_{\pi(l)}^r, \theta_l^h)$ between any two vertices $\theta_{\pi(l)}^r$ and θ_l^h whenever:

(i) if $l(>\bar{l}) \in \phi$ then $S_{\pi(l)}^r \subseteq S_l^h$

(ii) if $l(>\bar{l}) \in \psi$ then $S_{\pi(l)}^r \supseteq S_l^h$.

Set the 'cost' of arc $(\theta_{\pi(l)}^r, \theta_l^h)$ equal to $p_{\pi(l)} V_{\pi(l)}^r$. In addition add arcs, of zero cost, from θ^+ to θ_i^r for all corresponding $S_i^r \in \mathscr{F}_i'$, and arcs from θ_t^r to θ^-, of cost $p_t V_t^r$, for all corresponding $S_t^r \in \mathscr{F}_t'$, where t is the last operation involving liquid i.

Constraints (12.20) and (12.22) are now satisfied by any path from θ^+ to θ^- in G_i. Let the least-cost path from θ^+ to θ^- in G_i have cost $d_i(k)$. A lower bound to the complete problem at node L_k of the tree is then:

$$\text{LB}(L_k) = \sum_{i=1}^{n} d_i(k) + z(L_k). \qquad (12.34)$$

In addition to constraints (12.20) and (12.22), which are considered exactly (for a given liquid i), the re-definition of the families $\mathscr{F}_l'$ corresponds to a partial consideration of constraints (12.24).

An Improved Bound

Figure 12.8 shows the graphs G_1 and G_2 defined above for a hypothetical example involving two liquids. G_1 corresponds to liquid 1 and G_2 corresponds to liquid 2. The sets S_i^r corresponding to the vertices are also shown.

Loading Problems

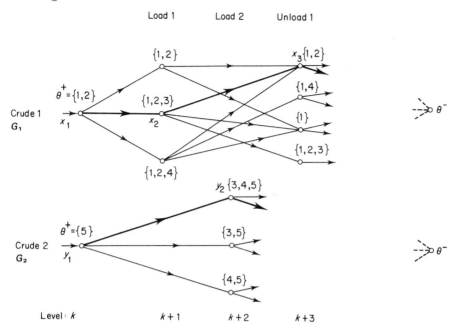

Figure 12.8 Example of improvement to the bound

Initially (i.e. as defined by L_k) liquid 1 is in tanks $\{1, 2\}$ and liquid 2 is in tank $\{5\}$.

Let us assume that the shortest path in G_1 derived during the calculation of the bound is $(x_1, x_2, x_3, \ldots)$ with cost C_1 and the shortest path in G_2 is $(y_1, y_2, \ldots)$ with cost C_2. The loading and unloading operations specified by these paths are infeasible since they violate constraint (12.24) for vertices x_2 and y_2, i.e. tank 3 is used for loading liquid 1 at level $k+1$ and is also used for loading liquid 2 at level $k+2$.

Let us now suppose that a penalty μ is placed on vertices x_2 and y_2 by adding μ to the cost of all arcs emanating from these two vertices. Let the shortest paths after the penalties are placed have costs $C_1(\mu)$ and $C_2(\mu)$. Quite clearly, constraint (12.24) requires that for any tank j at most one path can pass through a vertex θ_l^r for which $j \in S_l^r$. Thus, the maximum increase in the total cost of any two feasible paths (in G_1 and G_2) caused by the penalty is μ, and hence:

$$C_1(\mu) + C_2(\mu) - \mu$$

is also a lower bound to the problem.

In general, let W_h be a set of vertices of the graphs $G_1, \ldots, G_n$ (at most one from each graph) such that it violates constraint (12.24). Let μ_h be a penalty associated with W_h and $C(\mathbf{\mu})$ be the total cost of all shortest paths under the penalized costs. The expression

$$C(\mathbf{\mu}) - \sum_h \mu_h \qquad (12.35)$$

is a lower bound on L_k for all nonnegative vectors $\mathbf{\mu}$. Expression (12.35) can be maximized with respect to $\mathbf{\mu}$ to derive the best possible bound, in much the same way as was done for the travelling salesman problem (Christofides, 1975; Held and Karp, 1971; Held, Wolfe, and Crowder, 1974).

12.2.3.3 Dominance Tests

In Step 2 of the algorithm described in Section 12.2.3 a subfamily $\bar{\mathcal{F}}_k$ was produced by applying dominance tests. In this section we describe these dominance conditions.

Given two nodes a and b of the tree (at the same level k) which have the same 'parent' node, node a *dominates* node b if:

$$E_k(a) \doteq E_k(b) \qquad (12.36)$$

where a set A is equivalent to a set B (written as $A \doteq B$) if it is possible to map every $j \in B$ to an element of A, one-to-one and onto, by a function ρ so that

$$Q_j = Q_{\rho(j)} \text{ and } v_j = v_{\rho(j)}. \qquad (12.37)$$

Condition (12.36) above can be relaxed in the following case. If for liquid $\lambda(k)$, $\nexists l > k$ such that $\lambda(l) = \lambda(k)$ and $l \in \psi$, then condition (12.36) can be replaced by:

$$E_k(b) \dot{\subseteq} E_k(a)$$

where $\dot{\subseteq}$ is defined by conditions (12.37) above except that the mapping ρ need not be 'onto'.

12.2.4 Computational Results

The algorithm described in the previous section was coded and tested on a number of loading and unloading problems of varying size up to 45 operations, 11 liquids, and 20 tanks. It can be seen from Table 12.2 that the computational performance of the algorithm is good and that the algorithm is capable of solving problems of a practical size. Indeed, all problems except the third one in Table 12.2 are real problems. Problem

Loading Problems

Table 12.2 Computational performance of the algorithm

Problem	N	n	m	A		B		C	
				time	nodes	time	nodes	time	nodes
1	20	8	13	140.22	125 494	18.02	3 931	3.50	276
2	20	11	16	11.14	9 316	2.46	549	1.06	117
3	20	8	20	48.52	20 138	15.39	2 659	14.95	2 263
4	36	9	16	†	—	291.19	47 385	44.05	3 952
5	36	9	16	†	—	200.60	32 921	21.46	1 262
6	45	9	16	†	—	†	—	274.44	18 299

† No feasible solution found within the time limit (set at 512 s.)

number 3 is a problem specifically constructed to be hard and one on which heuristic procedures suggested for solving this type of problem fail to find even a feasible loading and unloading sequence. The data for this problem are reproduced in the Appendix.

Table 12.2 shows for each problem the computational time and number of nodes in the search tree for:

A. The algorithm without the bounds and dominance tests;
B. The algorithm without the bounds but including the dominance tests; and
C. The algorithm including both bounds and dominance.

All computing times are seconds on the CDC-6600 computer using a Fortran compiler.

Appendix

The data for Problem 3 in Table 12.2 is given below and a pictorial representation of the problem is shown in Figure 12.9.

$N = 20$, $n = 8$, $m = 20$, $T_f = 1761$

$t_l = (294, 296, 423, 528, 529, 687, 690, 792, 794, 826, 828, 975, 978, 1020,$
$1099, 1101, 1188, 1260, 1263, 1497)$

$T_l = (504, 387, 615, 675, 674, 705, 759, 816, 813, 946, 960, 1083, 1086,$
$1086, 1242, 1245, 1248, 1479, 1482, 1710)$

$q_l = (60, 80, 80, 75, 55, 10, 25, 12, 15, 50, 45, 60, 75, 15, 65, 75, 20, 75,$
$50, 140)$

$\lambda(l) = (4, 5, 1, 3, 4, 4, 7, 7, 4, 5, 7, 6, 3, 1, 2, 8, 3, 8, 3, 7)$

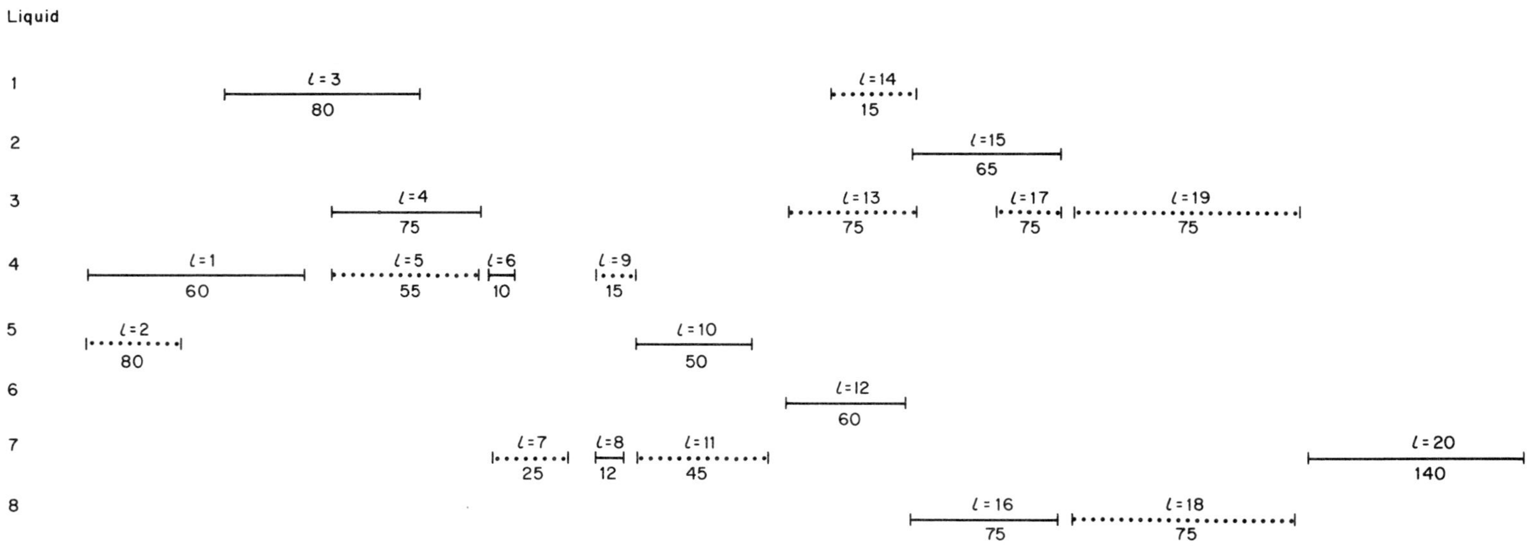

Figure 12.9 Data for problem 3. ——— Loading; ····· unloading. Amount to be loaded/unloaded at each operation is shown under the corresponding bar

$Q_j = (8, 8, 20, 20, 20, 20, 20, 20, 20, 20, 20, 45, 45, 45, 60, 60, 60, 75, 75, 75)$

$v_j = Q_j + 20 \qquad j = 1, \ldots, 20$

$y_0^j = (8, 8, 0, 12, 20, 20, 20, 20, 20, 20, 18, 45, 45, 45, 0, 0, 60, 75, 70, 35)$

$F_0^1 = \{1, 18\}$ $F_0^2 = \{4\}$ $F_0^3 = \{5, 6, 7, 19\}$ $F_0^4 = \{8, 12\}$ $F_0^5 = \{13, 20\}$
$F_0^6 = \{9, 10, 14\}$ $F_0^7 = \{2, 11, 17\}$ $F_0^8 = \varnothing$

$\phi = \{1, 3, 4, 6, 8, 10, 12, 15, 16, 20\}$

$\psi = \{2, 5, 7, 9, 11, 13, 14, 17, 18, 19\}$.

The weights p_l in the objective function (12.15) are defined from t_l, T_l, and T_f according to equations (12.15a).

References

Christofides, N. (1975). *Graph Theory—An algorithmic approach*, Academic Press, London.

Eilon, S., and N. Christofides (1971). The loading problem. *Man. Sci.*, **17,** 259.

Gilmore, P. C., and R. E. Gomory (1966). The theory and computation of knapsack functions. *Opns. Res.*, **14,** 1045.

Held, M., and R. M. Karp (1971). The travelling salesman problem and minimum spanning trees: II. *Math. Prog.*, **1,** 6.

Held, M., P. Wolfe, and H. P. Crowder (1974). Validation of subgradient optimisation. *Math. Prog.*, **6,** 62.

Johnson, D. S. (1974). Fast algorithms for bin packing. *J. Comp. Systems Sci.*, **8,** 272.

Lev, B. (1972). On the loading problem—A comment. *Man. Sci.*, **18,** 428.

Martello, S., and P. Toth, Branch and bound algorithms for the solution of the general unidimensional knapsack problem. in M. Roubens (Ed), *Advances in Operations Research*, North-Holland, Amsterdam.

CHAPTER 13

Minimizing Maximum Lateness on one Machine: Algorithms and Applications†

BEN J. LAGEWEG and JAN KAREL LENSTRA
Mathematisch Centrum, Amsterdam, The Netherlands

ALEXANDER H. G. RINNOOY KAN
Erasmus University, Rotterdam, The Netherlands

13.1 Introduction

We consider the following problem. Suppose we have n jobs $J_1, \ldots, J_n$, to be processed on a single machine which can handle only one job at a time. Job J_i ($i = 1, \ldots, n$) is available for processing at its release date r_i, requires an uninterrupted processing time of p_i time units and should ideally be completed by its due date d_i. Certain precedence constraints define a partial ordering $<$ between the jobs: '$J_i < J_j$' means that J_j cannot start before the completion of J_i. The index sets $B_i = \{j \mid J_j < J_i\}$ and $A_i = \{j \mid J_i < J_j\}$ indicate the jobs which are constrained to come before and after J_i respectively. Given a feasible processing order of the jobs, we can compute for J_i ($i = 1, \ldots, n$) a starting time $S_i \geq r_i$, a completion time $C_i = S_i + p_i$ with $C_i \leq S_j$ for all $j \in A_i$, and a lateness $L_i = C_i - d_i$. We want to find a schedule that minimizes the maximum lateness $L_{\max} = \max_i \{L_i\}$.

A number of special cases of this problem in which all r_i, p_i, or d_i are equal is studied in Section 13.2. In Sections 13.3 and 13.4, two branch-and-bound algorithms developed for the problem without precedence constraints are described and extended to the general case. Extensive computational experience is reported in Section 13.5. In Sections 13.6 and 13.7 we discuss two applications of the problem; one arises in the theoretical context of job-shop scheduling, the other occurred in a practical scheduling situation. Concluding remarks are contained in Section 13.8.

---

† This chapter was originally published in *Statistica Neerlandica*, and is reproduced by permission of the Managing Editor.

We will present an ALGOL-like description of several algorithms: the operation ':∈' in the statement '$s :\in S$' is defined to mean that s becomes an arbitrary element of S.

13.2 Special Cases

The problem to be considered is defined by n integer triples (r_i, p_i, d_i) and precedence constraints $<$.

To stress the symmetry inherent to the problem, it is useful to describe it in an alternative way. Let M_1 and M_3 be *nonbottleneck* machines of infinite capacity and M_2 a *bottleneck* machine of capacity one. Job $J_i (i = 1, \ldots, n)$ has to visit M_1, M_2, M_3 in that order and has to spend:

a *head* r_i on M_1 from 0 to r_i;
a *body* p_i on M_2 from $S_i \geq r_i$ to $C_i = S_i + p_i$ with $C_i \leq S_j$ for all $j \in A_i$;
a *tail* $q_i = K - d_i$ (for some constant $K \geq \max_i \{d_i\}$) on M_3 from C_i to $L_i' = C_i + q_i = L_i + K$.

We want to minimize the maximum completion time $L'_{\max} = \max_i \{L_i'\} = L_{\max} + K$.

The problem, now defined by n triples (r_i, p_i, q_i) and $<$, is clearly equivalent to its *inverse* problem defined by (q_i, p_i, r_i) and $<'$ with $J_i <' J_j$ if $J_j < J_i$; an optimal schedule for one problem can be reversed to obtain an optimal schedule for the other problem, with the same solution value.

Let us first assume that $A_i = B_i = \emptyset$ for all i. If all r_i are equal, an optimal schedule is provided by *Jackson's rule* (Jackson, 1955): $L_{\max}$ is minimized by ordering the jobs according to nondecreasing d_i.

If all d_i are equal, the problem is similarly solved by ordering the jobs according to nondecreasing r_i. This result can be interpreted as a consequence of the symmetry discussed above.

If all p_i are equal, such a simple solution method is usually not available, unless $p_i = 1$ for all i. In the latter situation, algorithm J below involving repeated application of Jackson's rule produces an optimal schedule (Baker and Su, 1974):

procedure algorithm J (n, r, q, S, C);
begin local N, N', t, i;
  $N := \{1, \ldots, n\}; t := 0$;
  while $N \neq \emptyset$ **do**
  begin $t := \max \{t, \min \{r_j \mid j \in N\}\}$;
    $N' := \{j \mid j \in N, r_j \leq t\}$;
    $i :\in \{j \mid j \in N', q_j = \max \{q_k \mid k \in N'\}\}$;
    $N := N - \{i\}; S_i := t; C_i := t := t + 1$
  end
end.

Minimizing Maximum Lateness on one Machine

The proof of this result is straightforward and depends on the fact that no job can become available during the processing of another one, so that it is never advantageous to postpone processing the selected job J_i. This argument does not apply if $p_i = p$ for all i and p does not divide all r_i; e.g. if $n = p = 2$, $r_1 = q_1 = 0$, $r_2 = 1$, $q_2 = 2$, postponing J_1 is clearly advantageous. However, algorithm J does solve the general problem if we allow *job splitting* (i.e. interruptions in the processing of a job); in this case we can interpret job J_i as p_i jobs with heads r_i, bodies 1, and tails q_i (Horn, 1974).

Let us now examine the introduction of precedence constraints in the problems discussed so far. As a general principle, note that we may set

$$r_i := \max\{r_i, \max\{r_j + p_j \mid j \in B_i\}\},$$
$$q_i := \max\{q_i, \max\{p_j + q_j \mid j \in A_i\}\},$$

because in every feasible schedule $S_i \geq C_j \geq r_j + p_j$ for all $j \in B_i$ and $L'_j \geq C_i + p_j + q_j$ for all $j \in A_i$. Hence, if $J_i < J_j$, we will assume that $r_i < r_j + p_i \leq r_j$ and $q_i \geq q_j + p_j > q_j$.

It follows that the case in which all d_i are equal is again solved by ordering the jobs according to nondecreasing r_i. Such an ordering will respect all precedence constraints in view of the preceding argument.

If we apply this method to the inverse problem to solve the case in which all r_i are equal, the resulting algorithm can be interpreted as a special case of Lawler's more general algorithm to minimize $\max_i \{c_i(C_i)\}$ for arbitrary nondecreasing cost functions c_i (Lawler, 1973).

A similar observation can be made with respect to the case that $p_i = 1$ for all i. Algorithm J will produce a schedule respecting the precedence constraints.

In the general case, however, the precedence constraints are not respected automatically. Consider the example specified by the data in Table 13.1 and the precedence constraint $J_4 < J_2$; note that $r_4 + p_4 \leq r_2$ and $q_4 \geq p_2 + q_2$. If the constraint $J_4 < J_2$ is ignored, the unique optimal schedule is given by $(J_1, J_2, J_3, J_4, J_5)$ with value $L'_{max} = 11$ (cf. Figure 13.1). Explicit inclusion of this constraint leads to $L'_{max} = 12$.

So far all the methods presented have been *good* algorithms in the by now conventional sense that their number of steps is bounded by a polynomial in

Table 13.1 Data for the example

i	1	2	3	4	5
r_i	0	2	3	0	7
p_i	2	1	2	2	2
q_i	5	2	6	3	2

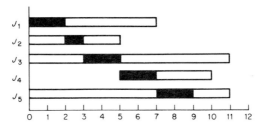

Figure 13.1 Schedule for the example

n (Edmonds, 1965). Such a method is unlikely to exist for the case in which r_i, p_i, and q_i may assume arbitrary values and $A_i = B_i = \varnothing$ for all i; in Lenstra, Rinnooy Kan, and Brucker (1977) this problem was proved to be *NP-complete*, which implies that a polynomial-time method for its solution would yield good algorithms for all other *NP*-complete problems as well. Because many notorious combinatorial problems such as the travelling salesman problem, job-shop scheduling, and graph colouring are *NP*-complete, the *NP*-completeness of the problem without precedence constraints serves as a formal justification to use enumerative solution methods such as *branch-and-bound*. Algorithms of this type have been proposed in Dessouky and Margenthaler (1972), Bratley, Florian, and Robillard (1973), Baker and Su (1974), and McMahon and Florian (1975). The first of these algorithms is not stated very clearly; the second one is surpassed by the fourth one both in elegance and efficiency (McMahon and Florian, 1975). In the following two sections, the algorithms from Baker and Su (1974) and McMahon and Florian (1975) will be described and extended to the general case.

13.3 The Algorithm of Baker and Su

The branch-and-bound algorithm to be discussed now has been developed for the problem without precedence constraints. It will be referred to as algorithm BS.

The *branching rule* generates all *active schedules* (Conway, Maxwell, and Miller, 1967) according to algorithm AS below.

procedure algorithm AS (n, r, p, S, C);
begin local i;
    procedure node (N, t);
    if $N = \varnothing$ **then comment** an active schedule has been generated **else**
    begin local N';
        $N' := \{j \mid j \in N, \ r_j < \min \{\max \{t, r_k\} + p_k \mid k \in N\}\}$;
        while $N' \neq \varnothing$ **do**

begin $i: \in N'$;
$N' := N' - \{i\}$; $S_i := \max\{t, r_i\}$; $C_i := S_i + p_i$;
node $(N - \{i\}, C_i)$
end
end;
node $(\{1, \ldots, n\}, 0)$
end.

At the lth level of the recursion, jobs are scheduled in the lth position. If the first assignment to N' is replaced by $N' := N$, all $n!$ schedules are generated. By means of the current assignment, only active schedules are generated; if $r_j \geq \max\{t, r_k\} + p_k$ for some $j, k \in N$, J_j is no candidate for the next position in the partial schedule since it can be preceded by J_k without being postponed.

The *bounding rule* is based on the observation that the value of an optimal schedule will not increase if we allow job splitting. A *lower bound* on all possible completions of a partial schedule $(J_{\pi(1)}, \ldots, J_{\pi(l)})$ is produced by the use of algorithm J to schedule the remaining jobs from $C_{\pi(l)}$ onwards while allowing job splitting. If no job splitting occurs, this particular completion is an optimal one, and the value of the complete solution is an *upper bound* on the value of an optimal solution. A partial schedule can be *eliminated* if its lower bound is not smaller than the global upper bound.

The branch-and-bound algorithm is now completely described if we specify a *search strategy* indicating which partial schedule will be chosen for further examination (Lenstra, 1977; Rinnooy Kan, 1976). In Baker and Su (1974) a *jumptrack* scheme was used, selecting a partial schedule with minimum lower bound. We implemented the recursive *backtrack* scheme of algorithm AS, selecting the unscheduled jobs in the order in which they appear in the solution, produced by algorithm J. Experiments in which these descendant nodes were chosen in order of nondecreasing lower bounds showed a 50–60% increase in solution time.

The above algorithm can easily be adjusted to take precedence constraints $<$ into account. As noted previously, they are automatically respected during the lower bound calculation and the only necessary change is a replacement of the first assignment to N' by

$$N' := \{j \mid j \in N, B_j \cap N = \emptyset, r_j < \min\{\max\{t, r_k\} + p_k \mid k \in N, B_k \cap N = \emptyset\}\}.$$

Algorithm BS is fairly straightforward and its general principles can be extended to other NP-complete sequencing problems with non-equal release dates.

13.4 The Algorithm of McMahon and Florian

A more sophisticated branch-and-bound algorithm for the problem without precedence constraints is due to McMahon and Florian (1975). Algorithm

MF is based on algorithm s below, a heuristic method suggested by Schrage (1971) for generating a good solution.

procedure algorithm s (n, r, p, q, S, C);
begin local N, N', t, i;
    $N := \{1, \ldots, n\}$; $t := 0$;
    while $N \neq \emptyset$ **do**
    begin $t := \max\{t, \min\{r_j \mid j \in N\}\}$;
          $N' := \{j \mid j \in N, r_j \leq t\}$;
          $i \in \{j \mid j \in N', p_j = \max\{p_k \mid k \in N', q_k = \max\{q_l \mid l \in N'\}\}\}$;
          $N := N - \{i\}$; $S_i := t$; $C_i := t := t + p_i$
    end
end.

The schedule $(J_{\pi(1)}, \ldots, J_{\pi(n)})$ produced by algorithm s can be decomposed into blocks. $J_{\pi(h)}$ is the last job in a block if $C_{\pi(h)} \leq r_{\pi(i)}$ for $i = h+1, \ldots, n$, i.e. if no job is delayed when $J_{\pi(h)}$ is completed. A set of jobs $\{J_{\pi(g)}, \ldots, J_{\pi(h)}\}$ forms a block if

a. $g = 1$ or $J_{\pi(g-1)}$ is the last job in a block;
b. $J_{\pi(i)}$ is not the last job in a block, for $i = g, \ldots, h-1$;
c. $J_{\pi(h)}$ is the last job in a block.

It follows that $J_{\pi(g)}$ is the first job in a block if $S_{\pi(g)} = r_{\pi(g)} \leq r_{\pi(i)}$ for $i = g+1, \ldots, n$.

With respect to J_i in block $\{J_{\pi(g)}, \ldots, J_{\pi(h)}\}$, we define

$$P_i = \{j \mid S_{\pi(g)} \leq S_j \leq S_i\}, \quad q_i^* = \min\{q_j \mid j \in P_i\}, \quad Q_i = \{j \mid j \in P_i, q_j = q_i^*\}.$$

We claim that lower bounds on the value of an optimal schedule are given by

$$\text{LB}_i' = r_i + p_i + q_i,$$

$$\text{LB}_i'' = \begin{cases} C_i + q_i^* & \text{if } i \in Q_i, \\ C_i + q_i^* + 1 & \text{if } i \notin Q_i. \end{cases}$$

LB_i' requires no comment, but the justification of LB_i'' is actually rather subtle. Defining C_{ji} as the minimum completion time of J_j if this job is scheduled as the last one of $\{J_k \mid k \in P_i\}$, we note that $C_{ji} \geq C_{ii} = C_i$ for all $j \in P_i$. A valid lower bound is now given by

$$\min\{C_{ji} + q_j \mid j \in P_i\}.$$

In the case that $i \in Q_i$, it is obvious that for all $j \in P_i$

$$C_{ji} + q_j \geq C_{ii} + q_i = C_i + q_i^*. \tag{13.1}$$

Suppose next that $i \notin Q_i$. If $j \notin Q_i$, we have

$$C_{ji} + q_j \geq C_i + q_i^* + 1. \tag{13.2}$$

Minimizing Maximum Lateness on one Machine

Consider finally the case that $i \notin Q_i$ and $j \in Q_i$. If we move J_j to the last position of $\{J_k \mid k \in P_i\}$, a gap of at least one unit idle time is unavoidable, unless a J_k with $r_k \leq S_j < S_k$ can be moved forward to start at S_j. From algorithm s we know that, if such a job exists, then $k \in Q_i$ and $p_k \leq p_j$. Thus, a gap now threatens to occur between S_k and $S_k + 1$. Repeating this argument as often as necessary, we conclude that $C_{ji} \geq C_i + 1$, and therefore

$$C_{ji} + q_j \geq C_i + q_i^* + 1. \tag{13.3}$$

Inequalities (13.1), (13.2), and (13.3) establish the validity of LB$_i''$.

At every node of the search tree, application of algorithm s yields a complete solution $(J_{\pi(1)}, \ldots, J_{\pi(n)})$ with value $L'_{\max}$ and a *lower bound* LB $= \max_i \{\max \{$LB$_i'$, LB$_i''\}\}$. We may decrease the *upper bound* UB on the value of an optimal solution by setting UB $:= \min \{$UB$, L'_{\max}\}$. If LB $\geq$ UB, the node is *eliminated*; otherwise, we apply the *branching rule* described below.

Let the *critical job* J_i be defined as the first job in the schedule with $C_i + q_i = L'_{\max}$. The schedule can only be improved if C_i can somehow be reduced. The set of solutions corresponding to the current node can now be partitioned into disjoint subsets, each characterized by the job J_j which is to be scheduled last of $\{J_k \mid k \in P_i\}$. However, jobs J_j with $j \in P_i$, $q_j \geq q_i - L'_{\max} +$ UB need not to be considered, since in that case $C_{ji} + q_j \geq C_i + q_i - L'_{\max} +$ UB $=$ UB. Therefore, only for each J_j with $j \in P_i$, $q_j < q_i - L'_{\max} +$ UB is a descendant node actually created.

We can effectively implement the precedence constraints $\{J_k < J_j \mid k \in P_i - \{j\}\}$ by adjusting r_j and q_k ($k \in P_i - \{j\}$) as described in Section 13.2. During the next application of algorithm s, J_j will then be scheduled last of $\{J_k \mid k \in P_i\}$. To maintain disjointness at deeper levels of the tree, we would have to update r_k and q_k for $k \notin P_i$ as well in view of previous choices. This would lead to the time-consuming administration of a continually changing precedence graph. Dropping the requirement of disjoint descendants, we will force J_j to follow the critical job J_i rather than the whole set $\{J_k \mid k \in P_i - \{j\}\}$. This can be done by putting r_j equal to any lower bound on $C_{ji} - p_j$ not less than r_j, such as $\max \{r_k + p_k \mid k \in P_i - \{j\}\}$, $C_i - p_j$ or simply (McMahon and Florian, 1975) r_i. Computational experiments have shown that the choice of a specific new r_j has only a minor influence on the performance of the algorithm; in our implementation, we put $r_j := \max \{r_i + p_i, C_i - p_j\}$.

The *search strategy* used in McMahon and Florian (1975) is of the jumptrack type, selecting a node with minimum lower bound. Again, our implementation is of the recursive backtrack type, choosing the descendant nodes in the reverse of the order in which the corresponding jobs J_j appear in the solution produced by algorithm s.

Algorithm MF is easily adapted to deal with given precedence constraints $<$. Since we may assume that $r_i < r_j$ and $q_i > q_j$ if $J_i < J_j$, they are respected by algorithm s. Obviously, the lower bound remains a valid one. With respect to the branching rule, descendant nodes have to be created only for jobs J_j with $j \in P_i$, $q_j < q_i - L'_{max} + \text{UB}$, $A_j \cap P_i = \emptyset$. We could branch by adding the precedence constraints $\{J_k < J_j \mid k \in P_i - \{j\}\}$; many heads and tails would then have to be adjusted. If, however, we drop the requirement of disjoint descendants and aim to preserve only the original precedence constraints, we may just as well restrict ourselves to adjust r_j in the way described above and update r_k for all $k \in A_j$. Since the tails still reflect the original precedence constraints, new solutions produced by algorithm s will respect those. Again, more extensive adjustments turn out to result in additional computing time.

13.5 Computational Experience

Algorithms BS and MF were coded in ALGOL 60 and run on the Control Data Cyber 73-28 of the SARA Computing Centre in Amsterdam.

For each test problem with n jobs, $3n$ integer data r_i, p_i, q_i were generated from uniform distributions between 1 and r_{max}, p_{max}, and q_{max} respectively. Here, $r_{max} = Rp_{max}$ and $q_{max} = Qp_{max}$. In the precedence graph, each arc (J_i, J_j) with $i < j$ was included with probability P. Table 13.2 shows the values of (n, p_{max}, R, Q, P) during our experiments; the values used in previously reported tests are also given. For each combination of values with $R \geq Q$ five problems were generated; inversion of these problems provided test problems with $R \leq Q$ (cf. Section 13.2). Significant and systematic differences between the solution times of a problem and its inverse would indicate advantages to be gained from problem inversion.

Tables 13.3 and 13.4 show the computational results for problems without precedence constraints, i.e. with $P = 0$. Algorithm BS solves 294 out of 300 problems with up to 80 jobs within the time limit of ten seconds. The limit is never exceeded for problems of the type on which the method has been tested previously (Baker and Su, 1974). Inspection of the results revealed no

Table 13.2 Values of parameters of test problems

Parameter	Baker and Su	McMahon and Florian	This paper
n	10, 20, 30	20, 50	20, 40, 80
p_{max}	2000/n	25	50
R	0.5n	0.5n, 2n	0.5, 2, 0.5n, 2n
Q	0.75n, 0.875n, n†	0.4, 1, 3	0.5, 2, 0.5n, 2n
P	0	0	0, 0.05, 0.15, 0.45

† In this case, the q_i are not distributed uniformly.

Table 13.3 Computational results for $P=0$: a survey†

		Median solution time			Maximum solution time		
n	P	Alg. BS	Alg. MF	Alg. FM	Alg. BS	Alg. MF	Alg. FM
20	0	0.05	0.02	0.03	>10:2	0.99	0.11
40	0	0.09	0.06	0.06	1.09	>10:1	0.17
80	0	0.23	0.16	0.15	>10:4	>10:3	0.57

† See legend to Table 13.5.

Table 13.4 Computational results for $P=0$: the influence of R and Q†

$n=80$	Maximum solution time											
$P=0$	Algorithm BS				Algorithm MF				Algorithm FM			
$R\downarrow Q\rightarrow$	0.5	2	0.5n	2n	0.5	2	0.5n	2n	0.5	2	0.5n	2n
0.5	0.26	0.25	5.54	5.75	0.19	0.21	1.64	>10:1	0.19	0.25	0.15	0.14
	0.25				0.19				0.25			
2		0.25	>10:1	4.84	0.18	>10:1	>10:1			0.22	0.19	0.16
	0.24	0.27			0.19	0.17			0.20	0.25		
0.5n			3.43	3.67			0.33	0.47			0.57	0.17
	>10:1	>10:2	3.60		0.10	0.12	0.52		0.08	0.11	0.49	
2n				2.54				0.11				0.17
	0.10	0.11	2.51	2.55	0.09	0.07	0.13	0.13	0.09	0.08	0.12	0.19

† See legend to Table 13.5.

obvious rule according to which problem inversion might take place and hence this additional feature was not incorporated into algorithm BS. Even better results were obtained with algorithm MF. It turns out that this method has been tested in McMahon and Florian (1975) on the very easiest types of problems. In general, algorithm MF performs especially well on problems with $R>Q$. Accordingly, we also tested algorithm FM, which inverts a problem if $\max_i \{r_i\} - \min_i \{r_i\} < \max_i \{q_i\} - \min_i \{q_i\}$ before applying algorithm MF. The remarkable quality of algorithm FM is clear from Tables 13.3 and 13.4.

Table 13.5 shows the effect of precedence constraints, which was investigated only with respect to algorithms MF and FM. For problems with $P \geq 0.15$, most of the solution time is spent on adjusting the r_i and q_i in accordance with the precedence constraints, as described in Section 13.2; this takes 0.06 seconds for $n=20$, $P=0.15$ and 0.70 for $n=80$, $P=0.45$. For each positive value of P which we tested, the median number of generated nodes is equal to one; for $P=0.45$ branching never occurs. Inversion according to the rule given above leads to some improvement, albeit not so spectacular as in the case without precedence constraints.

Table 13.5 Computational results: the influence of P

n	P	Median solution time		Maximum solution time	
		Algorithm MF	Algorithm FM	Algorithm MF	Algorithm FM
20	0	0.02	0.03	0.99	0.11
	0.05	0.06	0.05	0.41	0.43
	0.15	0.07	0.07	0.14	0.15
	0.45	0.07	0.08	0.12	0.11
80	0	0.16	0.15	>10:3	0.57
	0.05	0.36	0.33	>10:6	>10:4
	0.15	0.47	0.42	0.85	0.57
	0.45	0.73	0.75	0.81	0.80

Each entry in Table 13.3 (Table 13.4, 13.5) represents 100 (5, 100) test problems. In Table 13.4 the entries below the staircases represent the results for problems with $R \geq Q$; the remaining entries represent the inverted problems with $R \leq Q$.
Solution times: CPU seconds on a Control Data Cyber 73-28.
$>l:k$: the time limit l is exceeded k times.
Algorithm BS: see Section 13.3.
Algorithm MF: see Section 13.4.
Algorithm FM: algorithm MF with problem inversion if $\max_i \{r_i\} - \min_i \{r_i\} < \max_i \{q_i\} - \min_i \{q_i\}$.
n: number of jobs.
R: relative range of r_i.
Q: relative range of q_i.
P: expected density of precedence graph.

Summing up our computational experience we conclude that, NP-complete though the problem may be, algorithms BS, MF, and FM (especially the last) are able to solve problems of reasonable size fairly quickly. In view of the applications which are to be discussed in the following sections, this is a hopeful result.

13.6 A Theoretical Application

The one-machine problem can find application in the theoretical context of the *general job-shop scheduling problem*. This classical combinatorial problem can be formulated as follows (Conway, Maxwell, and Miller, 1967; Rinnooy Kan, 1976).

Suppose we have n jobs $J_1, \ldots, J_n$ and m machines $M_1, \ldots, M_m$ which can handle at most one job at a time. Job $J_i (i = 1, \ldots, n)$ consists of a sequence of n_i operations O_r $(r = \sum_{j=1}^{i-1} n_j + 1, \ldots, \sum_{j=1}^{i} n_j)$ each of which corresponds to the processing of job J_i on machine $\mu(O_r)$ during an uninterrupted processing time of p_r time units. We seek to find a processing order on each machine such that the maximum completion time is minimized.

Minimizing Maximum Lateness on one Machine

The above problem is conveniently represented by means of a *disjunctive graph* $G = (\mathcal{V}, \mathcal{C} \cup \mathcal{D})$ (Roy and Sussmann, 1964) where $\mathcal{V}$ is the set of vertices, representing the operations, including fictitious initial and final operations:

$$\mathcal{V} = \left\{0, 1, \ldots, \sum_{j=1}^{n} n_j, *\right\};$$

$\mathcal{C}$ is the set of directed *conjunctive arcs*, representing the given machine orders of the jobs:

$$\mathcal{C} = \left\{\left(0, \sum_{j=1}^{i-1} n_j + 1\right) \,\bigg|\, i = 1, \ldots, n\right\}$$

$$\cup \left\{(r, r+1) \,\bigg|\, r = \sum_{j=1}^{i-1} n_j + 1, \ldots, \sum_{j=1}^{i} n_j - 1; i = 1, \ldots, n\right\}$$

$$\cup \left\{\left(\sum_{j=1}^{i} n_j, *\right) \,\bigg|\, i = 1, \ldots, n\right\};$$

$\mathcal{D}$ is the set of directed *disjunctive arcs*, representing the possible processing orders on the machines:

$$\mathcal{D} = \{(r, s) \mid \mu(O_r) = \mu(O_s)\};$$

a *weight* p_r is attached to each vertex r, with $p_0 = p_* = 0$.

The disjunctive graph for a problem with $n = m = 3$ is drawn in Figure 13.2. Processing orders on the machines can be defined by a subset $D \subset \mathcal{D}$ such that $(r, s) \in D$ if and only if $(s, r) \in \mathcal{D} - D$. D contains the *chosen* or *settled arcs*: if $(r, s) \in D$, then O_r precedes O_s on their common machine. The resulting schedule is *feasible* if the graph $G(D) = (\mathcal{V}, \mathcal{C} \cup D)$ contains no directed cycles. The value of such a schedule is given by the weight of the

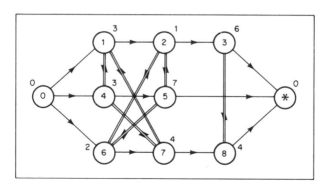

Figure 13.2 Disjunctive graph $G = (\mathcal{V}, \mathcal{C} \cup \mathcal{D})$ for the example

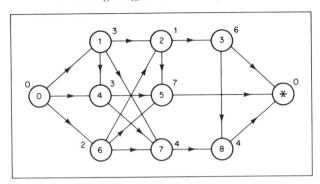

Figure 13.3 Directed graph $G(D) = (\mathcal{V}, \mathcal{C} \cup D)$ for the example

maximum-weight path (also called 'longest' or 'critical' path) in $G(D)$. As to our example, the graph $G(D)$ corresponding to processing orders (O_1, O_4, O_7) on M_1, (O_6, O_2, O_5) on M_2 and (O_3, O_8) on M_3 is drawn in Figure 13.3; the value of the schedule is equal to 14.

The general job-shop scheduling problem has been proved to be NP-complete (Lenstra, Rinnooy Kan, Brucker, 1977) and several branch-and-bound algorithms have been developed in the past. In these algorithms, any node of the search tree corresponds to a partial solution which is characterized by a subset $D \subset \mathcal{D}$ of chosen arcs such that $G(D)$ is acyclic.

Let us select an arbitrary M_k. For each O_r with $\mu(O_r) = M_k$ we can determine

a *head*, i.e. the maximum weight of a path in $G(D)$ from 0 to r;
a *body*, i.e. the processing time p_r;
a *tail*, i.e. the maximum weight of a path in $G(D)$ from r to $*$ minus p_r.

Furthermore, for O_r and O_s with $\mu(O_r) = \mu(O_s) = M_k$ we have a *precedence constraint* $O_r < O_s$ if $G(D)$ contains a path from r to s.

It follows that on each machine we can define a one-machine problem of the type discussed in this paper and, since the arcs in $\mathcal{D} - D$ are neglected, that the maximum of the m optimal solution values yields a lower bound on all extensions of D.

We conclude with some remarks on the combination of this lower bound and the two branching schemes that have been applied to the job-shop scheduling problem. We refer to Lageweg, Lenstra, and Rinnooy Kan (1977) for a more extensive discussion.

One branching scheme generates all active schedules (Giffler and Thompson, 1960) just as algorithm AS generates all active schedules for the one-machine problem (see Section 13.3). The use of the above lower bound within such a scheme has led to the best job-shop scheduling algorithms developed so far (Bratley, Florian, and Robillard, 1973; McMahon and

Minimizing Maximum Lateness on one Machine 383

Florian, 1975), although also non-active schedules may be generated by the branching scheme that is actually used (Florian, Trépant, and McMahon, 1971) and some precedence constraints may be neglected in the one-machine problems.

In the second scheme branching takes place by adding either (r, s) or (s, r) to D (Charlton and Death, 1970a, 1970b). The use of the above bound within this type of scheme requires a one-machine algorithm that is able to handle precedence constraints. The latter branching scheme seems very flexible in the sense that early branching decisions may settle essential conflicts within the problem, but computational experiments as reported by Lageweg, Lenstra, and Rinnooy Kan (1977) have revealed it to be inferior to the former one.

13.7 A Practical Application

A practical scheduling situation in which the one-machine problem occurs arose in the context of the production of aluminium aeroplane parts. In a certain section of the factory in question, the production is centred around a rubber press. The metal pieces are first processed either by a cutting or by a milling machine. They next have to pass a fitting shop and subsequently have to spend a full working day in an annealing furnace before being pressed into their proper shape by the rubber press. After passing the fitting shop for a second time they are completely finished. The processing time of each operation is known in advance.

There are nine operators available to process the jobs. One of them operates the cutting and milling machines, six are working in the fitting shop, and two handle the rubber press; the annealing furnace requires no attention and can be assumed to have an infinite capacity, i.e. it can handle any number of jobs at the same time.

Since the rubber press is a relatively costly machine, the objective was to choose processing orders in such a way that the total completion time is minimized while idle time on the rubber press is avoided as much as possible.

If we denote the operations of job J_i by O_{ir} with processing times p_{ir} ($r = 1, \ldots, 5$), typical data for a week's production look like those presented in the left-hand part of Table 13.6. Note that some jobs, which are left over from last week, have completed some of their initial operations.

We can model the above situation as a job-shop with four machines:

M_1 represents the cutting and milling machines and has capacity 1;
M_2 represents the fitting shop and has capacity 6;
M_3 represents the annealing furnace and has capacity ∞;
M_4 represents the rubber press and has capacity 1.

Each job has the same machine order $(M_1, M_2, M_3, M_4, M_2)$.

Table 13.6 A practical scheduling problem: data and results

i	p_{i1}	p_{i2}	p_{i3}	p_{i4}	p_{i5}	$\dfrac{p_{i1}}{p_{i4}+p_{i5}}$	C_{i1}	C_{i2}	C_{i3}	C_{i4}	C_{i5}
1	4	4	8	2	6	0.50	32	36	48	50	56
2	2	2	8	1	4	0.40	22	24	32	33	38
3	1	1	8	1	2	0.33	6	7	16	27.5	32
4	—	2	8	0.5	6	—	0	2	16	22	32
5	6	2	8	2	7	0.67	50	52	64	66	73
6	1	1	8	1	3	0.25	4	5	16	24	33
7	1	1	8	0.5	4	0.22	2	3	16	22.5	31
8	—	5	8	0.5	7	—	0	5	16	18.5	27
9	1	1	8	0.5	4	0.22	3	4	16	23	34
10	—	—	8	2	6	—	—	0	8	14	20
11	1	1	8	0.5	2	0.40	23	24	32	33.5	36
12	1	1	8	0.5	2	0.40	24	25	40	45	47
13	1	1	8	1	2	0.33	7	8	16	28.5	34
14	2	1	8	1	4	0.40	26	27	40	43	47
15	—	5	8	1	9	—	0	5	16	17	26
16	7	10	8	2	16	0.39	20	30	40	42	58
17	—	—	—	1.5	6	—	—	—	0	6.5	12.5
18	1	1	8	1.5	6	0.13	1	2	16	20	30
19	—	—	8	2	8	—	—	0	8	10	18
20	6	6	8	3	6	0.67	56	62	72	75	81
21	—	—	—	2.5	12	—	—	—	0	2.5	15
22	—	—	8	2	7	—	—	0	8	12	19.5
23	—	—	—	2.5	11	—	—	—	0	5	16
24	2	2	8	1.5	3	0.44	28	30	40	44.5	47.5
25	—	—	—	—	4	—	—	—	—	0	6
26	6	3	8	2.5	7	0.63	44	47	56	61.5	68.5
27	—	—	8	1	5	—	—	0	8	15	20
28	1	1	8	1	3	0.25	5	6	16	26.5	34
29	—	—	—	1.5	6	—	—	—	0	8	14
30	1	1	8	1	2	0.33	8	9	24	29.5	35
31	1	1	8	1	2	0.33	9	10	24	30.5	36
32	—	3	8	1	7	—	0	3	16	18	25
33	6	6	8	3	7	0.60	38	44	56	59	66
34	—	6	8	1.5	6	—	0	6	16	21.5	31
35	4	2	8	1.5	10	0.35	13	15	24	25.5	41

Approaching the problem in a heuristic way, we note that

$$\sum_{i=1}^{35} p_{i1} = 56, \quad \sum_{i=1}^{35} p_{i2} = 70, \quad \sum_{i=1}^{35} p_{i4} = 48.5, \quad \sum_{i=1}^{35} p_{i5} = 202.$$

Clearly, all jobs cannot be processed on M_1 and M_4 within one week of 40 hours and some overflow will result. It seems quite possible to schedule O_{i2} and O_{i3} directly after the completion of O_{i1}, but some waiting time for the jobs before the processing of O_{i4} and O_{i5} seems unavoidable. It is expedient to schedule O_{i1} in such a way that many jobs are quickly available for further processing, thereby taking p_{i4} and p_{i5} into account.

These intuitive considerations led to the following heuristic method, in which C_{ir} stands for the completion time of O_{ir}.

1. Schedule O_{i1} on M_1 according to nondecreasing $p_{i1}/(p_{i4}+p_{i5})$, thereby minimizing the weighted completion time $\sum_{i=1}^{35} (p_{i4}+p_{i5})C_{i1}$ (Smith, 1956).

2. Schedule O_{i2} as early as possible on M_2 according to nondecreasing C_{i1}.

3. Schedule O_{i3} on M_3 according to $C_{i3} := 8\lceil C_{i2}/8 \rceil + 8$ ($\lceil x \rceil$ is the smallest integer not less than x).

4. Schedule O_{i4} on M_4 by solving the one-machine problem as discussed in this paper, defined by heads C_{i3}, bodies p_{i4} and tails p_{i5}.

5. Schedule O_{i5} as early as possible on M_2 according to nondecreasing C_{i4}.

A schedule resulting from application of this heuristic to the problem data is given by the completion times in Table 13.6; the corresponding Gantt-chart is shown in Figure 13.4. The one-machine problem on M_4 was solved by algorithm MF (see Section 13.4); the first application of algorithm s yielded an optimal solution.

The approach described above seems to be more generally applicable. Basically, it involves the determination of *critical machines* in the production process, i.e. the machines that are important from a cost minimizing point of view and on which the processing orders have a crucial influence on the quality of the schedule as a whole. The problem is then decomposed into problems involving one or more of those critical machines; these problems may be solved by methods inspired by sequencing theory. The resulting schedules are concatenated by suitable processing orders on the other machines leading to an overall schedule of reasonable quality.

Our experience with this heuristic approach has been limited to the small example above and our only conclusion would be that it seems to merit further experimentation. We feel that through this approach the models of machine scheduling theory, which may well correspond to an oversimplified

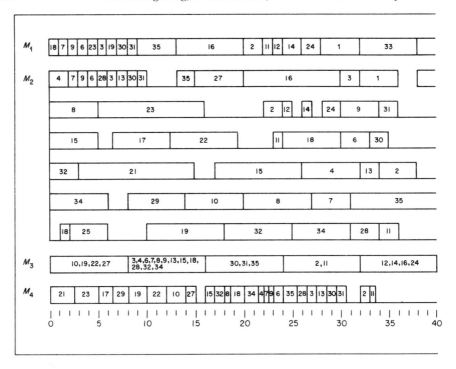

Figure 13.4 A practical scheduling problem: Gantt-chart

picture of reality, may find application in varying situations that do not fit the standard models.

13.8 Concluding Remarks

The computational experience reported in Section 13.5 leads us to conclude that the problem of minimizing maximum lateness on one machine can be satisfactorily solved by the algorithms described in Sections 13.3 and 13.4. If solution by implicit enumeration is indeed unavoidable, there seems to be little room for further improvement.

It might be worth investigating whether the ideas behind algorithms BS and MF could be applied to other machine scheduling problems. An interesting candidate is the problem of minimizing maximum completion time in a two-machine flow-shop with release dates. This problem can be interpreted as a variation on the three-machine model introduced in Section 13.2: a nonbottleneck machine M_1 deals with the release dates and two bottleneck machines M_2 and M_3 constitute the flow-shop. Again, the case in which all r_i

Minimizing Maximum Lateness on one Machine

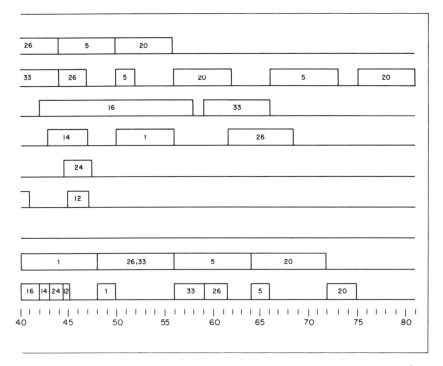

are equal can be solved in $O(n \log n)$ steps (Johnson, 1954), whereas the general problem is *NP*-complete (Lenstra, Rinnooy Kan, and Brucker, 1977). Similar remarks apply to the inverse problem, i.e. minimizing maximum lateness in a two-machine flow-shop.

The problem discussed in this paper finds application in theory and practice, as has been demonstrated in Sections 13.6 and 13.7. Especially the heuristic approach suggested in Section 13.7 deserves further examination. It might be a suitable response to the frequent complaint about the lack of successful practical applications of machine scheduling theory.

Acknowledgements

We are grateful to M. Florian and G. B. McMahon for valuable discussions concerning their algorithm. Also, we would like to thank J. H. Galjaard and J. S. Knipscheer of the Graduate School of Management, Delft, for providing the data for the practical scheduling problem.

References

Baker, K. R., and Z.-S. Su (1974). Sequencing with due-dates and early start times to minimize maximum tardiness. *Naval Res. Logist. Quart.*, **21,** 171–176.

Bratley, P., M. Florian, and P. Robillard (1973). On sequencing with earliest starts and due dates with application to computing bounds for the $(n/m/G/F_{max})$ problem. *Naval Res. Logist. Quart.*, **20,** 57–67.

Charlton, J. M., and C. C. Death (1970a). A method of solution for general machine scheduling problems. *Operations Res.*, **18,** 689–707.

Charlton, J. M., and C. C. Death (1970b). A generalized machine scheduling algorithm. *Operational Res. Quart.*, **21,** 127–134.

Conway. R. W., W. L. Maxwell, and L. W. Miller (1967). *Theory of Scheduling*, Addison-Wesley, Reading, Mass.

Dessouky, M. I., and C. R. Margenthaler (1972). The one-machine sequencing problem with early starts and due dates. *AIIE Trans.*, **4,** 214–222.

Edmonds, J. (1965). Paths, trees, and flowers. *Canad. J. Math.*, **17,** 449–467.

Florian, M., P. Trépant, and G. McMahon (1971). An implicit enumeration algorithm for the machine sequencing problem. *Management Sci.*, **17,** B782–792.

Giffler, B., and G. L. Thompson (1960). Algorithms for solving production scheduling problems. *Operations Res.*, **8,** 487–503.

Horn, W. A. (1974). Some simple scheduling algorithms. *Naval Res. Logist. Quart.*, **21,** 177–185.

Jackson, J. R. (1955). Scheduling a production line to minimize maximum tardiness. *Research Report 43, Management Science Research Project, University of California, Los Angeles.*

Johnson, S. M. (1954). Optimal two- and three-stage production schedules with setup times included. *Naval Res. Logist. Quart.*, **1,** 61–68.

Lageweg, B. J., J. K. Lenstra, and A. H. G. Rinnooy Kan 1977. Job-shop scheduling by implicit enumeration. *Management Sci.*, **24,** 441–450.

Lawler, E. L. (1973). Optimal sequencing of a single machine subject to precedence constraints. *Management Sci.*, **19,** 544–546.

Lenstra, J. K. (1977). *Sequencing by Enumerative Methods*, Mathematical Centre Tracts 69, Mathematisch Centrum, Amsterdam.

Lenstra, J. K., A. H. G. Rinnooy Kan, and P. Brucker (1977). Complexity of machine scheduling problems. *Ann. Discrete Math.*, **1,** 343–362.

McMahon, G., and M. Florian (1975). On scheduling with ready times and due dates to minimize maximum lateness. *Operations Res.*, **23,** 475–482.

Rinnooy Kan, A. H. G. (1976). *Machine Scheduling Problems: Classification, Complexity and Computations*, Nijhoff, The Hague.

Roy, B., and B. Sussmann (1964). Les problèmes d'ordonnancement avec contraintes disjonctives. *Note DS no. 9 bis, SEMA, Montrouge.*

Schrage, L. (1971). Obtaining optimal solutions to resource constrained network scheduling problems. Unpublished manuscript.

Smith, W. E. (1956). Various optimizers for single-stage production. *Naval Res. Logist. Quart.*, **3,** 59–66.

CHAPTER 14

The Crew Scheduling Problem: A Travelling Salesman Approach

FRANCO GIANNESSI
Department of Mathematics, University of Pisa, Italy
and
BERNARDO NICOLETTI
Operations Research Center, Alitalia, Rome, Italy

14.1 The Crew Scheduling Problem

'Crew rostering' (determining turns of duty for a crew) can be seen essentially as the assignment of the flight and training activities scheduled for the following months to each crew member. The assignment must be done within the constraints given by the union regulations in such a way as to reach the objective of an equal distribution of the workload and of a satisfactory utilization of the personnel.

The problem becomes complicated because of the number of people and activities involved (in the case of Alitalia Airlines, there are roughly 50 000 activities to be assigned to 3000 crew members, each month). In addition, the union regulations are rather complicated. For this reason the problem is usually split in two phases:

1. Construction of 'rotations', that is of flight sequences beginning and ending in the crews' 'home base', having a duration of a certain number of days and satisfying all the union regulations.

2. Construction of personal 'rosters', by assigning the rotations, prepared in advance, to each crew member. Of course, if there are preset activities, such as training courses, etc., they should not be changed.

This Chapter is concerned essentially with the first problem, from now on designated as the 'crew scheduling problem' (CSP). As far as the second phase is concerned, many studies are available in the literature (Nicoletti, 1975).

A rotation is itself composed of sequences of flights called 'duty periods', each of which can be thought of as a day's work, interspersed by 'layovers', which are long rest periods, intended for sleeping. Thus, the group ⟨duty period, layover, duty period⟩ might form a rotation.

Of course, various union rules as well as government regulations govern the solution of rotations. These generally involve assurance of sufficient rest, sufficient time at the home base, and flight time limitations.

Almost all flights repeat weekly, and for many airlines most repeat daily (or nearly daily). In order to take advantage of this periodicity, the problem for medium or large airlines is usually first formulated on a daily basis—i.e. assuming complete daily periodicity and finding the minimum cost set of rotations covering each flight. All flights are then covered by repeating each rotation on each day. A separate phase then deals with weekly frequency exceptions, re-pairing the flights not actually covered by the idealized daily schedule, and introducing a number of costly 'dead-heads'† in order to produce a weekly schedule.

14.2 The Need for Problem Modelling

Airline crews are expensive. The crew cost is about 10% of total costs in Alitalia, and is the largest single cost item after fuel and oil. This means that an improvement of 1% could represent something like $600 000 per year.

The scheduling rules, which are partly imposed by legislation and partly negotiated between individual carriers and the crew unions are complex and this gives rise to high costs in several ways:

(a) the rules may lead to poor utilization;

(b) a lot of staff are needed to produce and control the roster;

(c) standby crews must be provided to cover for illegalities caused by delays;

(d) standby crews may need to be provided to cover illegalities in the published roster;

(e) in North America, the rules stipulate levels of payment, e.g. one hour flight pay for every two hours on duty, even if this level of utilization cannot be achieved (this is the so-called 'multiplier rule').

Thus, the crew scheduling problem is complex and therefore time-consuming and error-prone and determines the crew utilization the airline will achieve. Moreover, as airline planning departments equip themselves with computer models and timetable editors, they are able to change the timetable more and more frequently. This imposes extra pressure on the

† Every scheduled flight must have a crew assigned. If a flight is doubly covered, one crew will fly, while the other crew will be carried as passengers.

crew scheduling departments which are required to report back rapidly on crew cost implications every time the timetable is changed. This process also tends to go on longer so that when the timetable is finalized the crew schedulers have only a short time available before the rosters must be published.

These pressures have led most large airlines to investigate the possibilities of using computer-based techniques to aid in the crew scheduling process.

The approaches followed by other authors have been essentially (1) heuristic, (2) mathematical (Arabeyre et al., 1969; Cavanagh, 1975).

Purely heuristic approaches to the problem were common in the earlier years. Various rules of thumb were coded to attempt to construct 'good' pairings. However, the lack of methodology for systematic improvement often produced schedules inferior to manually produced ones.

There have been several formal mathematical formulations. The currently dominant one is a set covering (with deadheading), or a set partitioning (without deadheading), problem. If one generates some selection of legal rotations, and lets

$x_j = 1$, if rotation j is in the solution, 0 if not;
$a_{ij} = 1$, if flight leg i is covered by rotation j, 0 if not;
c_j = penalty cost of rotation j;

then the nondeadhead problem can be formulated as

$$\min \sum_j c_j x_j$$

$$\text{subject to } \sum_j a_{ij} x_j = 1 \quad \forall i.$$

Deadhead can be permitted by relaxing the preceding constraints to

$$\sum a_{ij} x_j \geq 1 \quad \forall i$$

It should be pointed out that one area of imprecision is in the statement of work rules. Some of the parameters involved in pairing generation, such as the minimum time for a crew to change planes, are often flexible to some extent, and an in-context decision by a scheduler as to whether to permit relaxation of a rule must be contrasted with inflexibility of use of the parameter by a computer. In practice, none of these difficulties has proved insuperable, although an additional run with relaxed parameters, or manual adjustment, is often necessary due to the imprecision of the model.

There are several solution methods available for the set covering problem. Many of them have been tried on the problem under study (Arabeyre et al., 1969; Cavanagh, 1975). The difficulty of this approach lies in the fact that for any of the larger problems, involving a few hundreds or more daily flight legs, the number of legal rotations is astronomical, so that any practical selection of rotations is no more than a tiny percentage of the total.

For larger problems, the approach which seems to have been most successful is to search systematically for improvements of a given solution by concentrating on a relatively small section of the problem at a time. For example, if one chooses three rotations from a current valid solution covering twelve flights, say, one can set up a set partitioning problem in which the rows are the twelve flights and the columns are all those rotations covering only those twelve flights.

If a pairing generator is incorporated into the technique, all rotations for the twelve flights can be generated on the spot, thus avoiding the task of pregenerating an enormous matrix. If one has deliberately limited the pairing generation process in order to save time, then the three original rotations will of course appear in the matrix, but will not necessarily constitute the optimal arrangement of the twelve flights.

Solving the subproblem thus constructed may lead to an improved set of rotations covering the same twelve flights, which would then replace the original three rotations in the overall current solution.

There are a number of variations in the implemented versions of this technique.

The greatest source of variation is in the method of selecting subproblems. One approach is based on trying all possible combinations of n rotations at a time from the current solution, where $n = 1, 2, 3, \ldots$. For larger n, not only the number of subproblems but also the subproblem size increases, in which case a randomly selected sample of the above rotations is used.

14.3 A Travelling Salesman Approach

In the previous sections it has been pointed out that one of the most important aspects of a mathematical formulation of the crew scheduling problem consists of the possibility of handling side constraints (e.g. union regulations) without enlarging the model too much.

Now it will be shown how it is possible to obtain such a result by considering a travelling salesman problem (TSP) formulation of the crew scheduling problem. (See also Chapter 6 of this book)

To this end consider a directed graph, G, having $n+1$ vertices N_0, $N_1, \ldots, N_n$; N_0 denotes the crew home base (for the sake of simplicity, assumed to be only one), while the remaining vertices denote the flights to be executed; N_0 may be regarded also as a flight, namely a dummy one, which means rest. Arc (i, j) means that flight j can be executed after flight i; the corresponding distance c_{ij} is the lost time (or any cost function) associated with arc (i, j). G may be considered to be complete by defining $c_{ij} = +\infty$ when arc (i, j) does not exist (i.e. when flight j cannot be executed after flight i).

The Crew Scheduling Problem: A Travelling Salesman Approach

The pure (without side constraints) problem can be formulated as a TSP.

The CSP consists of finding a set of tours, each one containing N_0 and a subset of the remaining vertices, in such a way that every vertex be contained in at least (or exactly) one tour, and the total distance be a minimum.

It is easily realized that such a problem is a multiple travelling salesman problem (m-TSP): a salesman is now a crew, the cities to be visited are the flights to be executed, N_0 is again a dummy flight which means rest, and so on.

The m-TSP can be formulated as a TSP (Svestka and Huckfeldt, 1973; Gavish, 1976; Svestka, 1976) and therefore, as a consequence, the CSP can be formulated in terms of a TSP.

Assume that there are m available salesmen, with $m \geqslant 1$: now a tour is replaced by an m-tour, namely a sequence of $m+n+1$ integers, taken from $\{0, 1, \ldots, n\}$, in which 0 appears as first and last elements and $m-1$ further times, and any other integer appears exactly once. The problem consists of finding a tour with the minimum total distance. It will be shown how to formulate such a problem as a sequence of linear assignment problems.

Consider a complete directed graph G_m, having $n+m$ vertices

$$N_{01} \ldots, N_{0m}, N_1, \ldots, N_n.$$

Thus, n vertices of G_m are also vertices of G, m nodes of G_m replace vertex N_0 of G, and are in a certain sense a dummy 'partition' of it. To every arc (i, j) of G_m a nonnegative real number (distance) is associated, call it d_{ij}, $i, j = 1, \ldots, m+n$; call $D = (d_{ij})$ the corresponding distance matrix. Set

$$d_{ij} = \begin{cases} M & \text{if } 1 \leqslant i, j \leqslant m \\ c_{0,j-m} & \text{if } 1 \leqslant i \leqslant m; \ m+1 \leqslant j \leqslant n+m \\ c_{i-m,0} & \text{if } m+1 \leqslant i \leqslant n+m; \ 1 \leqslant j \leqslant m \\ c_{i-m,j-m} & \text{if } m+1 \leqslant i, j \leqslant n+m. \end{cases}$$

In other words, the matrix C is augmented with $m-1$ rows and columns, where each new row and column is a duplicate of the first row and column of C. The distance between two dummy vertices is assumed to be a large positive number M; in fact, all these vertices represent N_0, and no arc is desired in a feasible solution of the m-TSP.

Now it will be shown how it is possible to define a decomposition procedure for the TSP—and hence for the pure CSP—and as a consequence we will obtain a way of treating side constraints and answering some of the questions which have been posed in Section 14.2.

14.4 Travelling Salesman Tours as Penalized Assignments

When a TSP is formulated as a linear assignment problem, it is well known that to an optimal solution of the latter does not necessarily correspond a

Hamiltonian circuit of G. It can be shown that, either by suitably modifying the objective function, or by setting to zero some variables, a Hamiltonian circuit can be obtained as the optimal solution to the assignment problem.

Consider the real variables μ_i, $i = 1, \ldots, n$, associated with the vertices $N_1, \ldots, N_n$, and binary variables $\alpha_{ij} \in B \triangleq \{0, 1\}$, $1 \leq i \neq j \leq n$, associated with the arcs of G^*; set

$$\mu = (\mu_1, \ldots, \mu_n) \qquad \alpha = (\alpha_{12}, \ldots, \alpha_{1n}, \ldots, \alpha_{n,n-1})$$

and consider the set

$$\mu(\alpha) \triangleq \mu \in \left\{ R^n : \frac{n-1-(\mu_i - \mu_j)}{n} \geq 0, \ 1 \leq i \neq j \leq n; \text{ and} \right.$$

$$\left. \frac{n-1-(\mu_i - \mu_j)}{n} \geq 1, \text{ if } \alpha_{ij} = 1 \right\}.$$

Another $n(n-1)$-uple of binary values

$$\beta = (\beta_{12}, \ldots, \beta_{1n}, \ldots, \beta_{n,n-1}) \qquad \beta_{ij} \in B$$

will be considered. Let M be a large positive real, and consider the penalized distances†

$$c_{ij}(\alpha_{ij}) \triangleq \begin{cases} c_{ij} + (1-\alpha_{ij})M & \text{if } 1 \leq i \neq j \leq n \\ M & \text{if } i = j = 0, 1, \ldots, n \\ c_{ij} & \text{otherwise} \end{cases}$$

Finally the binary vector

$$b = (b_0 = 1, b_1, \ldots, b_n), \qquad b_i \in B$$

is introduced.

Now consider the problem

$$\min \left\{ z(x, \alpha) = \sum_{i,j=0}^{n} c_{ij}(\alpha_{ij}) x_{ij} \right\} \tag{14.1a}$$

$$\sum_{j=0}^{n} x_{ij} = b_i, \qquad i = 0, 1, \ldots, n \tag{14.1b}$$

$$\sum_{i=0}^{n} x_{ij} = b_j, \qquad j = 0, 1, \ldots, n \tag{14.1c}$$

$$x_{ij} \geq 0, \qquad i, j = 0, 1, \ldots, n \tag{14.1d}$$

$$x_{ij} \leq \beta_{ij}, \qquad i, j = 1, \ldots, n \tag{14.1e}$$

† It is assumed that G has no loops, otherwise trivial variants are necessary.

The Crew Scheduling Problem: A Travelling Salesman Approach 395

and call it $\mathcal{P}(\alpha, \beta, b)$; call the corresponding minimum and an optimal solution $Z(\alpha, \beta, b)$ and $x(\alpha, \beta, b)$, respectively. Moreover, denote by $R(\beta, b)$ the feasible region of (14.1), i.e. the polyhedron defined by (14.1b)–(14.1e).

We remark that α and β play the same role: $\alpha_{ij} = 0$ penalizes arc (i, j), so that one expects $x_{ij} = 0$ in an optimal solution of (14.1); the same fact is obtained from $\beta_{ij} = 0$. Thus, such two kinds of penalties will be used alternatively; their introduction is motivated by the need to obtain different results. Thus, if e_h denotes a vector of h unit components, the general problem (14.1) will be considered alternatively in the particular cases as:

$$\mathcal{P}(\alpha, e_{n(n-1)}, b) \text{ and } \mathcal{P}(e_{n(n-1)}, \beta, b).$$

We also note that $b_j = 0$ implies $x_{ij} = x_{ji} = 0$, $i = 0, 1, \ldots, n$, in a feasible solution of (14.1). This means that every path through N_j is considered infeasible, i.e. N_j is cancelled, and the TSP is considered on the remaining graph. Thus

$$h = \sum_{j=1}^{n} b_j \quad (h \geq 1)$$

is the number of vertices of G which it is legal to visit, so that a constraint of the form $\sum_{j=1}^{n} b_j \geq h$ imposes the need to visit at least h vertices in order to have a feasible circuit. When $h = n$ all the vertices of G must be visited, and the problem becomes the classical TSP.

First we will study problem† $\mathcal{P}(\alpha, e_{n(n-1)}, e_n)$ with the aim of characterizing a Hamiltonian circuit.

Remark that an optimal solution of $\mathcal{P}(\alpha, 1, 1)$ has $n+1$ elements equal to 1, and $n-1$ of them are x_{ij} having $1 \leq i \neq j \leq n$. If $\alpha \in B^{n(n-1)}$ has been chosen having more than $(n-1)^2$ elements equal to zero, then for at least an arc (i, j), with $1 \leq i \neq j \leq n$, we have $\alpha_{ij} = 0$ and in an optimal solution of (14.1a) is $x_{ij} = 1$, so that the minimum in $\mathcal{P}(\alpha, 1, 1)$ is $\geq M$, and it does not correspond to a total distance of a path of G. For this reason we assume that α satisfies the constraint (assuming that $n > 2$)

$$\sum_{1 \leq i \neq j \leq n} \alpha_{ij} = n - 1. \tag{14.2}$$

Even if (14.2) is satisfied, there is not necessarily a feasible solution to problem $\mathcal{P}(\alpha, 1, 1)$ with total distance less than M. In fact, if there are $n-1$ of the α_{ij} equal to 1, and if they have the same row index or the same column index, then there is no feasible solution of $\mathcal{P}(\alpha, 1, 1)$ with total distance $<M$. For this reason we are led to consider the set of all $\alpha \in B^{n(n-1)}$

---

† For the sake of simplicity, and without fear of confusion, $e_{n(n-1)}$ and e_n will be replaced symbolically by 1, when considered as arguments of $\mathcal{P}(*, *, *)$, $R(*, *)$, $Z(*, *, *)$, $x(*, *, *)$.

which satisfy (14.2) and which correspond to intersections of different rows and different columns. Let this set be H_1:

$$H_1 = \{\alpha \in B^{n(n-1)}: \alpha_{i_1 j_1} = \ldots = \alpha_{i_{n-1} j_{n-1}} = 1; j_s \neq i_1, \ldots, i_s; j_s = i_{s+1};$$
$$s = 1, \ldots, n-1; \alpha_{ij} = 0, \text{ otherwise}\}.$$

It is easy to show that $\mathcal{P}(\alpha, 1, 1)$ has feasible solutions with total distance less than M, iff $\alpha \in H_1$.

Unfortunately, the set H_1 does not characterize the set of problems $\mathcal{P}(\alpha, 1, 1)$ which have optimal solutions corresponding to Hamiltonian circuits. To obtain this we consider the set of all $\alpha \in B^{n(n-1)}$ and such that

$$\mu(\alpha) \neq \emptyset; \tag{14.3}$$

Theorem 14.1 Let $\alpha \in H_1$. *Every Hamiltonian circuit of G corresponds to an optimal solution of $\mathcal{P}(\alpha, 1, 1)$, iff $\alpha \in H_2$.*

Proof. Sufficiency. Let x^* be an optimal solution of $\mathcal{P}(\alpha, 1, 1)$ which always exists. (14.1b)–(14.1d) ensure that, for every vertex, there is only one arriving arc; (14.1c)–(14.1d) that there is only one departing arc. If follows that the set of arcs corresponding to the nonzero elements of x^* form a circuit, which is not necessarily Hamiltonian, i.e. it may be composed of subcircuits which do not contain N_0. Assume that there are at least two such subcircuits, so that there exists a circuit defined by $k < n$ vertices different from N_0. Call $\mathcal{I}$ the set of arcs (i, j) of indices of the k variables corresponding to such a circuit. We have

$$x^*_{ij} = 1 \quad (i, j) \in \mathcal{I}. \tag{14.4}$$

Since $\mathcal{P}(\alpha, 1, 1)$ has feasible solutions with total distance less than M iff $\alpha \in H_1$, the total distance corresponding to x^* is less than M, so that

$$\alpha_{ij} = 1 \quad \forall (i, j) \in \mathcal{I}.$$

These equalities and (14.3) imply the existence of μ, such that

$$\frac{n - 1 - (\mu_i - \mu_j)}{n} \geq 1 \quad \forall (i, j) \in \mathcal{I}.$$

The addition of these inequalities leads to the absurd inequality

$$k \frac{n-1}{n} \geq k.$$

Necessity. Let x^* denote an optimal solution of $\mathcal{P}(\alpha, 1, 1)$, which always exists. It is not restrictive to assume that the Hamiltonian circuit corresponding to x^* be $(N_0, N_1, N_2, \ldots, N_n, N_0)$. Set $\mu_s = s$, $s = 1, \ldots, n$. It follows that

$$\frac{n - 1 - (\mu_i - \mu_j)}{n} \geq 0 \quad \forall (i, j) \text{ such that } 1 \leq i \neq j \leq n$$

and moreover that

$$\frac{n-1-(\mu_i-\mu_j)}{n} = \frac{n-1-[s-(s+1)]}{n} = 1 \quad \forall(i,j) \text{ such that } x_{ij}^* = 1.$$

If follows that (14.3) is satisfied and this completes the proof.

The preceding theorem says that, by solving $\mathcal{P}(\alpha, 1, 1)$ for every α, i.e. by solving the problem

$$\min_{\alpha \in H} [\min_{x \in R(1,1)} z(x, \alpha)] \quad H \triangleq H_1 \cap H_2$$

one obtains the minimum and an optimal solution of the TSP at $b = e_n$.

It is easy to show that Theorem 14.1 has a corresponding proposition, which holds for $\mathcal{P}(1, \beta, 1)$:

Theorem 14.2 *Let $\beta \in H_1$. Every Hamiltonian circuit of G corresponds to an optimal solution of $\mathcal{P}(1, \beta, 1)$, iff $\beta \in H_2$.*

The proof of Theorem 14.2 is quite similar to that of Theorem 14.1, and is omitted.

Theorem 14.2 says that, by solving $\mathcal{P}(1, \beta, 1)$ for every β, i.e. by solving the problem

$$\min_{\beta \in H} \min_{x \in R(\beta, 1)} z(x, 1)$$

one obtains the minimum and an optimal solution of TSP at $b = e_n$.

Consider now the problems $\mathcal{P}(\alpha, 1, b)$ and $\mathcal{P}(1, \beta, b)$. Note that $\alpha \in H_2$ implies that a circuit must contain N_0, but does not imply that it contains all the vertices of G; this is implied by the equality $h = n$. The same fact is true for $\beta \in H_2$. If one considers less than n vertices of G, i.e. if $h < n$, then a Hamiltonian circuit through h vertices is characterized again by $\alpha \in H_1 \cap H_2$, where of course H_1 must be defined taking into account that now only h rows and columns have $b_j = 1$, i.e. now one must set

$$H_1(h) \triangleq \{\alpha \in B^{n(n-1)} : \alpha_{i_1 j_1} = \ldots = \alpha_{i_{h-1} j_{h-1}} = 1;$$

$$j_s \neq i_1, \ldots, i_s; j_s = i_{s+1}; s = 1, \ldots, h-1; \alpha_{ij} = 0, \text{ otherwise}\}.$$

Obviously the equality $H_1(n) = H_1$ holds. The preceding remark, together with Theorems 14.1 and 14.2, imply the following:

Theorem 14.3 *Let $\alpha \in H_1(h)$. Every Hamiltonian circuit of G corresponds to an optimal solution of $\mathcal{P}(\alpha, 1, b)$, iff $\alpha \in H_2$.*

Theorem 14.4 *Let $\beta \in H_1(h)$. Every Hamiltonian circuit of G corresponds to an optimal solution of $\mathcal{P}(1, \beta, b)$, iff $\beta \in H_2$.*

Theorems 14.3 and 14.4 say that, by solving $\mathscr{P}(\alpha, 1, b)$ for every α, i.e. by solving the problem

$$\min_{\alpha \in H(h)} [\min_{x \in R(1, b)} z(x, \alpha)] \qquad H(h) \triangleq H_1(h) \cap H_2 \qquad (14.5a)$$

or by solving $\mathscr{P}(1, \beta, b)$ for every β, i.e. by solving the problem

$$\min_{\beta \in H(h)} [\min_{x \in R(\beta, b)} z(x, 1)] \qquad (14.5b)$$

one obtains the minimum and an optimal solution of the TSP, when $h \leq n$ vertices must be visited.

14.5 A Decomposition Procedure for Solving the Travelling Salesman Problem

The general problem $\mathscr{P}(\alpha, \beta, b)$ will now be studied in order to obtain an optimality condition, which will be used for the particular problems $\mathscr{P}(\alpha, 1, b)$ and $\mathscr{P}(1, \beta, b)$: the former will enable us to find a Hamiltonian circuit shorter than a given one, or to state that this is optimal; the latter will let us determine a lower bound of the total distance of a given Hamiltonian circuit.

Denote by K a given subset of B^n (for instance, the set of elements of B^n, whose coordinates have sum equal to h if a circuit must go through exactly h nodes).

Assume that an optimal solution $x(\alpha, \beta, b)$, and the corresponding minimum $Z(\alpha, \beta, b)$, of $\mathscr{P}(\alpha, \beta, b)$ is available at $(\alpha, \beta, b) = (\bar{\alpha}, \bar{\beta}, \bar{b})$; set $\bar{z} = Z(\bar{\alpha}, \bar{\beta}, \bar{b})$.

Denote by A^r, A^c, and A^v the matrices of the coefficients of the unknowns of Equations (14.1b), (14.1c), and (14.1e), respectively; by c and $c(\alpha)$ the row-vectors of the c_{ij} and $c_{ij}(\alpha_{ij})$, respectively; and by I an identity matrix of order $n+1$.

We want to know if $x(\bar{\alpha}, \bar{\beta}, \bar{b})$ is also an optimal solution of the TSP, that is if $\bar{z}$ is the minimum total distance of the circuits of G which pass through h nodes. This happens iff x does not satisfy the system (14.1b)–(14.1e) and

$$\sum_{i,j=0}^{n} c_{ij}(\alpha_{ij}) x_{ij} < \bar{z} \qquad (14.6)$$

i.e. the system

$$\begin{cases} -c(\alpha)x > -\bar{z} \\ \begin{pmatrix} A^r \\ -A^r \\ A^c \\ -A^c \\ -I \\ A^v \end{pmatrix} x \leq \begin{pmatrix} b \\ -b \\ b \\ -b \\ 0 \\ \beta \end{pmatrix} \end{cases} \qquad (14.7)$$

whatever the parameters α, β, and b may be. A well-known alternative theorem (Mangasarian, 1969) (or, equivalently, the dual of (14.1): Benders, 1962; Mangasarian, 1969) states that (14.7) is impossible iff there exist vectors u, v, ω, which satisfy the system

$$\begin{cases} uA^r + vA^c - \omega A^v \leq c(\alpha) \\ (u+v)b - \omega\beta \geq \bar{z} \\ \omega \geq 0 \end{cases} \quad (14.8)$$

We have obtained the following:

Theorem 14.5 *Let $(\bar{u}, \bar{v}, \bar{\omega})$ be any solution of the system*

$$\begin{cases} uA^r + vA^c - \omega A^v \leq c(\bar{\alpha}) \\ (u+v)\bar{b} - \omega\bar{\beta} \geq \bar{z} \\ \omega \geq 0. \end{cases} \quad (14.9)$$

Then, to every $(\alpha, \beta, b) \in H^2 \times K$ and such that

$$\begin{cases} \bar{u}A^r + \bar{v}A^c - \bar{\omega}A^v \leq c(\alpha) \\ (\bar{u}+\bar{v})b - \bar{\omega}\beta \geq \bar{z} \end{cases} \quad (14.10)$$

does not correspond in $\mathcal{P}(\alpha, \beta, b)$ a minimum $Z(\alpha, \beta, b) < \bar{z}$.

This theorem enables us to answer the initial question. In fact, if every $(\alpha, \beta, b) \in H^2 \times K$ satisfies (14.10), then $x(\bar{\alpha}, \bar{\beta}, \bar{b})$ is also an optimal solution of the TSP. If there is any $(\tilde{\alpha}, \tilde{\beta}, \tilde{b}) \in H^2 \times K$ which does not satisfy either (14.10), or the systems like (14.10) (met in the preceding applications of Theorem 14.5), then we can solve $\mathcal{P}(\tilde{\alpha}, \tilde{\beta}, \tilde{b})$ and apply again Theorem 14.5 until every $(\alpha, \beta, b) \in H^2 \times K$ satisfies at least one of the systems of type (14.10). At every application Theorem 14.5 cuts off at least one element of $H^2 \times K$, and as H has a finite number of elements, the described algorithm is finite.

From Theorem 14.5 we obtain the following particular case, whose proof is trivial.

Corollary 14.1 *Let $(\bar{u}, \bar{v})$ be any solution of the system*

$$\begin{cases} uA^r + vA^c \leq c(\bar{\alpha}) \\ (u+v)e_h \geq \bar{z} \end{cases} \quad (14.9a)$$

Then, to every $\alpha \in H$ satisfying

$$\bar{u}A^r + \bar{v}A^c \leq c(\alpha) \quad (14.10a)$$

does not correspond in $\mathcal{P}(\alpha, 1, 1)$ a minimum $Z(\alpha, 1, 1) < \bar{z}$.

If $\bar{\gamma}_{ij}$ denotes the generic element of vector $\bar{\gamma} \triangleq \bar{u}A^r + \bar{v}A^c$, (14.10a) can be equivalently written as

$$\alpha_{ij} \leq 1 + \frac{c_{ij} - \bar{\gamma}_{ij}}{M} \qquad 1 \leq i \neq j \leq n \tag{14.11}$$

where $\bar{\gamma} = \bar{u}A^r + \bar{v}A^c$.

Remark that, for every (i, j) such that $c_{ij} - \bar{\gamma}_{ij} \geq 0$, (14.11) is redundant. If the basis associated to $x(\alpha, 1, 1)$ is a *free basis*, i.e. if it contains only columns (i, j) such that $\bar{\alpha}_{ij} = 1$, the inequality $c - \bar{\gamma} \geq 0$ is the well-known optimality condition for the assignment problem (14.1). Then, from Corollary 14.1 it follows that the α which have $\alpha_{ij} = 0$ (and then $x_{ij} = 0$) for every (i, j) such that $c_{ij} - \bar{\gamma} < 0$, may be cut off. If a free basis has been chosen to express $x(\alpha, 1, 1)$, then this result can also be obtained directly.

Another straightforward consequence of Theorem 14.5 is the following:

Corollary 14.2 *Let $(\bar{u}, \bar{v}, \bar{\omega})$ be any solution of the system*

$$\begin{cases} uA^r + vA^c - \omega A^v \leq c \\ (u+v)e_n - \omega\bar{\beta} \geq \bar{z} \\ \omega \geq 0 \end{cases} \tag{14.9b}$$

Then, to every $\beta \in H$ satisfying

$$\bar{\omega}\beta \leq (\bar{u} + \bar{v})e_n - \bar{z} \tag{14.10b}$$

does not correspond in $\mathcal{P}(1, \beta, 1)$ a minimum $Z(1, \beta, 1) < \bar{z}$.

Similar propositions are easily deduced from Theorem 14.5, when problems $\mathcal{P}(\alpha, 1, \beta)$ and $\mathcal{P}(1, \beta, b)$ are considered.

Now consider the dual of $\mathcal{P}(\alpha, \beta, b)$, which may be written

$$\max \left[(u+v)b - \omega\beta \right] \tag{14.12a}$$

$$uA^r + vA^c - \omega A^v \leq c(\alpha) \tag{14.12b}$$

$$\omega \geq 0 \tag{14.12c}$$

and which will be denoted by $Q(\alpha, \beta, b)$. Call $S(\alpha)$ its feasible region.

It is easy to remark that (14.9) is possible, iff $(\bar{u}, \bar{v}, \bar{\omega})$ is an optimal solution of (14.12), since in the contrary case the second inequality of (14.9) is violated.

14.6 A Lower Bound for the Minimum Total Distance in the Travelling Salesman Problem

In the preceding sections it has been shown how to decompose a TSP into a sequence of linear assignment problems. This has been accomplished by

The Crew Scheduling Problem: A Travelling Salesman Approach

means of a suitable penalization of the objective function, i.e. by means of $\mathcal{P}(\alpha, 1, b)$.

Now it will be shown that, by considering $\mathcal{P}(1, \beta, b)$, it is possible to obtain a lower bound for the TSP at every step of the decomposition procedure of Section 14.5. The main fact, which will enable us to obtain such a lower bound, is the independence of $S(1)$ from β; while $S(\alpha)$ depends on α, and thus $\mathcal{P}(\alpha, 1, b)$ does not enable us to derive such a bound.

Denote $S(1)$ simply by S and remark that $S \neq \emptyset$, as $R(\beta, b)$ is bounded and, $\forall \beta \in H_1(h)$, not empty. Moreover, $Q(1, \beta, b)$ has finite extremum, $\forall \beta \in H_1(h)$. As a consequence the following relations hold, where vert. S denotes the set of all vertices of S, and $\bar{S}$ any subset of vert. S, and where the first equality is due to (14.2.5b).

$$\min_{\beta \in H(h); x \in R(\beta, b)} z(x, 1) = \min_{\beta \in H(h)} [\min_{x \in R(\beta, b)} z(x, 1)]$$

$$= \min_{\beta \in H(h)} \{\max_{(u,v,\omega) \in S} [(u+v)b - \omega\beta]\}$$

$$= \min_{\beta \in H(h)} \{\max_{(u^i, v^i, \omega^i) \in \text{vert.} S} [(u^i + v^i)b - \omega^i \beta]\}$$

$$\geq \min_{\beta \in H(h)} \{\max_{(u^i, v^i, \omega^i) \in \bar{S}} [(u^i + v^i)b - \omega^i \beta]\}.$$

Such relations let us state the following:

Theorem 14.6 *The minimum of the following problem*

$$\begin{cases} \min(z) \\ z \geq (u^i + v^i)b - \omega^i \beta & (u^i, v^i, \omega^i) \in \text{vert. } S \\ z \in R; \beta \in H(h) \end{cases} \quad (14.13)$$

where the unknowns are z and β and where the constraining inequality must be written for every element of vert. S, equals the minimum of the TSP. If vert. S is replaced by any of its subsets, then the minimum of (14.13) is a lower bound on the solution of the TSP.

Every time the method of Section 14.3 is applied, an element of vert. S is determined (really, only (u^i, v^i); ω^i can be obtained by simple considerations), so that (14.13) is solved, where vert. S is replaced by the (u^i, v^i, ω^i) until now found. A sequence of nondecreasing lower bounds is determined in this way.

Of course a first lower bound is given by the minimum of $\mathcal{P}(1, 1, b)$. Note that vert. S may be replaced with any subset of S, and not necessarily of vert. S.

14.7 An Algorithm for the Crew Scheduling Problem

Consider the reals μ_i, $i = m+1, \ldots, m+n$, associated with the vertices $N_1, \ldots, N_n$, respectively; and the binary values $\alpha_{ij} \in \{0, 1\} \triangleq B$, $m + 1 \leq i \neq j \leq m + n$, associated with the arcs of G; set $\mu = (\mu_{m+1}, \ldots, \mu_{m+n})$, and $\alpha = (\alpha_{m+1,m+2}, \ldots, \alpha_{m+1,m+n}, \ldots, \alpha_{m+n,m+n-1})$. Consider the generalized distances

$$d_{ij}(\alpha_{ij}) \triangleq \begin{cases} d_{ij} + (1 - \alpha_{ij})M & \text{if } m+1 \leq i \neq j \leq m+n \\ M & \text{if } i = j = m+1, \ldots, m+n \\ d_{ij} & \text{otherwise} \end{cases}$$

where M is again a large enough real number. Define the set

$$\mu(\alpha) \triangleq \left\{ \mu \in R^n : \frac{n - 1 - (\mu_i - \mu_j)}{n} \geq 0, \, m+1 \leq i \neq j \leq m+n; \right.$$

$$\left. \text{and } \frac{n - 1 - (\mu_i - \mu_j)}{n} \geq 1, \text{ if } \alpha_{ij} = 1 \right\}.$$

Define H_1, H_2, and H as in Section 14.4. Consider the problem

$$\min \left\{ z_m(x, \alpha) = \sum_{i,j=1}^{m+n} d_{ij}(\alpha_{ij}) x_{ij} \right\} \tag{14.14a}$$

$$\sum_{j=1}^{m+n} x_{ij} = b_i, \, i = 1, \ldots, m; \quad \sum_{j=1}^{m+n} x_{ij} = 1, \, i = m+1, \ldots, m+n; \tag{14.14b}$$

$$\sum_{i=1}^{m+n} x_{ij} = b_j, \, j = 1, \ldots, m; \quad \sum_{i=1}^{m+n} x_{ij} = 1, \, j = m+1, \ldots, m+n; \tag{14.14c}$$

$$x_{ij} \geq 0, \, i, j = 1, \ldots, m+n. \tag{14.14d}$$

Problem (14.14) is formally the same as $\mathcal{P}(\alpha, 1, 1)$, so that the properties and the method described for it are also valid for (14.14). Then, when $\alpha \in H$, an optimal solution of (14.14) is a Hamiltonian circuit of G_m. If now the frst m vertices are regarded as the same vertex of G, such a circuit of G_m corresponds to a set of circuits of G, having in common only N_0, and such that every vertex of G belongs to a circuit. Thus, it is a feasible solution of the m-TSP.

The decomposition procedure of the preceding sections may then be applied in a straightforward way to the m-TSP.

Consider the problem of determining the tours of a given set of m crews so that a given set of n flights be realized, and in such a way that the total cost (distance) be a minimum (Gurel, 1969). As already pointed out, the

The Crew Scheduling Problem: A Travelling Salesman Approach

salesmen are now the crews; the vertices $N_1, \ldots, N_n$ are the flights to be executed; N_0 is a dummy flight, which means rest; arc (i, j) means that flight j is executed after flight i; d_{ij} is the cost (lost time†) implied by the execution of flight j after flight i. When it is impossible to execute flight j after flight i, then $d_{ij} = M$.

Consider now a more general problem. Let the number m of crews, who must execute the n flights, not be given, and assume that this number has to be determined together with the tours. Such a problem can be solved by introducing the b-parameter, as in Section 14.4. More precisely, the right-hand sides of the first m of equations (14.14b) and of (14.14c), corresponding to the vertices $N_{01}, \ldots, N_{0m}$, are replaced with the binary parameters $b_1, \ldots, b_m$, respectively. After this has been done, problem (14.14) is quite analogous to $\mathcal{P}(\alpha, 1, b)$ of Section 14.4, and the decomposition procedure described there can also be applied to (14.14) in searching for both the tours of flights and the number of crews to cover them in an optimal way. The number $\sum_{i=1}^{m} b_i$ of crews may, of course, be subjected to an upper bound or to any other constraint.

The first step for solving the CSP consists in finding a first feasible solution of the CSP, i.e. an optimal solution of any problem (14.14) which satisfies some additional given constraints, like union regulations or similar which define the real problem. The first feasible solution can either be obtained from the outside or one can compute it with the following method, based on the rule of the 'nearest unvisited vertex', used to find a first feasible solution to the TSP. In the TSP, $n+1$ vertices, $N_0, N_1, \ldots, N_n$, and the distances-matrix are given.

The above-mentioned rule consists in selecting N_0 (or any other) as the first vertex. When s vertices have been chosen, the $(s+1)$th vertex is selected among the remaining ones, as the one which has a minimum distance from the sth. This method can be improved by considering the matrix of the 'transformed' distances in place of the given matrix.

In the formulation of the CSP considered in this paper, vertex N_0 is considered as 'partitioned' in the dummy vertices $N_{01}, N_{02}, \ldots, N_{0m}$, depending on how many crews will be employed.

To find a first feasible solution to the CSP, one can proceed in this way: choose N_0 as a first vertex, N_{i_1} as the second one, N_{i_2} as the third one, and so on, by making use of the rule of the nearest unvisited vertex. But, every time one chooses a vertex, i.e. a certain flight must be executed after another one, it is necessary to check that this choice does not violate the union regulations; for instance that the total flying time is not larger than the allowed maximum. If no regulation is violated, one can proceed; otherwise the choice suggested by this method is not accepted, and N_{02} is considered

† The addition of flight times does not affect an optimal solution.

as the following vertex, i.e. one goes back to the crew home base and chooses another crew for which a rotation can be built up. From here on, the process is repeated until all flights have been covered.

Note that the above method does not guarantee a feasible solution to the CSP, when $\sum b_i$ (i.e. the number of crews that one wants to employ) is fixed or is subjected to an upper bound.

This difficulty is easily overcome by finding a feasible solution to the CSP without any constraint on the number of available crews. The decomposition procedure of Section 14.3 starts with such a feasible solution and varies the number of crews (i.e. $\sum b_i$), if necessary.

14.8 Software and Interactive Aspects

The major advantage of the proposed algorithm is the possibility of coding it on a computer in such a way as to solve the problem in an interactive mode: a crew scheduler sits at a terminal, either a typewriter or a display screen and interacts with a computer.

Particularly, the following mode of use can be hypothesized. A given set of possible rotations, not necessarily being feasible from the point of view of union regulations, is fed into the computer and their cost is computed. Within a given CPU time, the algorithm computes a set of rotations of lower (but not necessarily minimal) cost, feasible from the point of view of union regulations.

The crew scheduler can ask the computer to evaluate the set of rotations from a cost (or other performance figures) point of view, as a function of the number of crews employed. He can fix some subset of rotations, and he can ask to evaluate how much this fact penalizes him. He can add some information or constraints which are difficult to code within the program. He can monitor the improvements with time of the performance figure. In other words, his productivity and, as a consequence, the economics of the process can be greatly improved. In particular, he retains control over the algorithm, adapting it to his needs and stopping and directing it according to his experience in the field.

Figure 14.1 shows, in a synthetic way, a possible procedure based on the previous algorithm using it in an interactive mode.

(A) is a 'direct approach', that is a procedure such that the user's intervention is restricted to the introduction of the initial conditions and to the recording of the results.

(B) shows the direction to follow for an interactive approach, and makes use of some aspects of the algorithm shown in the preceding sections and based on the m-TSP.

The method starts with the input of the flights and of the union regulations (see Section 14.1).

The Crew Scheduling Problem: A Travelling Salesman Approach

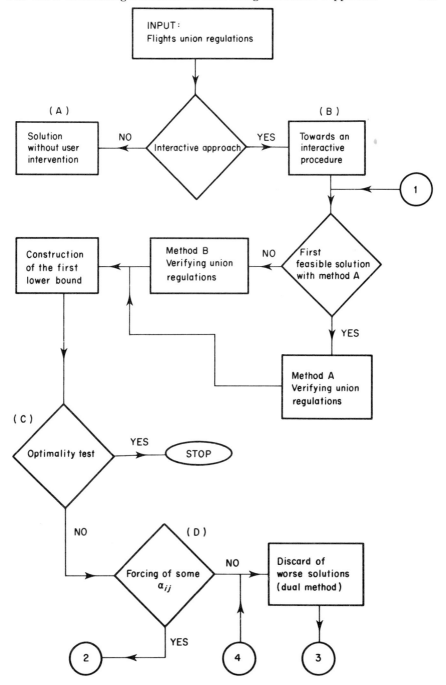

Figure 14.1(a)

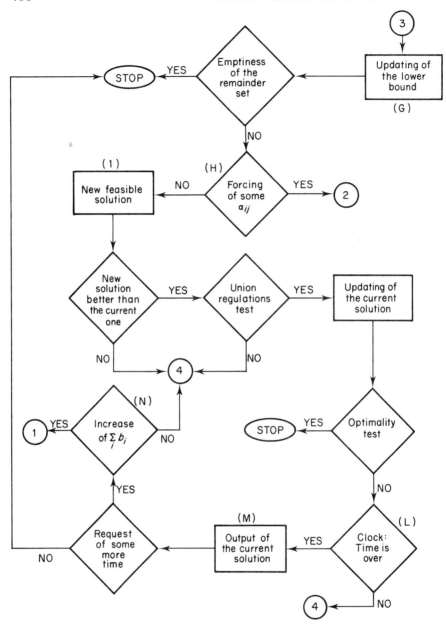

Figure 14.1(b)

The Crew Scheduling Problem: A Travelling Salesman Approach

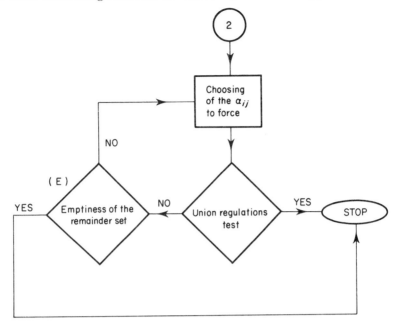

Figure 14.1(c)

The approach is made more flexible with the introduction of some tests, besides the optimality tests requested by any resolutive approach, which are careful and precise choices the user must operate to make the program more suitable to solve his specific problem.

As a second choice, the user must decide whether or not he wants to obtain a first feasible solution by the rule of the nearest unvisited vertex (utilizing the algorithm of Section 14.7).

If not, he can utilize another technique, but in both cases this step is followed by a check of the union regulations.

An initial lower bound is then computed which is useful for the optimality test (C).

At steps (D) and (H) appears one of the interactive variables of the approach of this paper, that is matrix α (see Section 14.4), where α is the matrix of the paths, that is $\alpha_{ij} = 1$ means flight j is paired after flight i, while $\alpha_{ij} = 0$ means the opposite. Hence the operator can force a sequence of flights, if he so wishes.

A positive action may be taken by the operator if he thinks that the cuts introduced by some prefixed values of α are better than the corresponding ones of the dual algorithm, or this action can be dictated for example by a specific need of substituting a certain crew. This step is followed by a test of the union regulations (E). If the regulations are satisfied, the procedure stops and gives the best solution found; otherwise a new cut is made if possible, and the process repeated.

The step marked with (F) is just the application of the algorithm of Section 14.3 (the dual method), for discarding solutions not better than the current one.

Every time this technique is utilized, one gets a reduction of the domain inside which one searches for the optimal solution (that is, the discarding of certain nonsuitable rotations), and a consequent updating of the lower bound (G).

In the case where the set of the discarded solutions coincides with the totality of the remaining solutions, the algorithm terminates with the best solution; otherwise a new possibility is offered to the user to force $\alpha(H)$. Then a feasible solution is chosen inside the new restricted domain, (I), and the choice acts in such a way that, if the objective function of the new solution is not better than the best one already found, this new solution is abandoned and control comes back to (F) where other solutions are discarded, till a feasible solution is found.

Next, there is a further check performed by the analysis of the union regulations. If the test is negative, control returns to (F), otherwise there is an updating of the current solution and an optimality check that stops the procedure and the displaying of the solution, in the case where the lower bound coincides with the best value of the solutions met till now.

In the opposite case, the program is interrupted, (L), by a programmed 'clock' of the computer, at intervals fixed in advance or from time to time.

In this way the user can see (M) the degree of optimality already reached. He must also decide whether to give some more CPU time or not.

References

Arabeyre, J. P., J. Feanley, F. C. Steiger, and W. Teather (1969). *The Airline Scheduling Problem: A Survey. Trans. Sci.*, **3,** 140–163.
Cavanagh, R. A. (1975). *Airline Crew Scheduling.* Presented at *XII TIMS International Meeting, Kyoto, Japan, July,* 1975.
Gavish, B. (1976). *A Note on 'The Formulation of the m-salesman Traveling Salesman Problem'. Manag. Science,* **22,** 704–705.
Gurel, O. (1969). *Algebraic Theory of Scheduling. IBM N.Y. Scientific Center Report* 320–2981, November, 1969.
Mangasarian, O. L. (1969). *Nonlinear Programming,* McGraw-Hill, New York.
Nicoletti, B. (1975). *Automatic Crew Rostering. Trans. Sci.,* **9,** 33–42.
Svestka, J. A. (1976). *Response to 'A Note on The Formulation of the m-salesman Traveling Salesman Problem'. Manag. Science,* **22,** 706.
Svestka, J. A., and V. E. Huckfeldt (1973). *Computational Experience with m-salesman Traveling Salesman Algorithm. Manag. Science,* **19,** 790–799.

CHAPTER 15

Graph Theoretic Approaches to Foreign Exchange Operations

NICOS CHRISTOFIDES, ROBIN D. HEWINS, and GERALD R. SALKIN
Department of Management Science, Imperial College, London

15.1 Introduction

The trading in currencies in order to obtain the best possible exchange rate is known as arbitrage, and can broadly be divided into three categories:

(1) *Space arbitrage* Transactions to take advantage of discrepancies between rates quoted at the same time in different markets.
(2) *Time arbitrage* Transactions to take advantage of discrepancies between forward margins for different maturities.
(3) *Interest arbitrage* Transactions to take advantage of discrepancies between yield on short term investments in different currencies. This form of arbitrage can be split into (a) covered and (b) uncovered (speculative) interest arbitrage. The former variety uses today's forward rate for forward conversion back into the holding currency, the latter allows the dealer to use the spot rate existing in the future.

Types (1) and (3a)—considered in detail here—are deterministic, while the other types are stochastic and require forecasts of future exchange rates to evaluate the possible outcomes.

This chapter considers deals for spot and forward delivery and gives graph theoretic formulations for the various types of space and covered interest arbitrage. Space arbitrage (and to a lesser extent interest arbitrage) transactions must be concluded with great speed and accuracy (due to the changeable rates) and any optimization method must have computational speed as a prime requirement. The problems discussed here could be formulated as linear programs but the graph theoretic method is superior in computational speed by several orders of magnitude and, furthermore, provides both an insight to the problems and additional useful information.

We assume that the set of transaction currencies is given, together with the prevailing cross exchange rates (both spot and forward). Unavailable, doubtful, or unwanted rates may be set to zero. For each currency pair (i, j) upper bounds on the amount of currency i that can be exchanged for currency j for delivery on a particular date are assumed known. Such bounds are the maximum transaction amounts which, in the existing circumstances, may cause the quoting dealer to alter his rate (Aliber, 1969). It is also assumed that the requisite interest rates (e.g. Einzig, 1966) are quoted.

For some arbitrage problems—discussed in the next section—some 'reference currency' or standard may be needed, together with 'notional rates of exchange' between the currencies and the reference, in order to evaluate a mixed holding of currencies. This reference will usually be one of the real currencies involved in the arbitrage, in which case these 'notional rates of exchange' may be the real future ones. Alternatively the reference may be some other standard such as gold or S.D.R.'s (Cutler, 1974; Harrod, 1965). The profit or loss resulting from the arbitrage transactions for those problems for which an absolute evaluation is impossible are then measured relative to this reference.

15.2 Some Arbitrage Problems

A. Arbitrage on Straight Rates

Type A1 arbitrage involves the most profitable exchange from a given single currency position to another currency position for the same delivery date.

(a) Given a certain amount α_k of currency k, what is the maximum amount of currency i that can be obtained?
(b) Given a currency k, what amount of this currency can be accumulated by circulating money flow amongst the other currencies?

Problem (b) is a pure arbitrage problem, i.e. all deals are made solely for profit purposes.

Type A2 arbitrage involves an extension of type A1:

(a) Given a certain amount α_k of currency k to be sold, what other currencies should be bought?
(b) Given a certain amount of currency i to be bought, find what amounts of other currencies should be sold.

Type A3 is the general type A arbitrage problem:

(a) Given amounts of money held in various currencies, find the optimal set of transactions to improve the holdings.

The above forms of arbitrage are possible for the spot and forward exchange rates.

Graph Theoretic Approaches to Foreign Exchange Operations

B. Arbitrage on Swop Rates

Whereas arbitrage types A1–A3 involve the switches of one currency into one or several others, all for delivery on the same date, in arbitrage of type B we may perform two deals which enable us to swop a currency holding into other currencies spot and then to return to the original holding of currencies after a given period at a rate fixed at the present. Such deals are called spot v forward swops and are an important part of the foreign exchange system since such deals are involved in both time and interest arbitrage.

Thus we have arbitrage of *Types B1, B2, B3* which correspond directly to A1, A2, A3 above but with the additional stipulation that all forward deals performed will entail a corresponding spot transaction in the reverse direction.

C. Interest Arbitrage

In the above we have assumed that currencies incur or accrue no interest—this is generally the case for the foreign exchange department of a bank. However, when foreign currency may be employed to yield interest in various centres we have the problem of deciding in which market (i.e. which currency) to place our money. We will only consider covered interest arbitrage here, i.e. where we return to our original currency holding(s) at a forward rate, fixed now, after the period of our placement.

Of particular importance in this form of arbitrage is the Euro-currency markets which are often the receivers of surplus funds within the foreign exchange markets. We will assume that surplus funds are placed in such Euro-currency markets.

Type C1. Given a certain amount α_k of currency k available for a certain time period, which currency i should we switch into and use for our investment? Since we assume, for covered arbitrage, that all currency i repayments are switched at today's forward rate back into currency k, we have no problems of a reference.

Type C2.

(a) Given a holding α_k of currency k, into which currencies should we switch for interest bearing deposit purposes?
(b) If we must meet a commitment to place a certain amount of currency i, what other currency holdings should we switch into i in order to maximize the resulting holdings of these currencies when reswitched from currency i in the forward market? This requires the use of a 'reference currency'.

Type C3. When we have money in various currencies available for deposit (placement) we encounter the general interest arbitrage problem of which currencies to use for our deposits. Once again, since we will return to our original currencies in the future, we have need of a reference currency.

Upper and lower bounds may again be placed on the amounts of certain transactions and holdings.

15.3 Graph Theoretic Formulations: Flows in Graphs with Arc Gains

A currency i is represented by a set of vertices x_{im}, $m = 0, 1, \ldots, n$ of a graph G, where n is the number of forward rates quoted in the market (e.g. for the usual quotes of 1, 2, 3, and 6 months we have $n = 4$). A possible transaction from currency i at time period p to currency j at time period r is represented by an arc (x_{ip}, x_{jr}). The graph G will be denoted by the triplet $\{X, A, N\}$ where X is the set of vertices (currencies), A the set of arcs (transaction possibilities), and N the set of forward dates with which we are concerned.

Each arc $(x_{ip}, x_{jr}) \in A$ has associated with it a gain g_{ijpr} usually (the exception lies in type C arbitrage when spot v forward swops are permitted) denoting:

(1) The exchange rate between currency i and currency j for forward date p when $r = p$. Hence the spot i to j rate is g_{ij00}.
(2) The value of a unit of currency i when held for period r when $i = j$ and $p = 0$.

The other arc possibilities are stochastic in nature or meaningless.

In addition, each arc (x_{ip}, x_{jr}) has an associated capacity q_{ijpr} representing the maximum amount of currency i possible for such a conversion/placement without affecting the corresponding exchange/interest rates. Possibilities of dealing beyond these limits may be included in our formulation with little difficulty.

By stipulating correct source, node s, and sink, node t, in the network described above we will see that a general algorithm for flows in graphs with arc gains (Christofides, 1975) can be used to solve the arbitrage problem. Let us consider type A in detail.

15.3.1 Model for Arbitrage of Type A1

Problem (a) is precisely the problem of finding the optimum flow of input value $v_s = \alpha_k$ between vertices $s = x_{kp}$ and $t = x_{ip}$ of a graph $G = \{X, A, N = (p)\}$ where vertices x_{kp} and x_{ip} correspond to forward p date positions of currencies k and i and all other vertices $x_{1p} \in X$ of the graph G correspond to the other currencies for that date.

Graph Theoretic Approaches to Foreign Exchange Operations

Table 15.1 Cross-exchange rates
BUY

	DM	$	Fr.	£	Sw. Fr.	
DM	—	0.3765	1.8400	0.1640	1.1756	
$	2.5925	—	4.8237	0.4253	3.0460	
Fr.	0.5422	0.2068	—	0.08826	0.6310	SELL
£	6.0775	2.3425	11.2600	—	7.1375	
Sw. Fr.	0.8485	0.3275	1.5740	0.1399	—	

This optimum flow pattern from $x_{kp} = s$ to $x_{ip} = t$ would provide the best sequence of transactions for converting the amount α_k of currency k into the largest amount of currency i on the forward date p.

Problem (b) is a special case of (a) where the source vertex s is the same as the sink vertex t, i.e. $s = t = x_{kp}$.

Example 15.1 Let us illustrate (b) above by means of an example. Five currencies will be considered, i.e. German Mark (DM), US Dollar ($), French Franc (Fr), Pound Sterling (£), and Swiss Franc (Sw. Fr). The spot cross-exchange rates between these currenices are g_{ij00} given in Table 15.1.

Further, we will take the maximum amount q_{ij00} of a currency i that can be sold spot for any other currency j without affecting the above exchange rates to be given by: for i representing Dollar—5 m; Sterling—1 m; Mark—5 m; French Franc—15 m; Swiss Franc—10 m. Thus $q_{1j00} = 5$ m, $j = 2, \ldots, 5$, etc.

We require to find the maximum amount of dollars that can be generated by circulating currencies within the spot market and the optimum transactions involved.

By applying the algorithm for flows in graphs with arc gains (Christofides, 1975) to the complete five-vertex graph with arc gains as shown in Table 15.1, arc capacities as given above, and $s = t = x_{20}$, the optimum flow pattern is found and is shown graphically in Figure 15.1.

The optimum set of transactions then involves seven deals as follows:

Sell $	3 109 646	Buy Fr	15 000 000
Sell Fr	15 000 000	Buy DM	8 133 000
Sell DM	5 000 000	Buy Sw Fr	5 878 000
Sell DM	3 133 000	Buy £	513 812
Sell Sw Fr	3 475 254	Buy £	486 188
Sell £	1 000 000	Buy $	2 342 500
Sell Sw Fr	2 402 746	Buy $	786 899

Net inflow of $ 3 109 646
Net outflow of $ 3 129 399
Net gain of $ 19 753 (i.e. 0.635% of input)

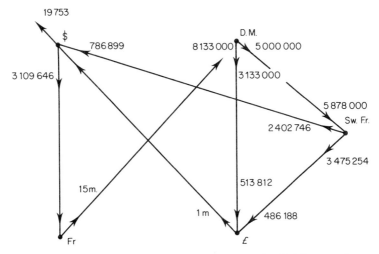

Figure 15.1 Optimal set of transactions for Example 15.1 (spot exchanges)

15.3.2 Model for Arbitrage of Type A2

Let us suppose that only some subset $S \subseteq X$ of the currency nodes is acceptable for the final currency holdings, but that the intermediate dealings can include the rest of the currencies.

Once again we form the graph $G = \{X, A, (p)\}$ corresponding to the arbitrage problem (Figure 15.2). In addition we now introduce a reference vertex x_{0p} and arcs (x_{ip}, x_{0p}) from every vertex $x_{ip} \in S$ to x_{0p}. The capacity of an arc (x_{ip}, x_{0p}) is set to β_{ip}, the maximum allowable final position of currency i that can be held at date p, and the gain g_{i0pp} is the 'notional' exchange rate for evaluation purposes.

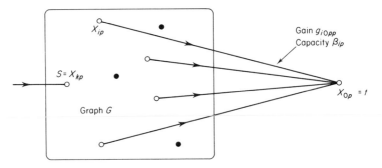

Figure 15.2 Sink vertex arcs for type A2 arbitrage. ○ Vertices $\in S$;
● vertices $\notin S$

We can now set $s = x_{kp}$, $t = x_{0p}$, and what is required is the optimum s-to-t flow in this graph with input value α_k, i.e. the input arc has capacity α_k with gain value 1.

The arbitrage problem of type A2b can be transformed into a network flow problem in an exactly analogous way to that described above for the problem of type A2a.

15.3.3 Model for Arbitrage of Type A3

Where a holding of a set $T \subseteq X$ of two or more currencies is to be changed to maximize the evaluation we introduce two extra vertices to the basic graph $G = \{X, A, (p)\}$; the reference vertex $x_{0p} = t$ and a source vertex s. We also introduce arcs from s to every other vertex $x_{ip} \in T$ and arcs from every x_{ip} to x_{0p} (Figure 15.3). If g_{i0pp} is the notional exchange rate between a currency i and the reference for date p, set the gain of an arc (x_{ip}, x_{0p}) to g_{i0pp} and the gain of an arc (s, x_{ip}) to g_{0i0p} (where g_{0i0p} is the present notional forward date p exchange rate from reference to currency i) for all vertices $x_{ip} \in T$.

If α_{ip} is the position of currency i initially held for date p we set the capacity of arcs (s, x_{ip}) to α_{ip}/g_{0i0p}. The capacity of arcs (x_{ip}, x_{0p}) is set to β_{ip}, where β_{ip} is the maximum allowable position of currency i for forward date p that can be held at the end of all the transactions.

Taking the inflow into s to be

$$\alpha_{sp} = \sum_{x_{ip} \in T} (\alpha_{ip}/g_{0i0p})$$

the optimum flow pattern from s to t will then correspond to the optimum sequence of transactions, and the net outflow from t will be the evaluation of the new currency holding.

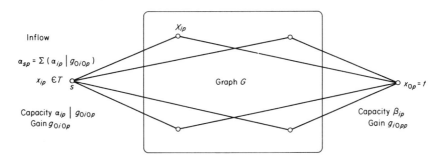

Figure 15.3 Network for type A3 arbitrage

15.3.4 Models for Arbitrage of Types B1, B2, B3

The problems B1, B2, and B3 can be formulated in the same way as for types A1, A2, and A3 respectively. It is usual practice in the market to quote for forward v spot swops by simply quoting the difference (margin) in the spot and forward rates; the forward rate may therefore be at a premium or discount with respect to the spot rate (Einzig, 1969). It can be shown that it is irrelevant (to first order) exactly which spot rate we base the forward margin on and hence it is only the margin which is important in quoting for spot v forward swops. Consequently, we can assume that all forward margins quotations are based on the same, consistent (in that we assume no spot arbitrage possibilities) spot rate and hence treat the spot v forward arbitrage problems simply by considering the forward part in the same way as Types A above.

15.3.5 Models for Arbitrage of Types C1, C2, C3

To permit spot v forward swop deals (the usual method of forward dealing) as well as outright spot and forward deals then we require several 'dummy nodes' denoting the currency held temporarily as the result of a spot v forward swop deal. Thus, if we may convert currency i into currency j spot with a reverse transaction in the future we denote this by including an additional node $x_{i,p}$ between the nodes x_{i0} and x_{ip}. Otherwise we construct two type A arbitrage network systems (one spot and one for the forward date) with connecting arcs between the corresponding nodes, the gains on which will be one plus the requisite interest rate, i.e. we have separate complete graphs $\{X, A, (0)\}$ and $\{X', A', (p)\}$ for the loan period date p and the nodes x_{i0}, x_{ip} are joined by arcs with gain $1 + I_i$ where I_i is the p period placement rate of interest. These 'interest' arcs will, in general, have an upper limit associated with them beyond which the requisite interest rates no longer hold (Aliber, 1969; Einzig, 1966). For the additional nodes we have $g_{ij,0p} = 1 + I_j$ (where I_j is the p period placement rate of interest for currency j),

$g_{j,ipp}$ = spot rate (i to j) × forward rate (j to i),
$q_{ij,0p}$ = maximum permitted placement of currency j in its market at the stated rate of interest,
$q_{j,ipp}$ = the maximum amount of currency i convertible into currency j spot v forward.

Hence we will have a network of the form shown in Figure 15.4 considering 3 currencies. Once again, the optimum s-to-t flow will be our desired solution obtained from the algorithm.

Graph Theoretic Approaches to Foreign Exchange Operations

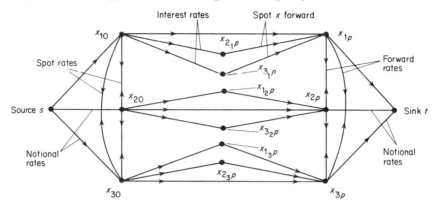

Figure 15.4 Typical network for type C2 arbitrage

15.4 Graph Theoretic Formulations: Shortest Path Models

By amending the problems slightly we can solve the space arbitrage problems for both spot and forward exchanges with a form of shortest path algorithm (Hu, 1967). The flows in arcs algorithm contains general contraints on the transaction amounts permissible between certain currencies, whereas the present shortest path method has the number of permissible deals as its central limiting factor. The latter has greater recognition of the need for speed and the difficulties in performing several 'simultaneous' deals; however, it lacks a certain amount of flexibility when compared with the flows with arc gains method above.

We develop all optimum routes between any two currencies either spot or forward, subject to visiting a given maximum number of intermediate nodes (or currencies). By so doing, ways of 'bettering' the straight swop rate are identified and all profitable loops (where we finish with the same currency held) are found.

Let us consider the spot rate matrix $\mathbf{R} = \langle r_{ij} \rangle$ where r_{ij} is the spot rate for converting currency i into currency j (i.e. the buying rate for currency j against i when rates are quoted as units of j per unit of i); and hence $r_{ii} = 1$ for all i.

If we now form the 'product' of $\mathbf{R}$ with itself we have

$$\mathbf{R}^2 = \mathbf{R} \otimes \mathbf{R}$$

where $\otimes$ is defined such that $\mathbf{R}^2$ has entries $r_{ij}^2 = \max(r_{ik} r_{kj})$ and gives the best rate from currency i to currency j for value on the spot date using at most one intermediary currency. Values in $\mathbf{R}^3$ give optimal rates with at most 2 intermediary currencies, and $\mathbf{R}^4 = \mathbf{R}^2 \otimes \mathbf{R}^2$ gives the best rates using at most 3 intermediary currencies, etc.

The limit on the number of intermediate currencies is governed solely by the time allowed to accept or reject all the quotes involved and the number of banks whom we can contact at one time.

By carrying a 'route matrix' along with the calculation, as defined by

$$\text{ROUTE}\,(i,j) = j \quad \text{if} \quad r_{ij} = r_{ij}r_{jj}$$
$$= k \quad \text{if} \quad r_{ij} = r_{ik}r_{kj}$$

we obtain the optimum route as well as its value.

These operations are obviously very simple and, in addition, the optimum route using any number of intermediate currencies can be obtained in only two matrix operations by using a cascading algorithm (Farbey, Land, and Murchland, 1967). However, we run into a little difficulty with this method because, when profitable arbitrage loops are discovered, an infinite profit can be made by continually exchanging money around such a loop. To overcome this we may place limits upon the amounts transferable at any one time between two currencies at the rate quoted.

The algorithm for forward rates is slightly harder because each forward transaction in the market is usually accompanied by a corresponding reverse spot transaction.

By arranging the currencies in decreasing order of unit currency value it can be shown that using the matrix $\mathbf{R}_t$ where

$$\mathbf{R}_r = \begin{bmatrix} 1 & \dfrac{r_{12}}{r_{12}+{}_tm_{12}} & \dfrac{r_{13}}{r_{13}+{}_tm_{13}} & \cdots \\ \dfrac{r_{12}+{}_tm'_{12}}{r_{12}} & 1 & & \cdots \\ \vdots & & & \end{bmatrix}$$

in the same fashion as $\mathbf{R}$ we can calculate the optimum routes and rates for all exchanges on forward date t using swop transactions. In the above notation ${}_tm'_{ij}$ and ${}_tm_{ij}$ are respectively the currency j buying and selling margins for currencies i and j for forward date t, i.e. the rate is quoted as ${}_tm'_{ij} \to {}_tm_{ij}$.

15.5 Conclusions

Many extensions to the basic algorithm above are possible and are easily incorporated into a general model (Hewins, 1977). It suffices to say here that the above techniques are adaptable to many similar problems in the financial markets. Furthermore, by defining two-way arcs to allow for borrowing as well as lending in the type C problem form, the general network flow approach, with its visual simplicity, is proving useful for analysis of the money market operations of international companies.

It has been demonstrated that all types of deterministic arbitrage can be formulated and solved by graph theoretic methods. These methods are computationallly very efficient (both in terms of memory requirements and in terms of computational speed and can be used to solve arbitrage problems involving tens of currencies in less than a second on a mini-computer. The technique can, therefore, be the basis of a practical system for optimizing arbitrage transactions, particularly if a Visual Display Unit is available to the dealers. The choice of which of the two basic methods above to use is dependent on the size of bank involved and the available staff, etc. In most situations a combination of methods is the most reasonable approach.

References

Aliber, R. Z. (1969). *The International Market for Foreign Exchange*, Praeger, New York.

Christofides, N. (1975). *Graph Theory—An Algorithmic Approach*, Academic Press, New York.

Cutler, D. (1974). The Valuation of the SDR. *Euromoney, August*, 1974, 27–31.

Einzig, P. (1966). *The Euro-Dollar System*, Macmillan, New York.

Einzig, P. (1969). *A Textbook on Foreign Exchange*, Macmillan, New York.

Farbey, B. A., A. H. Land, and J. P. Murchland (1967). The Cascade Algorithm for finding all Shortest Distances in a Directed Graph. *Man. Sci*, **14,** 16–28.

Harrod, R. (1965). *Reforming the World's Money*, Macmillan, New York.

Hewins, R. D. (1977). Management Science Models for Foreign Exchange. *Ph.D. Thesis, Imperial College, London.*

Hu, T. C. (1967). Revised Matrix Algorithm for Shortest Paths. *S.I.A.M. J. Appl. Math.*, **15,** 207–218.

Index

Abelian semi-group 40
Accuracy 9
Additive semi-group 25
Algorithm
  efficient 108–111
  greedy 109
Alternating subgraph 182
Arbitrage 409–419
  interest 411
  on straight rates 410
  on swop rates 411
  shortest path models for 417
Arborescence, *see* Directed spanning tree
Assignment problem 136–140
Augmenting sequence 110
Augmenting subgraph 182

Baker and Su algorithm for scheduling 374–375
Balanced matrix 159
Benders' partitioning method 21
Benichou partitioning strategy 9
Bin packing 119
Binary encoding 116
Bitroid 110
Bottleneck machine 372
Bivalent programs 93–106
  solving of 101–104
Bounded knapsack problem 275–276
Bounding function 74–78
Branch and bound 1–20, 28, 53, 133, 243–277
Branching 13–14
Breadth-first branch and bound 243

Canonical form 178

Capital budgeting 239
Cargo-loading problem 239
m-Centre problem 281–314
  absolute 281
  classification scheme for 283
  complexity of 285–287
  computational results for 304–307
  extensions of 307–313
  inverse 283
  on tree networks 283
  relaxation algorithm for 287–307
Change-making problem 277
Chord 162
Chordless cycle 158
Christofides' heuristic 145–146
Chromatic number 113–116, 213
Chvátal's operation 33
Clique 157, 159–162, 165, 170, 177, 201, 213,
  covering problem 151, 157
  number 213
  problem 112–116
Codes for mixed integer programming 17
r-Colourable 213
Colouring function 213
Column generation 90, 152
Comb inequalities 143
Complementarity theory 21
  conditions 48
Complexity 107–129, 285
  norm 108
Composite inequalities 178
Complexity 107–129, 285
  norm 108
Composite inequalities 178
Computational geometry 111

Concave function 73
Constraints
    contradictory 8
    facial 58, 59, 63
    tightening 63
Convex function 73
Convex span 59–61
Core knapsack problem 267–270
Crew scheduling 153, 389–408
Critical
    cutset 161–163
    edge 161, 169
    machine 386
Cuts 58
Cutset 161
    arc 132
Cutting planes 21–72, 152, 171–174, 178–179
    methods for travelling salesman 142–143
    strengthening of 26
    weakening of 26, 39, 48, 50, 55, 58
Cutting stock problem 239

Dantzig upper bound 239–240
Decomposable set 181
Degeneracy 7
Degree of infeasibility 11
Depth-first branch and bound 243
Directed spanning tree 134–136
Disjunctive arcs 381
Disjunctive graph 381
Disjunctive methods 47–71
Disjunctive normal form 171
Disjunctive programming 171
Disjunctive rule theorem 60
Dominance criteria 261
Dynamic programming 238–239, 274–277

Edge covering problem 151, 155
Edge matching problem 151, 155, 156
Efficiency of an algorithm 285
Elementary inequalities 178
Enumeration tree 3–5, 10–12
Eulerian subgraph 120
Expected case norm 108

Face 59

Facets 58, 167, 172–174
    of the set packing polytope 158–170
    producing graph 164–168, 177
Fathoming 3–5, 15–16
    premature 16
Finiteness 9
Flows in graphs with gains 41
Foreign exchange 409

Game, two-person 61–62
Generalized lagrangian method 238
Gomory's fractional row cut 32
Gradient 73
Graph colouring 122, 211–235
    algorithms
        colour sequential methods 223–224
        dichotomous search methods 221–223
        multiple method 228–233
        vertex sequential methods 217–221
    as a set covering problem 214
    formulations 214–216
    problems; size reduction 216–217
Greedy solution of the knapsack problem 260, 262, 269
Group knapsack problem 6
Group problem 25, 30–31, 39, 65

Hamiltonian circuit 131, 113–116, 131–149, 397
Heuristics 107–129
    approximation methods 118
    constructive methods 118
    decomposition methods 118
    deterministic evaluation of 118–122
    direct search 15
    for travelling salesman 143–146
    inductive methods 118
    probabilistic evaluation of 122–126
Homomorphisms 40
Hypergraph 158

Implicit enumeration 3, 152, 238, 274–277
Improper face 176
Independence number 160, 213
Independent set 213
Integer form 7
Intersection theorem (for binary integer programs) 59

Index

Intersection theorem (for graph colouring) 228–229

Jackson's rule for scheduling 372
Job-shop scheduling problem (general) 380–383

Knapsack problem 113–116, 186, 237–279, 340
   algorithms for, (dynamic programming) 250–257
   algorithms for, (implicit enumeration) 243–250
   dominance criteria for 261
   large size 266–274
   reduction procedure for 257–260
   unbounded 274–275
   upper bound for 239–243
   value independent 276
   zero–one 238–276

Lagrangean multipliers 131–149
Lateness of jobs on one machine 371–388
LIFO strategy 14
Linear combination of constraints 22–23, 35
Linear inequalities systems 74
Linear programming polytope 181–182
Lin's r-optimal heuristic 145
Loading problem 339–369
   computational results for 349, 367
   dominance for 344, 366
   dynamic 354–369
   feasibility test for 360–363
   formulation of dynamic 357–358
   formulation of static 344–345
   lower bounds for 346, 363–366
   minimal sets for 344
   sensitivity for 350–353
   static 343–354
   weak maximal sets for 356
   weak minimal sets for 356
Location theory 281–314

Machine scheduling 371–388
Matching problem 110, 113–116, 140
Matroid 109–110
McMahon and Florian algorithm for scheduling 375–378
Maximal r-colourable subset 224

Median 282
Mid-point property 287
Minimax criterion 282
Minimal inequalities 39
Minisum criterion 282
Mixed integer linear program 1, 29
Modular arithmitic 22–25, 33, 35
Monoid 25
   finite 31
   infinite 31
   mapping of 40
Monoidal cut strengthening 51–52
Multipliers 48, 50, 66

Nearest insertion rule for travelling salesman 144
Nearest neighbour rule for travelling salesman 144
Network flow problem 110
Node covering problem 151, 155
Node packing problem 151, 155, 157
   polytope 160
Nonsingular submatrix 159
Normal hypergraph 159
NP-complete problem 112–116, 286, 374
   strong 116
NP-hard problem 112–116

Odd anti-hole 160, 161, 165
Odd hole 160, 161, 165
ε-Optimality 16
Order relations 93–105

Partitioning 8, 9
   problem 113–116
n-Paths 140–142
Path-tree 113–116
Penality 6
   down 7
   up 7
Perfect graph 158, 159, 160
Perfect matching 155
Perfect matrix 159, 160
Polynomial bounded algorithm 108
Precedence constraints 371
Priorities 9
Projection method 14
Pseudo costs 9, 12–13
Pseudo-polynomial algorithms 116

Quasi-integer variables 9

m-Radius 281
  absolute 281
Reduced costs 136
Reduction—(*a priori*) for travelling salesman 143
Relaxation 3–4, 28, 40, 54, 58
  group theoretic 5–6
  lagrangean 5–6, 131–149
  linear programming 5–8
Round-off error 9
Route expansion criteria 323–324
  based on extra mileage 323
  based on radial position 323
  based on 'savings' 323
  composite 323
Routing methods 324
  Clarke and Wright 327–328
  computational results for 333–337
  heuristic 327–333
  heuristic tree search 331
  Mole and Jameson 328–329
  parallel 324
  sequential 324
  sweep 329–330
  two-phase 332–333

Scheduling problem 153
Scheduling timetables 211–212
Semi-group problem 31
Separation 3–5
Sequencing problem 113–116
Sequential lifting procedure 164
Set cover 113–116
Set covering 151, 154, 291, 297
Set packing polytope 151, 158–170, 175–182, 200
Set packing problem 113–116, 151–157, 197
Set partitioning 16, 151–210
  lower bound 186–192
  polytope 151, 171, 175–183, 188
  reduction rules 184
Set partitioning algorithms
  column generating 191–193
  cutting plane 181–191
  hybrid cutting plane/implicit enumeration 193–196
  implicit enumeration 184–188
  node covering 200–204

subgradient cutting plane 196–200
Shortest spanning tree 133–136
  constrained 113–116
Simultaneous lifting procedure 164
Slack variable 74
Special ordered set 16
Steiner Network problem 113–116
Stickstacking problem 276
Stopping rule 17
Strong intersection graph 177
Subadditive collection 40, 56
Subadditive function 25
  extension of 28
  generation of 28
Subadditive methods 21–47
Subdifferential 73–90
Subgradient method 73–90, 134, 141, 152
Subgraph, induced 132
Surrogate constraint 101, 102

Ternary relations 97–98
Tolerance 9
Transitive closure 95
Travelling salesman problem 120, 131–149, 392–407
  asymmetric 134–136
  bounds for 133–142, 400–401
  decomposition procedure for 398–400
  heuristics for 143–146
  soluable cases of 146–148
  symmetric 132–134
m-Travelling salesman problem 393
Tree-search (*see also* Branch and Bound) 29

Unilaterally connected graph 138
Unimodular matrix 159
Unloading only problem 353–354

Valid inequality 22
Value function 27
Vehicle routing problem 315
  direct tree search algorithm for 319–321
  extended 322, 323
  formulation of 317
  heuristics for 321–322

Index

set partitioning formulation of 318–319
Vertex covering problem 113–116
Voronoi diagram 111

Web 160–165

Weighted node covering problem 200, 203
Weighted node packing problem 200
Worst-case analysis of heuristics 143–146
Worst-case norm 107